LASER AND OPTOELECTRONIC ENGINEERING

Series in Electrical Engineering

Editor: S. A. Nasar

Bradley Molecules and Molecular Lasers for Electrical Engineers
de Jong AC Motor Design: Rotating Magnetic Fields in a Changing Environment
Nasar and Boldea Electric Machines: Steady-State Operation
Muncheryan Laser and Optoelectronic Engineering

in preparation

Nasar and Boldea Electric Machines: Dynamics and Control

LASER AND OPTOELECTRONIC ENGINEERING

Hrand M. Muncheryan, Sc.D.

President
Laserkinetics Corporation
Orange, California

HEMISPHERE PUBLISHING CORPORATION

A member of the Taylor & Francis Group

New York Washington Philadelphia London

LASER AND OPTOELECTRONIC ENGINEERING

1 2 3 4 5 6 7 8 9 0   E B E B   9 8 7 6 5 4 3 2 1

This book was set in Times Roman by WorldComp.
The editors were Carolyn Ormes and Mary Prescott. Cover design by Sharon M. DePass.
Printing and binding by Edwards Brothers.

A CIP catalog record for this book is available from the British Library.

Library of Congress Cataloging-in-Publication Data

Muncheryan, Hrand M.
Laser and optoelectronic engineering / Hrand M. Muncheryan.
p. cm.
Includes bibliographical references and index.
1. Lasers. 2. Optoelectronic devices. I. Title
TA 1675.M85 1990
621.36′6—dc20 90-37720
CIP

ISBN 1-56032-062-1

CONTENTS

PREFACE

The unprecedented advent of laser science and technology during the past two decades has wrought tremendous changes in the advancement of scientific and engineering knowledge and has given impetus to new techniques for the execution of this knowledge. With this in view, a concerted effort has been directed to the writing of this book to supply a uniform account of the present-day knowledge of laser science as applied to professional and industrial fields. The book is prepared for those who aspire to specialize in the theoretical and practical aspects of laser technology. For professionals engaged in the use of lasers, the book is designed to expand their knowledge and stimulate their thinking in keeping up with the trend of rapidly progressing knowledge of this technology to provide new directions to the technical growth of their organizations.

The material in this book is intended to provide the reader with a comprehensive understanding of the various disciplines encompassing radiation physics as applied to the design and development of laser systems for industrial uses, holographic applications, telecommunication, military weapon systems, medical surgery and treatment systems, nuclear transmutation to create controlled atomic power, and metrology of specialized measurement techniques. The organization of the text may permit the teaching of this technology, where applicable, at any academic level of study desired.

This work is particularly appropriate for use as an instructional sourcebook for professionals to guide them in designing systems of their own innovation. In this respect, the last chapter contains a categorized miscellany of ideas for laser systems and devices of all character to be improved or created. They will motivate the thinking of the scientist to design and develop laser-based products in demand in many disciplinary fields. The medical profession has repeatedly emphasized the need for versatile laser instrumentation systems and, to this end, the author has given numerous medical cases in the pertinent chapter to guide the reader to attain this accomplishment for this young and fast-moving laser-aided instrumentation field.

The text gives a complete explanation of design, construction, and operations of all kinds of laser systems and devices and their performance characteristics. These include ion lasers, molecular lasers, solid-state lasers, semiconductor diode lasers, electron-beam-aided lasers, metal-vapor lasers, liquid-dye lasers, x-ray lasers, and excimers. The descriptive specifications on laser systems and accessories extend from beam expanders and collimators, spatial filters, mode lockers, Q-switches, and cavity dumpers to beam stabilizers, all of which are fully discussed in the pertinent chapters. Metrological measurement systems, including aerial pollution-detection and measurement systems, sea-floor-mapping systems, precision automatic target tracking systems, security surveillance systems, weapon targeting systems, and medical diagnostic and treatment systems are some of the extensively explored and discussed subjects in this book. In short, the principles of laser theory on the latest state-of-the-art developments are covered in easy-to-read language consistent with scientific writing.

Some of the more complex but practical concepts are supplemented by applicable equations, many of which are exemplified by problems that are worked out for the benefit of the reader. The treatment of the subjects is pedagogical and offers the opportunity of self-study of the subject. The work combines an intensive theoretical background with a practical approach to support a thorough knowledge of the subject of coherent radiation. The review questions and problems at the end of each chapter further consolidate the reader's learning of the subject studied. Although the text is sufficient unto itself, each chapter terminates with a list of selected reference sources from which the reader may obtain specialized information, when desired.

The appendixes at the end of the book contain a table of chemical elements and their characteristics used in conducting laser research, electrical units of measurement and their equivalents, international units of measurements, and the names and addresses of laser and laser-oriented companies. Some of the important products of each manufacturer are also given to aid the laser worker in the selection of devices and components for purchase and building of equipment.

The author acknowledges his indebtedness to those many institutions, laboratories, and individual professionals who have contributed some of the scientific data on their most recent research and development in laser technology and related subjects.

Hrand M. Muncheryan

CHAPTER

ONE

FUNDAMENTAL INFORMATION

1-1 ELECTROMAGNETIC WAVES

The discovery of electromagnetic waves dates back to 1887–1888, at which time Hertz observed that an oscillatory electric discharge set up a train of electromagnetic waves in space and that the character of these wave trains was similar in many ways to that of light rays. The theory was exactly predicted by Maxwell in 1865, but it was never demonstrated experimentally in his time. The theory proposes that the electromagnetic wave trains are radiations from accelerated charges. According to this concept, an accelerated charge must radiate, an idea which, as we shall see later, must be accepted with reservation in view of Bohr's extension of the radiation hypothesis.

Radiations of all types propagate in space in the form of wave trains. A radiation consists of transverse vibrations set up in space by the interlinkage of the electric and magnetic vectors executing simple harmonic motions. Creating a harmonic motion of this character involves energy, the rate of travel of which is proportional to the product of the electric and magnetic vectors, as set forth in Poynting's theorem, which is derived on the basis of the energy per unit volume in an electric field of strength $F = eF^2/8\pi$ and that in a magnetic field of a strength $H = \mu H^2/8\pi$. What attribute the radiation energy possesses while speeding through space cannot be determined in a fundamental fashion except by its effects on recipient objects.

In accord with the electromagnetic wave theory, we may infer that the energy from any radiating source is given in ergs per cubic centimeter of the space traveled. Again, we would not be justified by our assumption, which confronts conflicting explanations when applied to the photoelectric phenomenon. The electromagnetic theory fails to adduce a solution to the phenomenon of expulsion of photoelectrons at the instant a flash of light becomes incident on an active material surface. We have to refer to

Einstein's equation, supported by the quantum theory, to find an adequate interpretation for the effect; this equation is given as

$$\frac{1}{2}mv^2 = hf - w_0 \tag{1-1}$$

where $\frac{1}{2}mv^2$ is the kinetic energy of the electron, hf is the incident photonic energy, and w_0 is the work function, which is the amount of energy required to remove an electron from the atomic quantum level.

We cannot neglect, however, the importance of the two fundamental tenets assigned to the electromagnetic theory, in that (1) an accelerated charge must radiate and (2) the wave trains of the radiation energy must be continuous. That is, the emission of energy from the radiating source should produce a continuous radiant wave front with an intensity varying in accordance with the inverse square law. The latter law states that a diverging radiation from a point source decreases in intensity as the square of the distance from the source.

The velocity of propagation of an electromagnetic wave train may be computed by use of the following relation:

$$v = \frac{c}{\sqrt{\mu D}} \tag{1-2}$$

where v is the velocity of the wave train; c is the ratio of the electrostatic and electromagnetic charges and is equal to 2.9979×10^{10} cm/sec, which is the speed of light; μ is the permeability; and D is the dielectric constant of the medium traveled. But for empty space (vacuum) these constants are each unity. Thus, v becomes equal to c.

All wave phenomena of electromagnetic character become related to each other by the expression

$$\lambda = \frac{c}{f} \tag{1-3}$$

where λ is the wavelength of the radiation, c is its velocity, and f is the frequency with which the radiation propagates. This relation holds for any electromagnetic wave from cosmic rays through visible rays to electric waves. In Table 1-1 the range of electromagnetic waves is given in order of their increasing wavelengths. There are various established methods by which the wavelengths of any range of radiation can be determined.

James Bradley (1692–1762), in his systematic observations of the seasonal displacement of a certain star in the zenith, discovered that the shift in the position of that star was due to the motion of the earth with respect to the velocity of light from the star. The full significance of this phenomenon was not realized until many years later, when further investigations of the effect led to the discovery of relativity theory, proposed by Albert Einstein in 1905. The latter theory assumes equivalence between accelerated and gravitational motions and denies the existence of any absolute velocity.

Table 1-1 Range of Electromagnetic Waves

Type	Wavelength (angstroms)	Generation	Detection
Cosmic rays	0.00008–?	Extraplanetary sources	Cloud chamber and similar schemes
Gamma rays	0.004–1.4	Disintegration of atomic nuclei	The same as for x-rays but more penetrating
X-rays	0.06–1100	Sudden stoppage of high-velocity electrons	a. Radiometry b. Photography c. Phosphorescence d. Ionization of gases e. Diffraction f. Erythema
Ultraviolet rays	136–3900	From hot bodies, ionized gases, dye lasers Nd-YAG	The same as x-rays, erythema reaction
Visible rays	3900–8200 –8500?	Radiation from hot bodies and gases; the sun	Sensation of light rays and heat rays; see ultraviolet above
Infrared rays	8200–100,000	Radiation from hot bodies and gases	Heating effect, photography, bolometry, radiometry; reflected, diffracted, refracted
Laser radiations	2000–100,000	Electric energy, light energy, magnetic flux and light energy, microwaves	a. Radiometry b. Photography c. Phosphorescence d. Ionization e. Refraction/diffraction f. Erythema reaction g. Melting, fusing, and cutting metals h. Biomedical effects
Microwaves	1×10^{6}–10^{11}	Spark gap discharge, oscillating triodes, transistors, CBs	Spark gaps, radiation as short radio waves
Radio waves	10^{11}–10^{14}	Triodes, transistor-type oscillations	Communication, music, data transmission, coherer
Broadcasting band	10^{11}–10^{13}	Same as above	Same as above
Electric waves	3×10^{14}–10^{16}	Conductor coil rotating in magnetic field	Mechanical effect, electrical effect, thermal effect, fatal effect, etc.

Note: Values given above are to be considered as nominal for practical purposes.

It further maintains that an accelerated body acquires an increase in mass in measure with its increased energy.

On the assumption of the electromagnetic origin of the electron mass, Sir J. J. Thompson showed experimentally that there is a measurable increase in the mass of an electron when its velocity approaches that of light, a fact that was to be expected on the basis of relativity. Lorenz further computed independently that the variation in the mass of an electron moving at a velocity v could be given as

$$m = \frac{m_0}{\sqrt{1 - (v/c)^2}} \tag{1-4}$$

which is an exact picture of what the relativity requirement of the variation of mass with velocity exhibits, and for our purpose we shall not extend this subject any further until we study the transmutation of atoms in Chapter 12.

Einstein's energy-mass relationship has shown that a photon with a momentum p has a mass equivalent expressed by

$$p = mv = \frac{hf}{c} = \frac{h}{\lambda} \tag{1-5}$$

where h is Planck's constant, equal to 6.55×10^{-27} erg-second, v is the velocity of the particle, f is the frequency of the photon, c is the speed of light, and λ is the wavelength of the photon.

By the hypothesis of de Broglie, all particles such as atoms, electrons, protons, and neutrons are associated with a wave mechanism characterized by a phase velocity given by

$$u = \frac{hf}{p} = \frac{E}{mv} = \frac{c^2}{v} \tag{1-6}$$

where m is the mass of the particle and E is its energy (including the photon as a particle). Thus, multiplying both sides of the terms in Eq. (1-4) by the velocity v, we obtain

$$p = mv = \frac{m_0 v}{\sqrt{1 - (v/c)^2}} \tag{1-7}$$

where m_0 is the rest mass of the particle under consideration. This relation with respect to a photon indicates that a photon can exist only when it is traveling with the speed of light c.

The lifetime of the electron-photon interaction is relatively short, dependent on the transition probability of the excited atom. If we assume the excited atoms to be dominant in a state, for instance, s, the reduction in the number of atoms from that state will proceed exponentially as $N(t) = N_0 e^{-pt}$. Accordingly, the average lifetime of a given atom in state s can be expressed as

$$T = \frac{1}{N_0} \int_0^\infty tp\, N_0 e^{-pt}\, dt \tag{1-8}$$

where T is the average lifetime of the atom in the selected state s.

The radiance of, for instance, N_0 atoms in a ruby laser, which is a three-layer system, is influenced by the so-called rate equations, which may be given as

$$\frac{dN_3}{dt} = W_{1.3}N_1 - (W_{3.1} + A_{3.1} + S_{3.2})N_3 \tag{1-9}$$

and

$$\frac{dN_2}{dt} = W_{1.2}N_1 - (W_{2.1} + A_{2.1})N_2 + S_{3.2}N_3 \tag{1-10}$$

where N_3 is the number of atoms at level 3, $W_{1.3}$ is the excitation energy to level 3 of N_1 atoms, $W_{3.1}$ is the radiation energy to ground level 1, $S_{3.2}$ is the probability of radiationless transition from quantum level 3 to quantum level 2, N_3 is the number of excited atoms at level 3, $A_{2.1}$ is the spontaneous radiation from level 2 to level 1, and the other quantities are similarly identified, as shown in Fig. 1-1.

In the radiation from atomic and molecular collisions, a line proceeding with intermission of the radiation activity occurs, which results in deviation from the monochromatic emission of the total radiation. This effect further gives rise to misproportional radiation trains of waves. Thus, an expression due to the collision-broadened frequency distribution may be given as

$$\nu_d = \frac{\Delta\nu}{2\pi} \frac{1}{(\nu - \nu_0) + (\Delta\nu/2)^2} + \frac{1}{(\nu + \nu_0)^2 + (\Delta\nu_2)^2} \tag{1-11}$$

where $\Delta\nu$ is the linewidth and ν_0 is the uninterrupted control frequency during atomic and molecular collisions with no change in phase of the radiation train of wave motion. Since the collision frequency is directly related to the density of the lasing gas in the resonant cavity, the linewidth would be proportional to this density. The second term in parentheses may be dispensed with when the magnitude of the control frequency far exceeds the linewidth. The subject of line broadening will be taken up in more detail when we discuss radiation interactions.

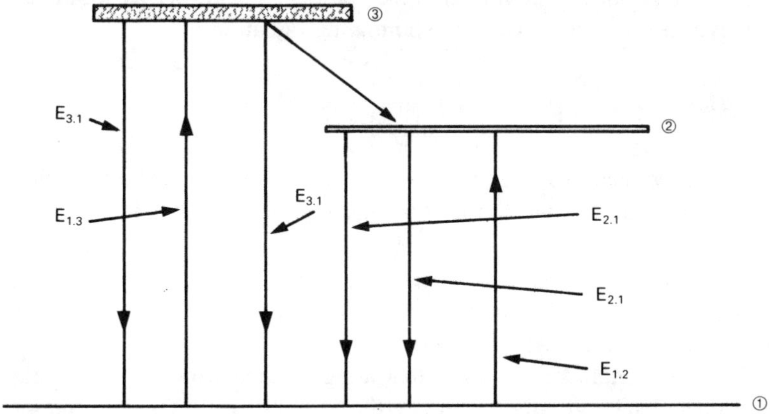

Figure 1-1 Diagram showing spontaneous and stimulated quantum transitions.

1-2 NATURE OF LASER RADIATIONS

The word "laser" is an acronym formed by the first letters in the phrase "light amplification by stimulated emission of radiation." To the uninitiated, this definition may fall short of an explicit meaning. Therefore, a simple definition for laser that is more definitive would be "a light-emitting body with a feedback for amplifying the emitted light." The word laser is generally used in referring to either the radiation or the device that produces it. Laser radiation can be produced in the spectral ranges from x-radiations through ultraviolet, visible, and infrared radiation. The laser generator is an elementary light emitter or generator, such as a ruby rod about 1 cm in diameter and 10 cm long; other sizes are also available.

The ruby rod is a highly optically transparent material, and its ends are highly polished for the transmission of light almost without absorption. A mirror is placed at each end of the rod and is aligned perpendicular to the principal axis of the rod; in recent years, the polished ends have been mirrored, thus eliminating the use of external mirrors. One end of the rod is 100% mirrored and the other end is mirrored (coated) only about 30 to 45% so that the light produced in the ruby rod can be transmitted to its exterior.

When the ruby rod is illuminated with an intense flashlight, it emits a fluorescent light that oscillates between the two mirrored ends and it becomes intensified. The process is known as amplification, the medium in which the oscillation occurs is known as an optical cavity, and the mirrored ends form the feedback mechanism. The emitted light from the optical cavity is known as laser radiation. As long as the flashlamp illumination continues, the laser emission persists, and when the illumination ceases the radiation ceases to be emitted. The amplified laser radiation exits from the lightly mirrored end of the ruby rod. The emitted light propagates through the atmosphere, with almost the same diameter as the ruby rod but with slight divergence, which effect will be discussed in a later section.

The propagating laser radiation may be considered as unidirectional light. In addition, the radiation is coherent because most of its optical spectrum is of the same wavelength and in-phase relation. The radiation thus produced is known as stimulated emission and its waves are in phase with the stimulating radiation.

Types of Laser Systems

Laser radiation can be produced in a large variety of materials, which can be classified, for convenience of identification, into the following categories:

Solid-state (rod-type) lasers. The radiation generators in lasers of this type are ruby, neodymium-doped glass, neodymium-doped yttrium-aluminum-garnet (Nd-YAG), alexandrite, and a number of others.

Gas-type lasers. This type includes all lasers that are generated from gas lasing elements in an electric discharge tube. They are the helium-neon laser, argon laser, krypton laser, carbon dioxide laser, xenon laser, nitrogen laser, and lasers using a number of other gaseous materials that have the characteristic properties to lase. As in solid-state lasers, mirrors are provided at each end of the discharge tube for the oscillation and amplification of the generated laser light.

Excimer lasers. These are also gas lasers, but because the lasing gas consists of an inert gas combined with a halogen (e.g., fluorine or bromine) in the presence of an electric field in a discharge tube and because of their toxic and corrosive nature, they are better classified separate from the noncorrosive gaseous laser materials. Examples of excimers are ArF, XeF, KrF, KrCl, XeBr, etc.

Liquid dye lasers. These are optically clear solutions of various organic dyes that are stimulated to radiate by illuminating them with another laser, such as a Nd-YAG, Nd-glass, or nitrogen laser, and similar higher-frequency laser wavelengths. The dyes used are oxazine, rhodamine 6, coumarin, and a great many others. Dye lasers are generally weak in light intensity but are important because of their easy tunability to various wavelengths

Semiconductor diode lasers. Diode lasers are generally made of gallium arsenide and its various derivatives. The radiation in this type of laser occurs at the *p-n* junction of the diode when forward-biased with a current. They have many applications, including uses in the military and communication fields.

Metal vapor lasers. In this type of laser, the active medium is an ionized metal vapor, which is mixed with one of the inert gases, such as helium or argon. Examples are the He-Cd lasers, which are operable continuously, and the copper-vapor laser, which is operable in the high-frequency pulsing mode with high efficiency. These lasers are optically pumped by an electric field.

Other laser sources. These sources are iodine vapor, erbium ions, lanthanum ions, and similar rare-earth active elements, in addition to a number of others that we will have occasion to study in later chapters of this book.

Since laser radiation is the same as any electromagnetic radiation in the spectrum between x-rays and microwaves, the same optical principles that govern visible light also apply to lasers. For instance, a laser beam can be reflected, refracted, polarized, diffracted, and split by means of beamsplitters. Also, the laser beam can be made to diverge and converge by means of lenses or spherical mirrors. Accordingly, whenever the applications of laser beams are discussed in the text, the reader should treat a laser beam optically in the same way as a light beam, with the exception that a laser beam is coherent and is many times more intense than the sun's rays incident on the earth. Therefore, one should be aware of the fact that both physical and optical dangers exist when handling laser beams and equipment.

Laser Emission and Threshold Energy

We have already stated that an atom at ground state (at rest) is devoid of active energy and must receive external energy, such as from a flashlamp illumination or an electric source, in order to change its quantum or energy state. In the diagram shown in Fig. 1-1, suppose an atom of helium at ground state 1 is impressed with a voltage energy in an electric discharge tube and makes a quantum transition to energy level 3. On returning to the ground state, the atom spontaneously emits radiation. However, in its excited state at level 3, if the atom collides with a photon, the radiation emitted from the atom would be known as stimulated emission because it now has the same wavelength and frequency as the stimulating photon, and it would be expected to radiate in the same direction as the photon. Amplification also occurs in this system due to oscillation of the photons between the feedback mirrors in the optical cavity and further collision and production of other photons from the excited atoms. This process ceases when the stimulating energy falls below the threshold condition, which represents the least amount of energy required to emit a photon from an excited atom.

1-3 OPTICAL ELEMENTS USED IN LASER SYSTEMS

The optical elements that are employed to control the projected beam profile are lenses, mirrors, prisms, etalons, optical fibers, beamsplitters, beam expanders, beam polarizers, and the like.

Optical Lenses

There is a large variety of optical lenses. Among them, those that are generally used in laser industry are biconvex lenses, planoconvex lenses, biconcave lenses, planoconcave lenses, spherical lenses, meniscus lenses, and others. Lenses used with ultraviolet laser beams are generally made of fluoride glass, quartz, and Pyrex for some wavelengths. Lenses that are used in the visible spectral range are made of crown glass, flint glass, Pyrex, and plastic materials. For infrared laser beams, the lenses are usually made of germanium, calcium fluoride, quartz, sapphire, dielectric-coated silver chloride, Lucite, etc., each having a certain range of beam transmission. Lenses are also manufactured with antireflection coatings. One generally employed coating material is magnesium fluoride (MgF_2) of about a quarter-wave optical thickness or its multiple, depending on the application of the laser lens system. Some of the popular lens designs are shown in Figs. 1-2 through 1-4.

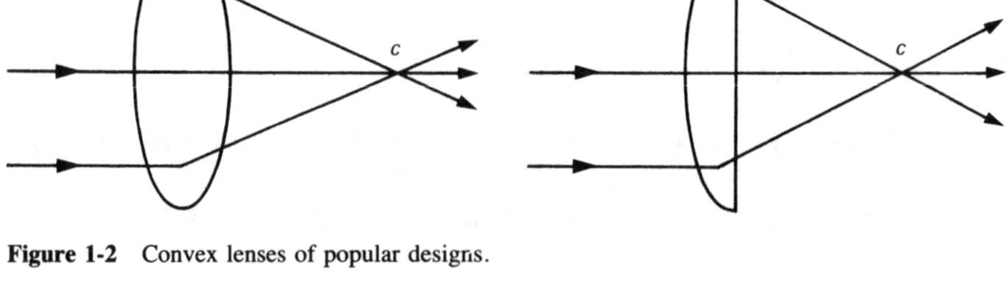

Figure 1-2 Convex lenses of popular designs.

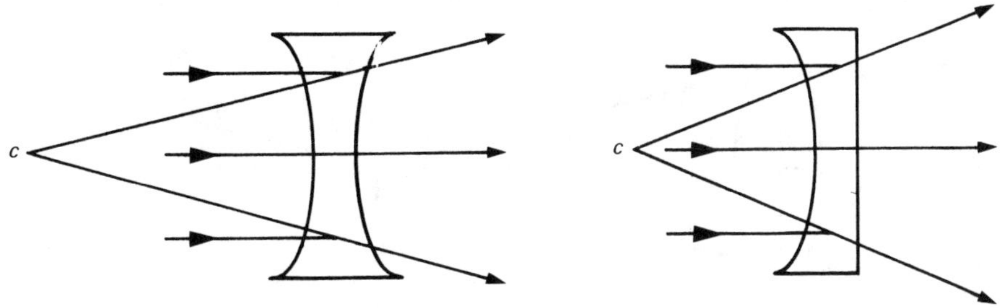

Figure 1-3 Concave lenses of popular designs.

Convex lenses are used in laser work requiring focusing of the laser beam to concentrate its energy level by reducing its incident beam diameter. Usually, biconvex and planoconvex lenses are used for this purpose. These lenses are employed for welding, cutting, drilling, scribing, and trimming thick- and thin-film resistors; they are also used in medical surgery. When using a planoconvex lens, the convex surface of the lens is usually positioned facing the oncoming laser beam. Biconcave and planoconcave lenses are used in the construction of laser beam collimators and beam expanders. For reducing coma and aberration to a minimum, a meniscus lens is used. A Fresnel lens has an aspherical surface made of many annular concentric rings, which permit the lens to have a short focal length with a very large diameter at low *f*-number; the lens is corrected for spherical aberration, thus yielding a better axial resolution than a planoconvex lens of equal *f*-number. The Fresnel lens is also used as the rear mirror in the Fabry-Perot cavity of a laser discharge tube; with this lens, a single wavelength can be selected within the range of the radiant emission.

Focal Length and Lens Magnification

In Figure 1-5, a paraxial parallel bundle of laser rays is incident on a biconvex lens with radii of curvatures r_1 and r_2 and index of refraction n. The emergent beam from the lens is focused at F, from which it propagates divergently in space. The point F is the principal focus of the beam, and the distance between the center of the lens and the focal point F is the focal length f of the lens.

The symbol f is positive for a convex lens and negative for a concave lens; it is a

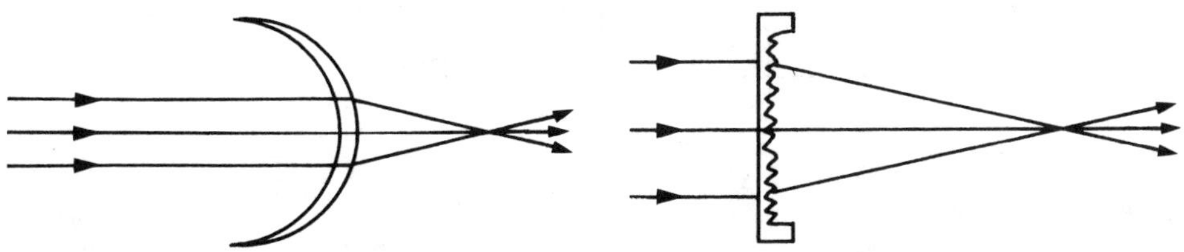

Figure 1-4 Meniscus and Fresnel lenses.

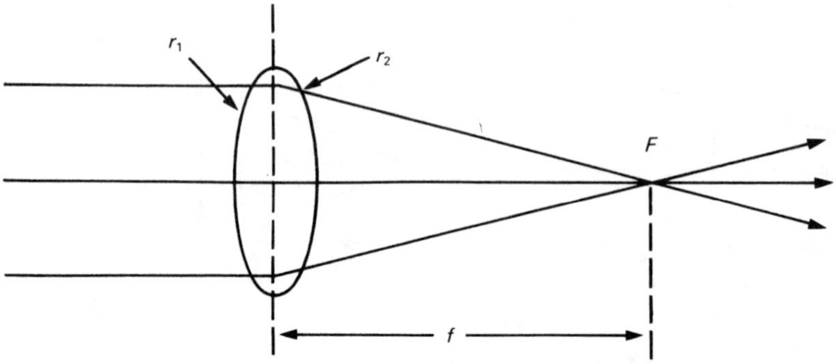

Figure 1-5 Focal length and radii of curvatures of a lens.

function of the radii of curvatures r_1 and r_2 of the lens and its refractive index n. Thus, we may express this relation in equation form as

$$\frac{1}{f} = (n-1)\left[\frac{1}{r_1} + \frac{1}{r_2}\right] + \left[\frac{(n-1)^2}{n} \times \frac{t}{r_1 r_2}\right] \tag{1-12}$$

where t is the thickness of the lens.

For thin or planoconvex lenses, such as those used in laser range finders, the second term of Eq. (1-12) vanishes and we may write the equation as

$$\frac{1}{f} = (n-1)\left[\frac{1}{r_1} + \frac{1}{r_2}\right] \tag{1-13}$$

When the focal length of a concave lens is considered, the positive sign in the brackets becomes negative.

The magnifying power of a lens is a function of its focal length and its index of refraction; it is generally expressed in diopters. A lens having 1 diopter power has a focal length of 1 m. Since the focal length of a lens depends on its curvatures, as shown in Fig. 1-5, the greater the curvatures of the lens, the shorter its focal length and the greater its magnifying power. Figures 1-6 and 1-7 show the focal length relationship of a biconvex lens and a biconcave lens.

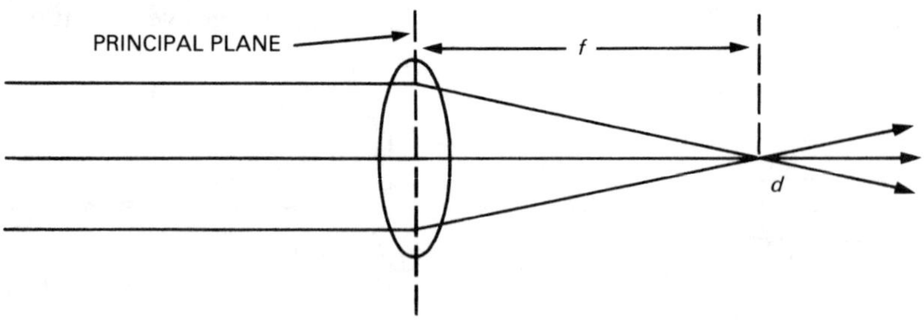

Figure 1-6 Focal length of a biconvex lens where d = focus diameter at $1/e^2$ point.

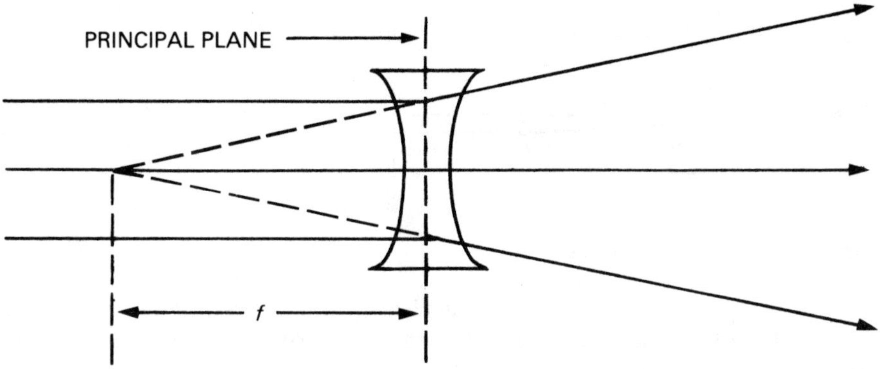

Figure 1-7 Focal length of a biconcave lens.

It is seen from Fig. 1-7 that the focal length of a biconcave lens will have a negative sign, in contrast to the focal length of a biconvex lens, which has a positive sign.

Example What will be the focal length of a lens that has an optical power of 0.125 diopter? What will be the magnifying power of a lens that has a focal length of 10 cm?

Solution

$$\text{Focal length} = \frac{100}{0.125} = 800 \text{ cm}$$

$$\text{Magnifying power} = \frac{100}{10} = 10 \text{ diopters}$$

For Gaussian laser beams, the focal spot diameter is given by the expression

$$d = \frac{8000}{3} \times \frac{1.22\lambda}{D} \times f \tag{1-14}$$

where f is the focal length of the leading optical element, λ is the laser beam wavelength, and D is the laser beam diameter, equal to 1.5 times the $1/e^2$ point, all quantities being expressed in millimeters.

Laser Beam Collimation

In collimating a diffraction-limited laser beam, illustrated in Fig. 1-8, usually a spatial filter of pinhole size d_m given by the following expression is used:

$$d_m = \frac{2f\lambda}{\pi r_e} \tag{1-15}$$

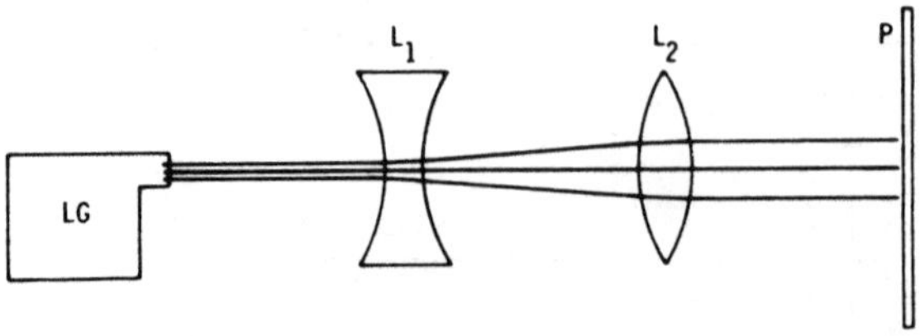

Figure 1-8 Collimation of a laser beam.

where f is the focal length of the leading optical element, λ is the laser beam wavelength, and r_e is the radius of the one-half irradiance profile at the first optical lens, all quantities being given in the same measurement unit.

In Fig. 1-8, a laser beam projecting from the laser generator, LG, is normally incident on biconcave lens L_1, from which the beam divergently projects to biconvex lens L_2, which collimates the beam into a parallel pencil beam and projects it to the plate, P. In Fig. 1-9, the laser beam, after passing through lens L_1 divergently, falls on lens L_2, which exaggerates the peripheral ray positions, as shown by the dashed lines. This effect causes peripheral blurring of the beam, and when the peripheral rays fall on the aluminum plate with a pinhole in its center, they are absorbed by the aluminum plate. The central beam, which is rectified of all blurring, is projected divergently on the screen, C; this process is known as spatial filtering of the laser beam.

The range over which the laser beam can be collimated is given by the expression

$$S = \left(\frac{\pi m^2}{4\lambda}\right)(2r_e)^2 \times 10^{-3} \text{ meters} \tag{1-16}$$

where m is the expansion ratio (5×, 10×, 20×, etc.), λ is the wavelength in millimeters, and $2r_e$ is equal to $1/e^2$ of the input beam diameter given in millimeters.

Optical Mirrors

Mirrors employed in laser work may be flat, concave, or parabolic, and their substrates may be made of glass, plastic, or metal. Concave or spherical mirrors are made in various focal lengths and are used in a resonant cavity to collimate or direct the laser beam to the output mirror in a single-mode operation. Glass and plastic mirrors are coated on their first surface to eliminate double reflection and blurring, which would

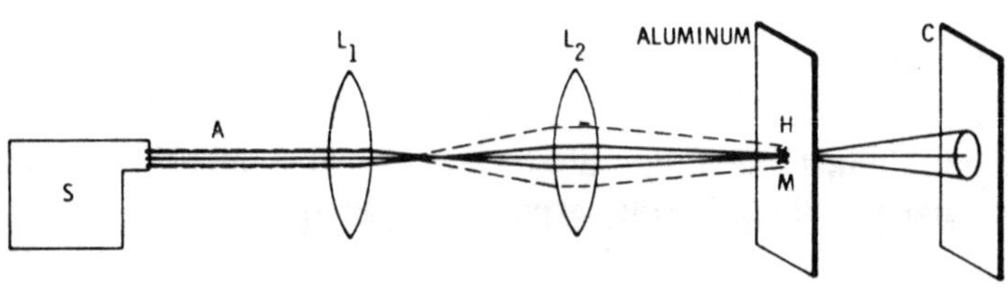

Figure 1-9 Spatial filtering of a laser beam.

occur if the mirrors were coated on their second or back surfaces. Front-surface coating also eliminates almost entirely the thermal effects arising from the back coated surface. The coating may be either metallic or dielectric. The metallic coating consists of a metal, such as aluminum, silver, gold, rhodium, or iridium, that is deposited by vacuum evaporation. The dielectric coating consists of magnesium fluoride (MgF_2) vacuum-coated at various quarter-wave thicknesses, depending on the degree of reflectance desired from the mirror.

Flat mirrors, usually coated on the front surfaces, are used to produce laser feedback in a Fabry-Perot resonant cavity. When lightly coated, the front-surface mirror may be used as a beamsplitter. The mirrors are also used in aerial communication systems to reflect the laser beam from the transmitting site to the receiving site. Second-surface-coated mirrors are rarely used in laser work. Mirrors that have a metallic coating are usually overcoated with a film of dielectric material to preserve the metal film from oxidation and scratching during handling.

Figures 1-10 and 1-11 illustrate the construction of a front-surface (first surface) mirror and a back-surface (second surface) mirror.

Optical Prisms

Optical prisms are generally triangular optical elements that are used to disperse a laser beam into its component wavelengths. They are made of the same material as lenses for different spectral transmissions. They operate by total reflection, and for this reason they are preferable to mirrors in many applications. The Brewster-angle prism, for instance, is used in a Fabry-Perot cavity for tuning a dye or ion laser beam that may contain several harmonic wavelengths in its bandwidth. An example of a Brewster prism is shown in Fig. 1-12*a*; its principal angular surface is about 70° with respect to the vertical passing through its base.

A right-angle prism (Fig. 1-12*b*) is an equilateral prism with two equal sides, a 90° included angle, and a hypotenuse that forms the base of the prism. The equal sides of the prism are used for dispersing the laser beam and the hypotenuse is used for

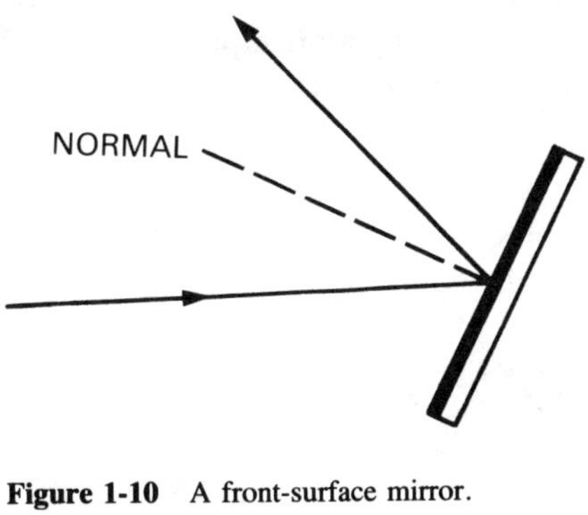

Figure 1-10 A front-surface mirror.

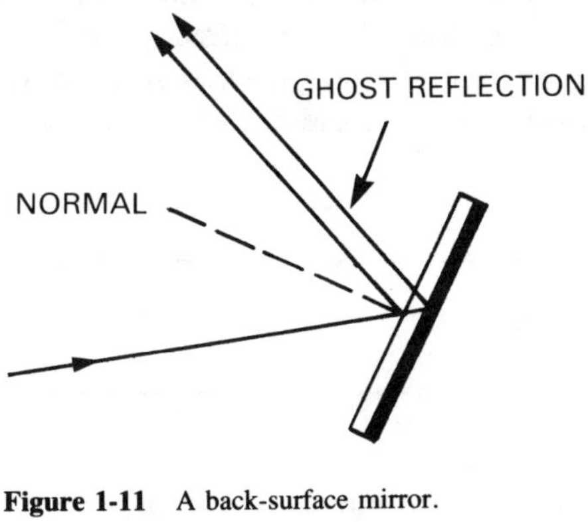

Figure 1-11 A back-surface mirror.

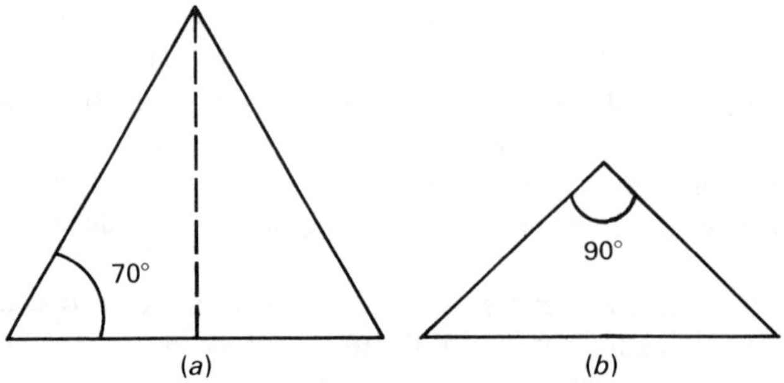

Figure 1-12 (*a*) Brewster-angle prism and (*b*) right-angle prism.

reflecting the incident beam. When the two equal sides are coated with a reflecting material, such as magnesium fluoride or silver, they can be employed as beam reflectors, as illustrated in Fig. 1-13. This type of prism is used for distance measuring by retroreflecting the laser beam from the prism.

Another type of prism, known as a Littrow prism, is made by splitting a Brewster prism at its apex down to the base in a vertical aspect into two Littrow prisms of equal size, each having an apex angle of 34° 21′. This prism is used in a resonator cavity to tune the laser beam by tilting the prism in the cavity. Tuning may be accomplished over a wide wavelength range if the laser beam is of the multiline type, that is, if the beam has harmonic wavelengths. A Littrow prism is illustrated in Fig. 1-14.

Prism Dispersion

A beam-dispersing prism is an optical element that is highly useful in the Fabry-Perot cavity of a tunable laser for selecting any wavelength within the constituents of the laser plasma generated in the cavity. Dispersion in a prism occurs because of the refraction of the laser beam in passing from the prism (of high index of refraction) into air (of lower refractive index), the degree of refraction being dependent on the wavelength of the beam. The ray having the shortest wavelength in a bundle of laser light is most refracted, as illustrated in Fig. 1-15.

The variation of refractive index n with wavelength is known as the optical dispersion of a prism. The amount of dispersion achieved by a prism is a function of its vertex angle and refractive index n, as given by the expression

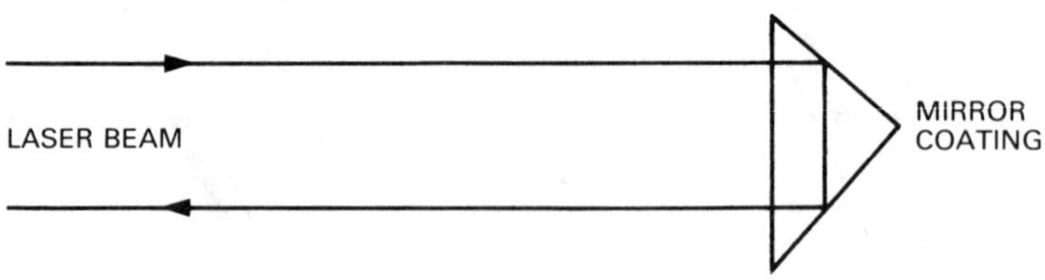

Figure 1-13 A Poro (back-reflecting) prism.

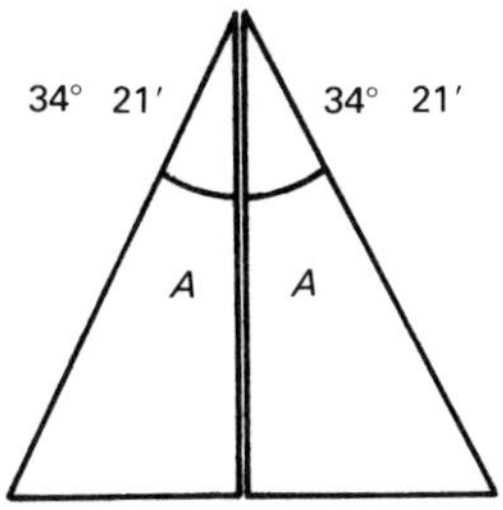

Figure 1-14 A Littrow prism of 34° 21′ angle.

$$n = \frac{\sin[(D + A)/2]}{\sin A/2} \qquad (1\text{-}17)$$

where n is the refractive index of the prism, D is the angle of deviation, and A is the vertex angle of the prism, as shown in Fig. 1-16.

When the angle A of a prism is small (up to 30°), the sines become equal to the angles and we may write an equation in the form

$$n = \frac{D + A}{A}$$

or

$$D = A(n - 1) \qquad (1\text{-}18)$$

The dispersion between any two wavelengths may be calculated by using Eq. (1-18) to determine the respective deviation values of the wavelengths dispersed. The amount of dispersion in a prism is a function not only of the angle of the prism (A) but also of the dispersive power of the prism material. Although in laser work one does not often encounter any usage of dispersive power, we shall study the application of such a prism in tunable ionic lasers and liquid dye lasers.

In Fig. 1-16, D_r is the minimum angle of deviation for the red wavelength and D_v

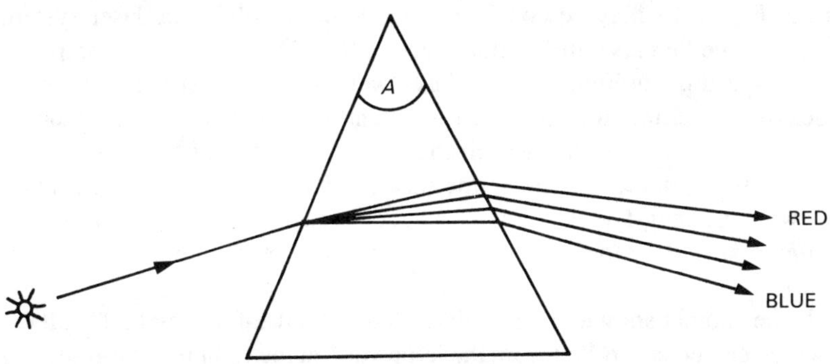

Figure 1-15 Light spectrum formed by refraction of light beam through a prism.

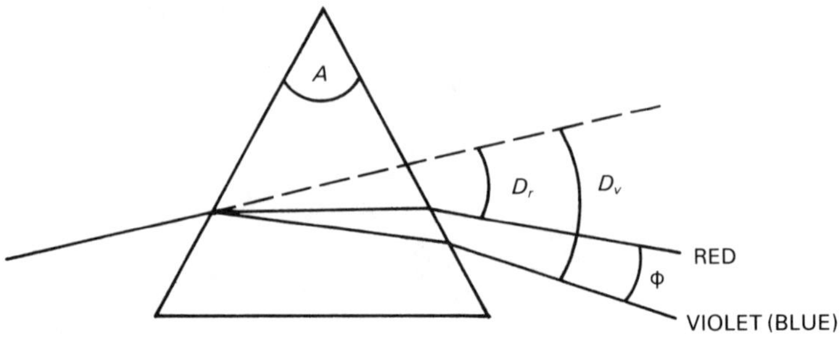

Figure 1-16 Dispersion of a multiline laser beam through a prism.

is the minimum angle of deviation for the blue or violet wavelength. The angle ϕ is the dispersion between the red and blue wavelengths, which can be calculated from the equation $\phi = D_v - D_r$.

1-4 LASER SYSTEMS AND CIRCUIT DIAGRAMS

There are numerous types and designs of laser systems to perform a variety of different functions in industry, medicine, and military work. Principally, they each contain a laser-generating element, an optical pump, a source of high-voltage power, and a system of optical elements to control the beam profile. In addition, each laser system with variable output or wavelength is provided with a control panel that has knobs for adjusting the power and directing the beam through the optical system to the area of interest. With high-power laser systems there is also a cooling system, either built into the same console or separate from it. The cooling system keeps the temperature of the laser generator at a given normal value, thus maintaining the output quality of the laser radiation at a uniform and constant value. Although each of the laser types will be discussed in detail in later chapters, the ruby laser system invented in 1960 will be discussed here as preliminary information. This system is shown in Fig. 1-17.

The system in Fig. 1-17 may be considered as the basic solid-state laser system containing a ruby rod, the first laser model manufactured by Hughes Aircraft Company. The author improvised this equipment into a laser welding system at the Autonetics Division of Rockwell International Corporation. Hundreds of welds were made in various types of metals employed in microelectronics. Thousands of holes were also drilled by means of this system, in spite of the fact that its power output was limited to 1.4 J. Since it was a pulsing-type laser, the capacitor network could be charged fully about twice a minute and, therefore, two pulses per minute could be obtained from it during operation.

In Fig. 1-17, the cabinet shown on the right contains a pulse-forming high-voltage power supply, which can be controlled from the large knob at the right in the top section of the instrument panel. The voltage could be adjusted up to 1250 V, which is indicated on the kilovoltmeter at the upper left side of the panel. The capacitors could be adjusted by closing the various switches inside the cabinet, after opening the top lid. Thus, both the voltage and the capacitor values could be adjusted up to the power rating of the system. The setup as shown in the figure is also provided with a binocular microscope to view the specimen and to locate the focus of the incident laser beam.

To operate the equipment, the main switch that transmits the 115-V AC to the laser system is turned on. The voltage is adjusted by means of the potentiometer shown in Fig. 1-18, which is the schematic circuit diagram of the system. Then the capacitors are switched in as required to bring the power to the required value. The combination of the voltage and capacitance values provides the input energy in joules to the laser generating section of the circuit. The specimen is then placed beneath the optical system, and the lens distance is adjusted to a given focus size. If the equipment is provided with a cooling system, that system is also turned on so that the coolant flows through the laserhead containing the flashlamp and the laser rod. In the diagram in Fig. 1-18, the cooling system may use either water or air as the coolant. In either case, the coolant is circulated continuously through the laserhead.

After positioning of the specimen under the laser optical system, the laser generating circuit is energized by means of a hand switch or a foot switch, as marked in the drawing. This operation produces a single electrical pulse from the capacitor network, which pulse is applied to the flashlamp. A momentary flash of light illuminates the laser rod, ruby in this case, which in turn produces a pulse of laser radiation. The laser

Figure 1-17 The first laser system manufactured by Hughes Aircraft Company.

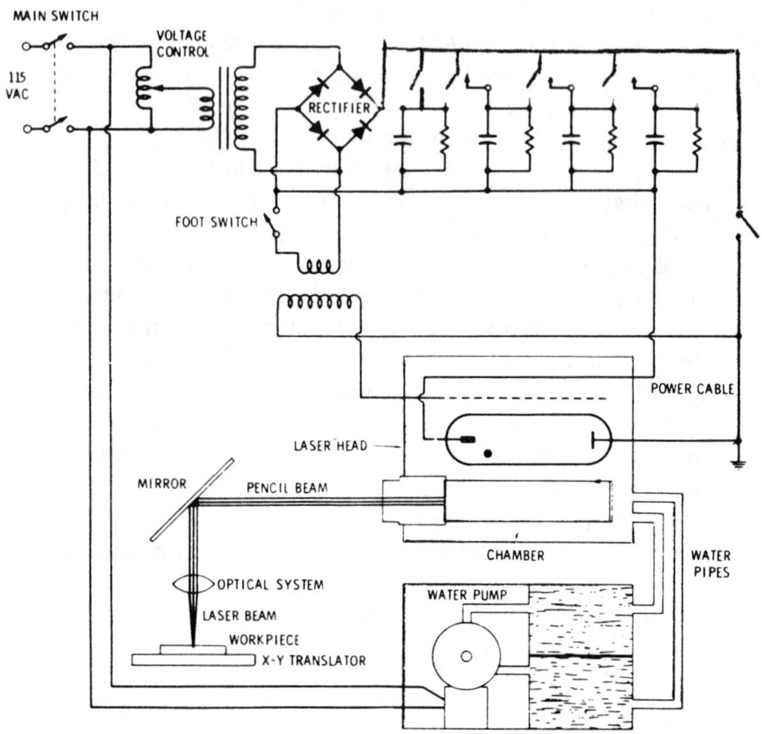

Figure 1-18 Circuit diagram of a typical solid-state laser system.

pulse performs the welding or drilling. The capacitor network bank immediately begins to charge for the next operation. It takes about 30 seconds to fully charge the capacitor network bank. In modern pulsed laser systems, the charging of the capacitor bank takes a fraction of a second, so multiple pulses per second can be obtained from the system.

For the pulsing system shown in the circuit diagram, the laser output can be calculated to a relatively close approximation by knowing the power applied to the flashlamp. An equation expressing the energy is

$$J = \frac{CV^2}{2} N \tag{1-19}$$

where J is the electrical energy in joules generated by the pulse-forming network, C is the charge in microfarads in each capacitor section, V is the charging voltage in kilovolts, and N is the number of pulse-forming sections, four in the diagram.

To illustrate the equation with a problem, let us assume that a laser power supply contains five pulse-forming sections and all sections, being equal in capacitance, are closed. Each capacitor section contains a 100-μF capacitance sustained at 1500 V (1.5 kV). The total energy produced in joules will be

$$J = \frac{100 \times (1.5)^2 \times 5}{2}$$

$$= 562.5 \text{ joules}$$

To calculate the laser emission from this input energy, we may write

$$J = (562.5 - 320) \times 0.004 \text{ (average)}$$

$$= 0.97 \text{ joules of laser, or simply } 1 \text{ J}$$

The last equation indicates that it takes about 1 J of flashlamp energy to produce 0.003 to 0.005 J of laser radiation above the threshold energy varying from 300 to 340 J. Thus, the arbitrary system chosen above can be rated to produce 1 J of radiation at the set electrical parameters.

1-5 SAFETY PRECAUTIONS

Two types of safety precautions should be taken in handling laser equipment: precautions against electrical shock and precautions against laser radiation. The first type involves avoidance of contact with uninsulated wires or electrical contact points while operating or repairing the laser system. The second type involves exposure to laser radiation.

Electrical Precautions

Almost all laser systems operate at dangerous voltage levels (kilovolts), with the exception of diode lasers. Therefore, the following precautions should be observed before attempting to work with the system circuitry.

- No one unfamiliar with the system circuitry should be allowed to work with a laser system.
- The main switch should be turned off before handling the circuit connections.
- A jumper with an insulated handle should be used to discharge the capacitors before work is performed on the circuits.
- A bare wire should never be touched while the current is on.
- The circuitry should be covered up when no one is working with it.
- For maximum safety, all laser systems should have grounded outlets.
- Safety interlocks should be in good operating condition; if they are not, they should be repaired immediately.
- The laser tube should be handled carefully, as it may implode if damaged or cracked by rough handling.
- Liquid dye lasers are toxic and in some cases may be corrosive; particular precautions should be taken to prevent manual contact with the liquid or to breathe the fumes from the solution.

- Excimer lasers are both corrosive and toxic when the vapors from them are breathed. When a leak is observed, it should be repaired immediately. On occasion, proper gas masks may be used, especially when making gas connections to the laser system. The expended gases from the system should be either expelled to the atmosphere remote from the operating room or transmitted into a storage tank. Electrical precautions are the same for this laser.

Laser Radiation Precautions

As stated in a previous section, laser beams are many times brighter than the rays of the sun reaching the earth and are much more hazardous. When a laser beam is incident on living tissue and is absorbed by it, the extent of damage caused is dependent on several factors: the energy level of the radiation, the type of tissue irradiated, the wavelength of the laser radiation, and the time of exposure to the radiation. If the energy of the radiation is sufficiently high, it interacts with the body tissue and coagulates or destroys the tissue cells.

For a given level of laser energy, certain tissues in the body are more vulnerable to laser radiation than others. For example, the human eye is more vulnerable to radiation than other human tissues. The retinal tissue is particularly susceptible to the radiation because (1) it has a dark color and readily absorbs the radiation and (2) the lens of the eyeball concentrates and focuses the laser beam on the retina, the focus being the hottest point in the beam.

Different laser wavelengths interact differently with body tissues. For instance, the retinal tissue is affected in decreasing intensity by ruby, argon, and helium-neon laser beams, in that order. The carbon dioxide laser is more effective in removing skin blemishes, such as warts, cancerous tissues, and other growths on the skin. Since the He-Ne laser does not damage the skin to the same extent as the ruby or Nd-YAG laser, it is used for radiotherapy (acupuncture) procedures and a technique for removing wrinkles from the dermal tissue. The effects of laser beams on different parts of the body will be discussed in the chapter on medical instrumentation.

As the laser beam impinges on the eyeball, part of the beam is prevented from entering the eye by the iris, a colored disk behind the cornea. The iris functions as an automatic photographic shutter and constricts when a high-intensity light impinges on the eye. The shutter action prevents part of the light from reaching the retina, where the image patterns are formed, and from injuring the retinal tissue when it is exposed for a short time. The retina is surrounded by a thin dark-brown choroid membrane containing arteries, veins, and pigment cells. Because of its dark color, the choroid membrane readily absorbs the radiation and must be protected by artificial means, such as goggles.

The retina is sensitive to all wavelengths, from 3000 to 9000 angstroms. It is also sensitive to some degree to infrared wavelengths beyond 9000 angstroms. Although the infrared and invisible ultraviolet rays are not detectable by the eye, the rays will affect the eye. In fact, the invisible ultraviolet rays are more dangerous to the eye than the visible laser rays because of the absorption of ultraviolet rays by the corneal and conjunctival tissues of the eyeball. If the eye is exposed to extreme amounts of ultraviolet rays, a cataract may form over the lens of the eye. The rays of the sun and welding arcs contain copious amounts of ultraviolet rays. Figure 1-19 shows the part of the laser beam that is absorbed by the corneal and conjunctival tissues.

In Fig. 1-19, the lens causes the laser beam to converge and focuses it on the fovea of the retina. From laser welding principles, it is well known that the focus of the beam is the hottest point. The laser energy density at the fovea is about 10^4 to 10^6 times more concentrated than that received by the cornea and the lens. The theoretically calculated retinal damage threshold is less than 10^{-6} J per square centimeter. It should be readily apparent now why a laser beam is hazardous to the eye.

Laser Safety Standards

The laser safety standards established by the U.S. Bureau of Radiological Health (BRH) are based on the potential hazards of various types of laser emissions and on the manner in which these hazards can be controlled. All lasers manufactured and sold in the United States must conform to these standards, which are assigned for five general laser classifications depending on the laser power emission. The standards further specify safety measures for the protection of the laser operator from dominant hazards. These classifications are as follows.

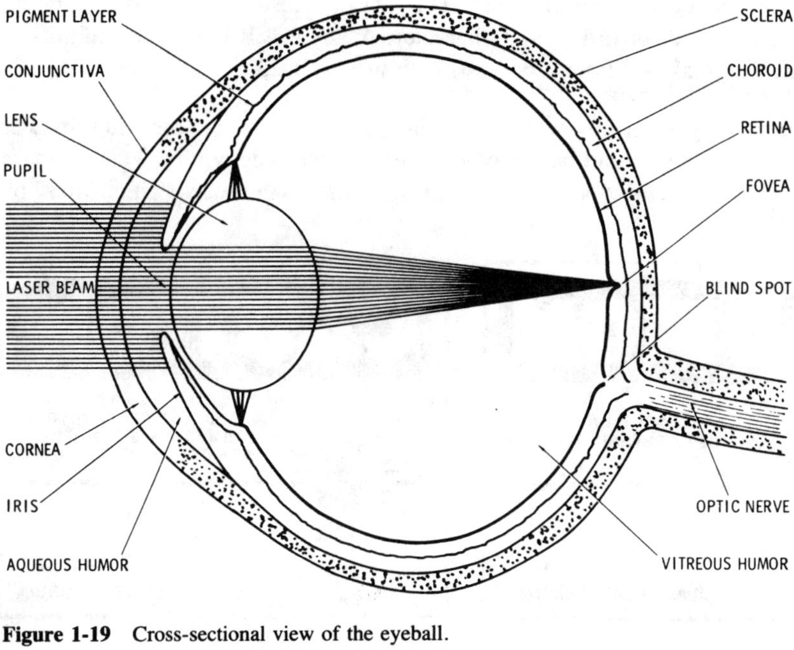

Figure 1-19 Cross-sectional view of the eyeball.

Class I: Exempt laser systems. The exempt laser systems are those that emit laser radiation that is not significantly hazardous to the eye, and when the total radiation is focused to a 1 mm^2 area on the skin for 24 hours no damage is incurred by the skin. Such lasers are gallium arsenide types, and there are no federal safety requirements for these radiations. The only requirement is common sense.

Class II: Low-power lasers. Class II laser systems are those that do not produce laser emissions in excess of 1 mW, and they include certain He-Ne lasers. The system must contain a pilot light, which must be automatically turned on when the laser emission is in progress. It must also have a shutter to turn off the laser beam when necessary. These laser radiations may produce retinal injury. The system must have a CAUTION label affixed on it, such as the label shown in Fig. 1-20.

Class III: Medium-power lasers. There are two subclasses in this classification: class IIIa and class IIIb. Class IIIa pertains to lasers with visible radiations (He-Ne type) that do not emit more than 4 mW of laser power. This type of laser can injure the eye when viewed directly. A CAUTION label is required on the system. Class IIIb lasers are those that do not exceed 50 mW of laser power and can produce injury to the eye. A system of this type must contain a pilot light, a beam shutter, and a DANGER label. These labels are shown in Figs. 1-21 and 1-22.

Class IV lasers: High-power lasers. These high-power lasers emit up to 50 W of radiation. Examples of this type are the ruby laser, Nd-YAG laser, Nd-glass laser, carbon dioxide laser, alexandrite laser, and the like. There are other lasers that emit in this range and are included in this category. These lasers present hazardous radiations, whether direct, reflected, or diffused in character. A DANGER label must be affixed on the system. The class IV laser label is shown in Fig. 1-23.

Class V lasers: Very high-power lasers. Class V lasers are enclosed high-power lasers, whose enclosures should be periodically examined to determine whether there is any leakage of laser emission. The system must have interlocks, which must be

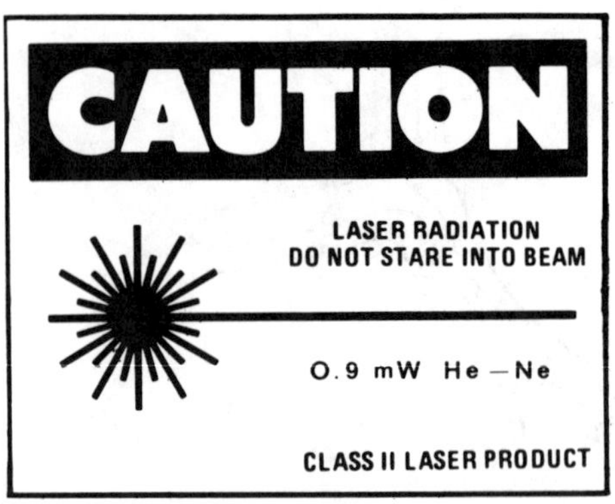

Figure 1-20 Class II laser label.

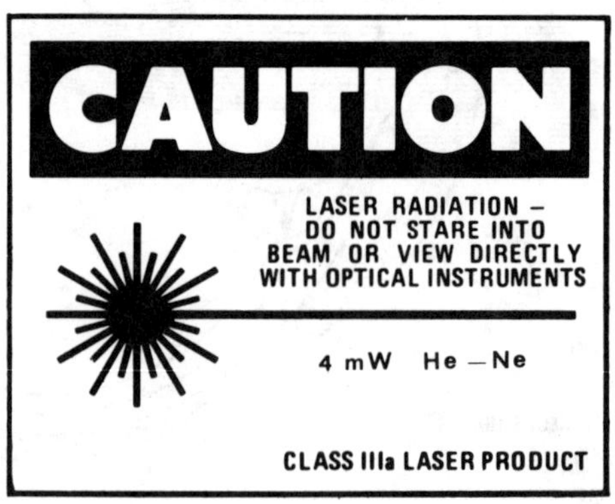

Figure 1-21 Class IIIa laser label.

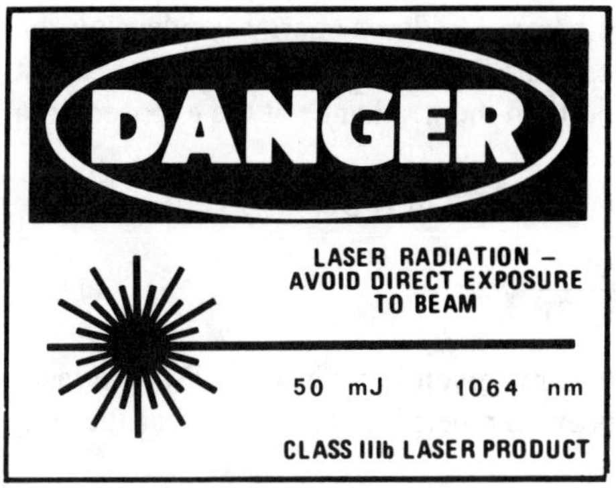

Figure 1-22 Class IIIb laser label.

Figure 1-23 Class IV laser label.

examined periodically to see that no accidental bypasses exist and the interlocks are firmly fitted when the system is assembled for operation. A HIGH-POWER LASER sign may be placed on the system.

A sixth classification strongly suggested by the author is CAUTION at all times regardless of the power rating of the laser system. There is always a possibility of human error leading to wrong marking on the system, fallen off or worn-out labels, and inadvertently fitted interlocks where leakage of the laser beam is a possibility. NEVER WORK WITH A LASER SYSTEM WHEN YOU DON'T KNOW ABOUT ITS POWER RATING AND WHETHER THE SAFETY CONTROL AND INTERLOCK RULES HAVE BEEN PROPERLY PROVIDED AND PRACTICED. When in doubt, always wear safety goggles of approved quality. Procure the latest safety rules and regulations from the BRH or LIA, study their contents, and put them into practice.

The laser system labels approved by BRH can be procured from the following firms:

Advanced Technology Publications, Inc.
119 Russell Street
Littleton, Massachusetts 01460

Rockwell Associates, Inc.
Post Office Box 43018
Cincinnati, Ohio 45243

Laser Hazard Calculations

To determine the hazard dominant in laser radiation, the manufacturer's specification must be consulted first. This specification lists the laser power output at the exit aperture, the beam divergence, and the beam diameter at the exit port, among many other parameters. Since laser damage is due to the amount of laser energy incident on the vulnerable part of a person (the eye being most vulnerable to the radiation), one must calculate first the irradiance at the exit port and then the irradiance at the exposed area at the distance considered.

The irradiance at the exit aperture of the laser system is given by

$$P_a = \frac{P}{\pi(D_a/2)^2} \tag{1-20}$$

where P_a is the irradiance at the laser exit aperture, given in watts/cm^2, P is the laser output power in watts, and D_a is the laser beam diameter in centimeters at the exit aperture.

The irradiance at a distance S centimeters from the exit aperture may be given as

$$P_S = \frac{4P}{\pi(D_S)^2} \tag{1-21}$$

where P_S is the irradiance in watts/cm^2 at the distance S in centimeters and D_S is the beam diameter in centimeters at distance S and is equal to

$$D_S = S\phi + D_a \tag{1-22}$$

where ϕ is the beam divergence in radians.

To illustrate these equations, let us assume that a certain laser system specification gives the laser power output as 1 mW, laser beam diameter as 1.5 mm at the exit aperture, and beam divergence as 1 milliradian. Answer the following:

1. What will be the beam diameter at the exposure area at a distance of 120 cm from the exit aperture?
2. What will be the irradiance at the exit aperture?
3. What will be the irradiance at the area 120 cm from the exit aperture?
4. What precautionary safety label is to be used?

The laser beam diameter at the exposure area at 120 cm from the exit port will be

$$\begin{aligned} D_S &= S\phi + D_a \\ &= 120 \times 0.001 + 0.15 \\ &= 0.27 \text{ cm} \end{aligned}$$

The irradiance at the exit aperture will be

$$\begin{aligned} P_a &= \frac{4\text{P}}{\pi(D_a)^2} \\ &= \frac{4 \times 0.001}{3.14(0.15)^2} \\ &= 5.66 \times 10^{-2} \text{ W/cm}^2 \end{aligned}$$

The irradiance at the area 120 cm from the exit aperture will be

$$\begin{aligned} P_S &= \frac{4P}{\pi (D_S)^2} \\ &= \frac{4 \times 0.001}{\pi (0.27)^2} \\ &= 1.75 \times 10^{-2}\ \mathrm{W/cm^2} \end{aligned}$$

Since the irradiance at 120 cm from the exit port is 17.5 $\mathrm{mW/cm^2}$, which is greater than 4 $\mathrm{mW/cm^2}$ as given for Class IIIa, a Class IIIb safety label must be used on the equipment. It would definitely be unsafe to look directly at the beam emission or at the reflections from any reflecting surface, recalling that the safe viewing level for any laser beam is $5 \times 10^{-5}\ \mathrm{W/cm^2}$, as given by the American Conference of Industrial Hygienists. Again, the author emphasizes that no one should work in an area where laser radiation is imminent without using safety goggles. Furthermore, no one should be admitted to the radiation area unless that person is wearing safety goggles approved by BRH.

1-6 LASER SYSTEM MAINTENANCE

Cleanliness and a dust-free environment are essential in a laser operating room. All equipment surfaces should be dusted off with a dust-catching soft cloth, and all fingerprints, to which dust and dirt adhere, should be removed to avoid charge accumulation. If necessary, the laser system should be covered with cloth when not in use. The main electric switch should be turned off when the laser system area is left. Other maintenance instructions are:

- All optical lenses, prisms, wedges, etalons, mirrors, and reflectors should be cleaned with lens tissue paper. They should then be wrapped in lens paper and stored in a box with a label to identify the lens diameter, lens power, and focal length. For the prisms and etalons, the vertex angle should be given on the label for each item.
- All lens holders, plates, and fixtures should be cleaned with soft cloth and stored in dust-proof drawers or shelves.
- Since for experimental work either glass or plastic lenses may be used, it is safer not to use any solvent, such as methanol or a similar solvent, to clean the surfaces of these elements whether they are glass or plastic. There is always a risk of using the solvent on the plastic lens or prism without knowing its constituent material. Use only lens paper to clean all optical surfaces.
- Never leave the location where the laser system is operating without first turning off the laser system; a better habit is to turn off the main power switch to the system.
- The system cord plug should be three-pronged, since the third prong (ground connection) will in most cases discharge the stray high-voltage components to the ground to make the system safe for the next operation or repair.
- Never let another operator make any change on the system you are using, or on your

setup, without your knowledge. For best results in this case, disconnect all parts, if practicable, before leaving the room, and put away each part on the particular shelf or in the drawer as assigned by the laser chief or operator.

- When any part of the system or optical component is not operable, always replace it with a new part in order to obtain reliable results from your experiment.
- The laser room illumination should be as bright as possible so that the pupils of the operator's eyes will be constricted at all times when the laser equipment is in operation in case a stray beam is accidently projected at the operator.

QUESTIONS AND PROBLEMS

1-1 What are electromagnetic waves? In what range of the electromagnetic waves are laser radiations? If the wavelength of a certain laser beam is 5000 angstroms, what is its frequency? What is Einstein's theory? State Lorenz's equation that relates the increase of mass with its velocity approaching the speed of light.

1-2 What is the meaning of "laser"? How is this word coined? What is a ruby rod? Describe the structure of a ruby rod. Why does light created in a ruby rod oscillate in the rod? What is a feedback mechanism with respect to laser emission? If both ends of a ruby rod are mirrored, how does a laser beam emerge from the rod?

1-3 In what variety of materials can laser radiation be produced? What are solid-state lasers? What are gas-type lasers? What are excimer lasers? What are dye lasers? What are semiconductor lasers? Can laser beams be treated as light waves through lenses, prisms, and other optically transparent materials? In what respect are laser beams different from light rays? Discuss.

1-4 If laser beams are made of light rays, why should their use require any safety precautions? Since the laser system operates from an electric source of high voltage, give several examples of electrical dangers against which one should be cautioned. Why one should take special precautions in handling the excimer material?

1-5 What radiation precautions should be taken when handling a laser system? How do different laser beam wavelengths interact with different body tissues? Which body tissue is more vulnerable than the other body tissues? Why do some laser beams damage the skin when they are incident on it while others don't? Discuss.

1-6 If ultraviolet rays are invisible, why do they injure the retina of the eye when they are incident on it? What is the laser density at the fovea, compared with that of the eye pupil, when a beam becomes incident on the eye? What is the threshold damage energy for the eye? Where is the hottest point of a laser beam? Why? Discuss.

1-7 Relatively speaking, how many classifications are there for laser safety standards? Describe them briefly. Does a laser beam diverge as it emerges from the exit port? How does one calculate the laser intensity at a particular distance from the exit port, if the initial intensity at the exit port is given? Give the equation for this.

1-8 A certain laser system has an output power of 2 mW and a beam diameter of 2 mm. What will be the intensity of the beam incident on a target 2 m from the exit port, ignoring the beam divergence? If the beam divergence is 1.5 mrad, what will be the intensity at the target?

1-9 How does one maintain the optical elements of a laser system that are in daily use? After completing work with a laser system, what procedures should be followed before leaving the operating room? What two main precautions should be taken before leaving the laser room to go home? Discuss.

1-10 What optical components are used in a laser system? What types of materials are used to transmit ultraviolet radiation? What types of materials are used to transmit visible radiation? What types of materials are used in transmitting infrared radiation? What materials can be used to transmit both ultraviolet and infrared radiations? What antireflection material is used for coating optical lenses? In what direction does a laser beam bend in a lens?

1-11 What types of optical mirrors prevail in laser systems? What is a Fabry-Perot cavity? What type of mirror is used as a rear mirror in a Fabry-Perot cavity? Can one employ a concave mirror for the output mirror? Which mirror is used for a single-mode operation? What are the mirrors made of? What coating is used on their surfaces? Is the coating of a mirror used on the first or second surface?

1-12 What is an optical prism? What types of prisms are used in a laser system? What is a Brewster-angle prism? What is its function when used in connection with a resonant cavity? What is a right-angle prism used for? What is a Littrow prism, and what is it used for? Why should there be a cooling system in a laser unit? Discuss.

1-13 Describe a typical, simple laser system, giving its various parts and their functions. What company manufactured the first ruby laser? Approximately what energy rating did it have? Could it be used for materials processing? Was the first laser system a pulsing or continuous type? What parameters constituted its electrical power? What percentage of the input power was converted into laser energy?

1-14 A certain laser system has six capacitor sections operating at 2 kV. If each capacitor section in the pulsing network provides 175 microfarads, what will the power output be in joules? How much laser energy can such a laser system produce? If this system operates for 1 second, what will be its power rating? If the system operates for 4 seconds, what will be the energy produced in joules?

1-15 A pencil laser beam is projected on a target and the point of incidence is marked on the target, which is positioned perpendicular to the part of the beam. The system is slightly rotated and another mark is made on the target where the beam is incident. If the angle through which the system is rotated is 20° and the distance between the two marks is 2 m, what is the distance between the target and the exit port of the laser system?

1-16 What is meant by the phrase focal length? If the focal length of a convex lens and a concave lens is the same, what is the difference between the two focal lengths? What will be the focal length of a convex lens that has an optical power of 0.10 diopter? What will be the magnifying power of a convex lens whose focal length is 6 centimeters? In Eq. (1-10), if the plus sign is changed to minus what type of lens focal length is being measured?

1-17 Can a photon travel with less than the speed of light? Is there any difference in speed between a ray with a blue wavelength and one with red wavelength, if both are traveling in water? If the wavelength of a ray of light passing through a prism is 4000 Å, what will its frequency be? What will be the momentum of a photon that collides with an electron at rest and imparts one-half its energy to the electron?

1-18 What is a photon emission threshold? At what quantum level with respect to its threshold can a laser photon be emitted? What is spontaneous emission? Can spontaneous emission occur below the threshold energy state? How does amplification of laser radiation occur in a Fabry-Perot cavity?

1-19 If a lens has radii of curvatures 25 cm and 30 cm and an index of refraction of 1.544, what will be its focal length? What will be its magnification power if it has a diameter of 10 mm?

1-20 A Gaussian laser beam of 8-mm diameter passes through a lens of 300-mm focal length and becomes focused on a target. If the focus diameter of the incident beam is 1.2 mm, what is the wavelength of the laser beam?

SELECTED SOURCES OF RELATED DATA

Cheo, P. K., ed. 1987. *Handbook of Molecular Lasers*. New York: Dekker.

Chester, A. N., ed. 1984. *Integrated Optics: Physics and Applications*. New York: Plenum.

Coherent Laser Systems. 1984. Technical Publications. Palo Alto, Calif.: Coherent, Inc.

Continental Optical Corporation. 1988. Laser and Optical Systems, technical brochure. Hauppauge, N.Y.

Dainty, J. C., ed. 1984. *Laser Speckle and Related Phenomena*. New York: Springer-Verlag.

Hora, H. 1981. *Physics of Laser-Driven Plasmas*. New York: Wiley.

Kompa, K. L., and J. Wanner, eds. 1984. *Laser Applications in Chemistry*. New York: Plenum.

Marathay, A. S. 1982. *Elements of Optical Coherence Theory*. New York: Wiley.

Penn, W. M., Assigner to P.R.C. Ltd. 1985. Gas Laser Tube Assembly. U.S. Patent No. 4,617,667.

Shannon, R. F., and J. C. Wyant, eds. 1983. *Applied Optics and Optical Engineering*. New York: Academic Press.

Simons, B. S., Itek Corp., Lexington, Mass. 1982. Laser Light Detection Systems. U.S. Patent No. 4,366,378.

Smith, B. 1986. Lamps for pumping solid-state lasers and characteristics. *Laser Focus* September, pp. 58–73.

Stark, H. 1982. *Applications of Optical Fourier Transforms*. New York: Academic Press.

Title, A. 1988. Infrared Polarizing Beamsplitters. U.S. Patent No. 4,733,926.

Wilson, J., and J. F. B. Hawkes. 1983. *Optoelectronics: An Introduction*. Englewood Cliffs, N.J.: Prentice-Hall.

Yoder, P. R., Jr. *Optomechanical Systems Design,* Vol. 9. New York: Dekker.

Young, M. 1986. *Optics and Lasers,* Vol. 8, Optical Science Series. New York: Springer-Verlag.

CHAPTER

TWO

PRINCIPLES OF LASER RADIATION

2-1 QUANTUM MECHANICS

Following Planck's introduction of the quantum theory, it became recognized that radiant energy had both an undulatory and a corpuscular nature. It behaved both as a wave motion and as energy quanta, now known as photons. From this latter theory arose the new quantum mechanics, which we will present with a brief introduction related to the quantum theory of radiation with special respect to the mechanism of laser radiation. The radiation energy can be exemplified by the use of Planck's quantum theory in which the energy of a photon is given by the relation

$$E = hf = \frac{hc}{\lambda} \tag{2-1}$$

where E is the photon energy in ergs, h is Planck's constant equal to 6.627×10^{-27} erg-sec, f is the frequency of the photon, c is the speed of light equal to 2.9979×10^{10} cm/sec, and λ is the wavelength in angstroms (cm $\times 10^{-8}$).

On the basis of Einstein's relativity theory, the mass m of a moving body is equivalent to the energy E, as shown in the relation

$$E = mc^2 \tag{2-2}$$

and the inertial characteristic of mass m is given by Eq. (1.4) as

$$m = \frac{m_0}{\sqrt{1 - (v/c)^2}}$$

In a discharge system, the kinetic energy of the recoil electron will be

$$E_r = \frac{mv^2}{2} = m_0c^2 \frac{1}{\sqrt{1-(v/c)^2}} - 1 \tag{2-3}$$

and the momentum of the electron in motion will be

$$mv = \frac{m_0 v}{\sqrt{1-(v/c)^2}} \tag{2-4}$$

in which mv is the momentum of the electron, given in gram-centimeters.

In order to elucidate where the electrons come from and how they produce laser radiation, we must review the atomic structure and its physical state under various electric field conditions. An atom consists of a nucleus containing neutrons and protons. The nucleus is surrounded by hypothetical orbits that contain definite numbers of electrons equal to the number of free protons (positive charges) in the nucleus. These electrons are known as extranuclear or orbital electrons. The electrons in the outermost orbit of the atom are conducive to combining one or more atoms together in a chemical reaction. When an atom dissociates in an electric field, the outermost orbital electrons are the first to be disrupted from the atom, resulting in the ionization of that atom.

An atom may dissociate to form equal numbers of negative and positive charges, known as ions. Such a dissociation occurring in an electric conducting medium can produce a flow of electricity. In metallic conductors, most of the electrons are projected at relatively great distances from the nuclei of the atoms. These electrons are known as free electrons; they occupy the interstices between the adjacent atoms and are shared by random atoms at random intervals. They are easily removed from the parent atom, or atoms, by attraction or repulsion in an electric field. Under these conditions, the electron acquires an average velocity of drift in the direction of the electric force gradient (voltage), thus producing a current of electricity; this force is depicted by the familiar electromotive force or electric potential. Since the electrons bear negative charges and are the only motive entities of an atom in a metallic conductor, it follows that the flow of current in metals is formed by the displacement of negative charges axially to the conductor.

According to the free electron structure of the atoms, substances may be classified as conductors or nonconductors (insulators), depending on the number of free electrons they contain. All substances contain free electrons to some extent, and those that contain a comparatively large number of these electrons are called conductors, whereas those that are devoid of free electrons, or contain only a few electrons per given volume, are known as insulators or dielectric substances. Conductors are usually metals and can readily transmit an electric current, whereas nonconductors are fibers, glass, Lucite, mica, rubber, etc., which cannot carry any current to a significant degree.

Electrical Units

The reader is expected to have some knowledge of electrical units, but for the benefit of those who have not had occasion to use this knowledge for a number of years we shall briefly review these units. The practical electrical units by which the production and transmission of an electric current are measured are the volt, the ampere, and the ohm. The volt is the unit of electromotive force behind the flow of electric current. The ampere is the unit amount of current driven by 1 volt in a unit resistor; an ampere is equal to 6.3×10^{18} or simply 10^{19} electrons moving per second in an electric conductor. The ohm is the unit electrical resistance offered by an electric conductor to the flow of 1 ampere of current driven by an electromotive force of 1 volt. The unit of electric power is a watt and is equal to 1 ampere times 1 volt; a watt expresses the rate of electric power flow per second. The electrical work performed in 1 second by a power of 1 watt is a joule, which also may be expressed as a watt-second. The electrical units may also be expressed in their multiples or in submultiples. For instance, 1 milliwatt of power, which is ordinarily used to define the power of ion laser radiations, is $1/1000$ of a watt. One kilowatt is 1000 watts; this unit is usually employed to specify the rating of higher-power systems. One kilovolt is 1000 volts and is used in multiples to energize an ion laser system. A microampere is one-millionth of an ampere. A megohm is a million ohms. For additional units of electrical measurements, the reader should refer to Appendix II at the end of this book.

Laser Radiation

In order to understand the mechanism of laser emission from atoms, we shall assume the Bohr model of orbital quanta, because of its ease of explanation, and discuss the behavior of atoms under various electric field conditions. According to Bohr concept, which is old but tenable, all atoms in the normal state exist at their lowest energy levels, which are variously known as stationary states, ground states, lowest quantum states, and at-rest states. In the stationary state, there is no energy change in the atom, since no energy is received or given off by the atom. For emission of laser radiation (or any electromagnetic radiation), an energy transition must occur in the quantum states. The energy may be imparted to the atom in several ways. For instance, if the atom is a gas, the gas can be confined in an electron discharge tube to which a high voltage has been applied. The voltage energy is transferred to the atom, causing it to become excited or ionized. Another way to transfer energy is to irradiate the atom by some electromagnetic radiation or by optical pumping (illumination with a high-intensity flashlamp). Before the atom can radiate, it must be excited or raised to a higher quantum state. The energy required to excite the atom is expended during its return to a lower quantum state. The wavelength of this radiation is inversely proportional to the energy expended.

All orbital electrons are confined to definite orbits or quantum states. Loss of one or more electrons from the orbits causes the atom to become ionized with a positive charge. All laser radiation occurs in the extranuclear quantum states; this involves an energy transition of the orbital electrons and their radiant behavior within the quantum states.

An atom at ground state, with its orbital electrons revolving in their normal, privileged orbits, possesses the lowest energy. When energy is applied to the atom, it becomes excited and the electrons move farther away from the nucleus and revolve in higher quantum levels. The process of transition from a state of lower energy to one of higher energy is known as excitation. The minimum voltage required to raise a normal atom to an excited state is called the excitation voltage. When the atom is energized beyond the excitation state, whereby one or more of its extranuclear electrons are completely removed from it, the atom becomes ionized positively. This process is called ionization. The nucleus then contains the same number of free protons as the number of electrons expelled from the atom. Such a process ideally occurs in an electron discharge tube impressed by a relatively high voltage, such as 500 V or higher. The process of ionization in a discharge tube is illustrated in Fig. 2-1.

In Fig. 2-1, when a high voltage is applied across the discharge tube filled, for instance, with neon gas, the orbital electrons move farther from the atomic nuclei and finally leave the influence of the nuclear electrostatic charge; the atoms that have thus lost electrons become positively ionized. The ionized atoms then are propelled to the negative electrode, the cathode, and the electrons move to the positive electrode, the anode. This process continues as long as the voltage is sustained across the discharge tube. During the operation of the tube, the neon gas in the tube glows with a characteristic red color. Since the atom has a strong tendency to return to its ground state, when the voltage is removed the atom regains its ground state and ceases to emit further radiation. The radiating photon possesses a frequency and wavelength dependent on the transition energy as given by Eq. (2-1).

Since the gas in our discussion was neon, the emitted radiation has a wavelength of 6328 Å, which is within the visible spectral range. Had the atom been argon, the radiation would have had a wavelength of typically 4500 Å in the blue spectrum; at this wavelength, the argon atom would have a higher frequency than that of neon and, therefore, higher photonic energy. Thus, taking the visible radiation as a reference spectrum, the low frequencies will emit in the red or infrared region of the spectrum and the higher frequencies will emit in the blue or near the ultraviolet region. An expression for the photonic energy hf may be given as

$$hf = \frac{Ve}{300} = \frac{1}{2}mv^2 \tag{2-5}$$

where h is Planck's constant, f is the frequency of the photon, V is the voltage applied to the laser tube, e is the charge of the electron equal to 4.8×10^{-10} esu (electrostatic units), m is the mass of the electron equal to 9.0×10^{-10} g, and v is the velocity of the electron in centimeters per second. The quantity 300 is the conversion factor of V to practical volts.

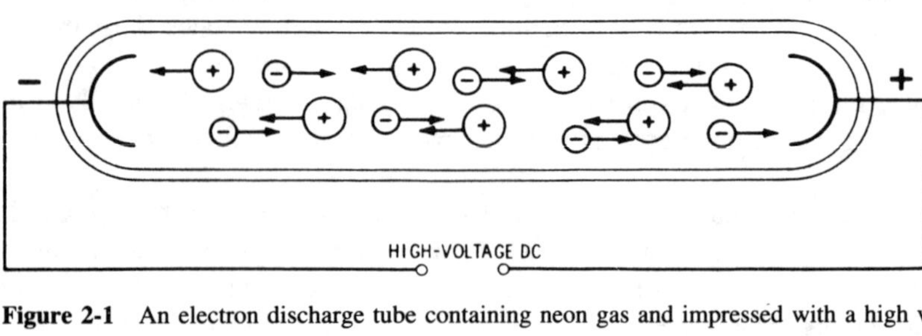

Figure 2-1 An electron discharge tube containing neon gas and impressed with a high voltage.

It must be noted that these three quantities, hf, $Ve/300$, and $\frac{1}{2}mv^2$, are respectively the photonic energy, the voltage energy applied to the electron, and the kinetic energy gained by the electron in the discharge tube. We may now write the photonic energy hf in terms of its wavelength:

$$hf = \frac{hc}{\lambda}$$

and

$$f = \frac{c}{\lambda} \tag{2-6}$$

where c is the velocity of light in centimeters per second and λ is the wavelength of the photon in centimeters. To convert the wavelength from centimeters to angstroms, the quantity is multiplied by 10^8.

2-2 ENERGY TRANSITIONS

In Eq. (2-5) it was assumed for practical purposes that the voltage energy applied to the electron was completely converted to photonic energy. In practice, this is not entirely true; some energy from $Ve/300$ is used up in exciting or ionizing the atom. Therefore, the minimum energy required to remove an electron from the outermost orbit of a given atom may be written as θ, which is equal to $V_1e/300$ and is known as the work function, where V_1 is the ionization potential in volts for that quantum state. With an applied voltage of $Ve/300$ on the electron in a discharge cavity, the kinetic energy with which the disrupted electron from the atom will move in the cavity may be given as

$$\frac{1}{2}mv^2 = \frac{Ve}{300} - \theta = \frac{Ve}{300} - \frac{V_1e}{300} \tag{2-7}$$

The velocity with which the electron will traverse the discharge cavity is given by

$$v = \sqrt{\frac{2(Ve - V_1e)}{300m}} \tag{2-8}$$

In the preceding discussion, the kinetic energy of the electron is given in ergs; 1 erg is equal to 1 dyne-centimeter, the dyne being the smallest force in physical relations. However, in practice, an erg is too small to express the laser energy. Therefore, the term generally used is either a joule or a watt-second, a joule being equal to 10^7 ergs. It is convenient to use the term joule for pulsed laser energy and watt for continuous-wave laser power. The relationship between the two terms may be given as follows:

$$\text{joules} = \text{watts} \times \text{seconds} \tag{2-9}$$

$$\begin{aligned} \text{watts} &= \text{joules/seconds} \\ &= \text{amperes} \times \text{volts} \end{aligned} \tag{2-10}$$

or, using the letter P for power in watts, J for energy in joules, and s for time in seconds, the relation can be written as

$$P = \frac{J}{s} \tag{2-11}$$

For example, if a laser system is rated at one-half kilowatt and operates at a pulsing rate of 80 pulses per second, what will be the energy in joules per pulse? What will be the total energy in joules per second?

The energy in joules per pulse will be

$$\begin{aligned} J &= Ps \\ &= 0.5 \times 1000 \times \frac{1}{80} \\ &= 500 \times \frac{1}{80} = 6.25 \text{ joules} \end{aligned}$$

The total energy in joules per second will be

$$\begin{aligned} J &= \frac{0.5 \times 1000 \times 80}{80} \\ &= 500 \text{ joules} \end{aligned}$$

If only two pulses were used (out of 80 pulses) for welding a certain metal, what would the welding time be in milliseconds?

$$\begin{aligned} s &= \frac{1}{80} \times 2 \text{ seconds} \\ &= \frac{1}{80} \times 2 \times 1000 \text{ milliseconds} \\ &= 25 \text{ milliseconds} \end{aligned}$$

2-3 CHARACTERISTICS OF BEAM PROPAGATION

A laser beam can be converged and diverged, reflected from reflecting surfaces, refracted in media of various densities, polarized by means of polarizing materials, diffracted by diffraction plates or slits, and collimated by means of optical lenses.

Convergence and Divergence of a Laser Beam

A laser beam can be converged to a point by means of a biconvex or planoconvex lens or by means of a spherical mirror. The beam can be focused to a point sharply because it is coherent and monochromatic. In contrast, a light beam, which has many wavelengths, does not focus sharply, and each wavelength travels at a different velocity through a given transmitting medium. Figure 2-2 illustrates a laser beam propagating

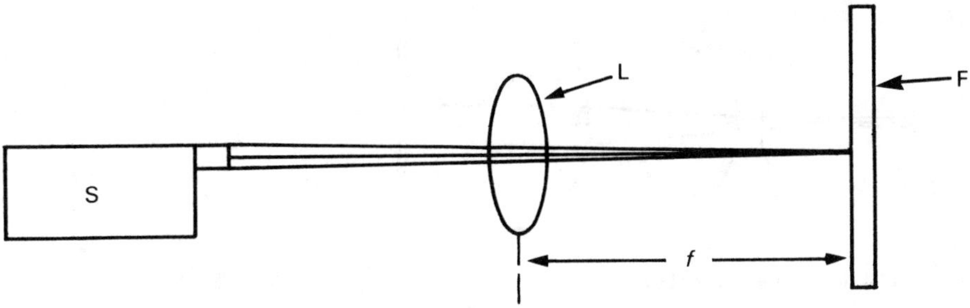

Figure 2-2 Laser beam convergence to a focus.

from a laser source S and after passing through the biconvex lens L it becomes focused on a screen F. The point of incidence on the screen will be sharp because of the coherent nature of the laser beam.

The principal purpose of focusing a laser beam is to increase the intensity of the beam by concentrating it into a small area. For instance, 1 joule of laser energy emanating from a ruby laser rod of ¼-inch diameter, when projected on a thin metal film, produces no effect on the metal. However, when the same beam is concentrated by means of a convex lens to a focus of, for instance, 5 mil diameter, the beam intensity increases more than 2000 times. The beam is now capable of drilling a hole of 5 mil diameter in the metal—for instance, in a kovar film 1 mil thick. Accordingly, by employing a convex lens, a laser beam can be used to cut, weld, drill, and remove metal from metallic objects positioned at the beam focus. In the medical field, such a focused beam can be used to cauterize, cut, and sterilize human tissue; in dentistry, it can be used to glaze human teeth in restorative treatment.

In Fig. 2-3, a biconcave lens is shown causing a collimated laser bean to diverge as it leaves the lens. The optical principle involved here, as well as in the biconvex lens illustration, is the bending (refraction) of the laser ray toward the thickest portion of a lens. In the case of a convex lens, the thickest portion of the lens is located at its principal axis; in a biconcave lens, the thicker part is located from its midpoint to its periphery, so the transmitted beam diverges away from the center of the lens (toward the lens periphery). Thus, a convex lens is called a converging lens and a concave lens is known as a diverging lens.

In Fig. 2-3, the collimated laser beam projecting from the laser generator falls on the concave lens, from which it divergently propagates and becomes incident on a

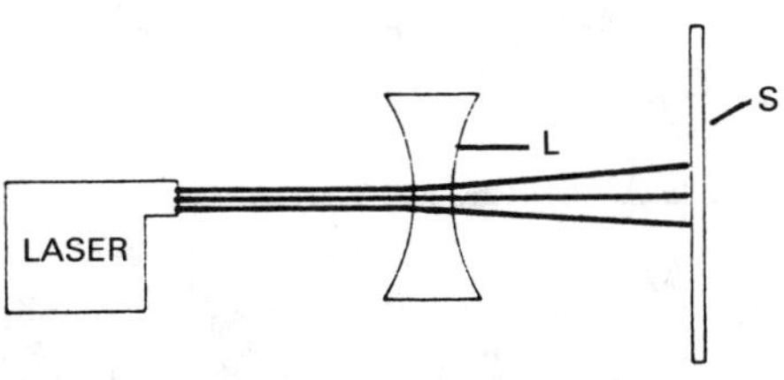

Figure 2-3 Divergence of a laser beam.

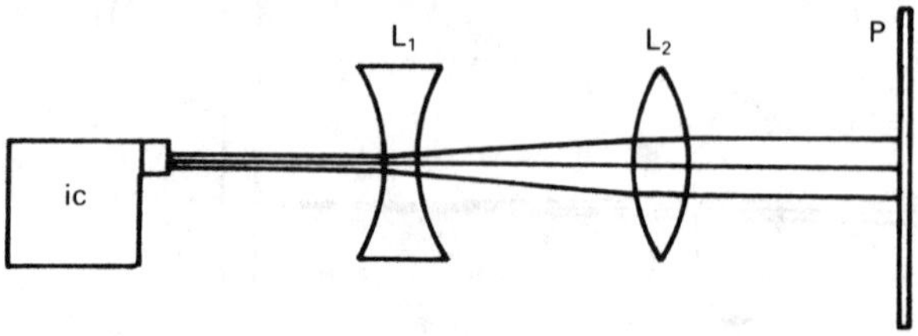

Figure 2-4 Collimation of a laser beam.

screen, S. A concave lens is rarely used by itself in laser work except in illuminating a large area with a laser beam, such as in viewing holograms or the like. This type of lens is generally used in combination with a convex lens for collimating a laser beam, such as that shown in Fig. 2-4.

Collimation of a Laser Beam

Figure 2-4 illustrates the arrangement of a concave lens and convex lens combination for collimating a laser beam. The laser beam in the figure projects from the laser source, LG, and falls perpendicularly on a biconcave lens, L_1, preferably at its center. As the beam leaves lens L_1 it diverges and becomes incident on lens L_2, which is a biconvex lens placed at a distance one focal length from the principal plane of lens L_1. The laser beam emerges from lens L_2 in a collimated format (all rays are parallel to each other) and is projected on a screen, P.

The true parallelism of the beam can be demonstrated by holding the graph paper P at right angles to the projecting laser beam. As lens L_2 is displaced to and from lens L_1, the area covered on the graph paper by the projected beam becomes larger or smaller, or fuzzier or sharper. A sharp delineation of the beam on the graph paper indicates that the distance between lens L_1 and lens L_2 is exactly right for producing 100% collimation of the laser beam. If the radii of curvatures of both lenses are numerically equal, the distance between them is one focal length.

The applications of a laser beam collimated in the manner described are many. In industrial work, the laser beam must first be collimated before it is focused on any surface for welding, drilling, scribing, or resistor trimming. A sharp focus is necessary in order to obtain a desired focus diameter. To produce a focus from the collimated beam, a biconvex or planoconvex lens is required, as illustrated in Fig. 2-5.

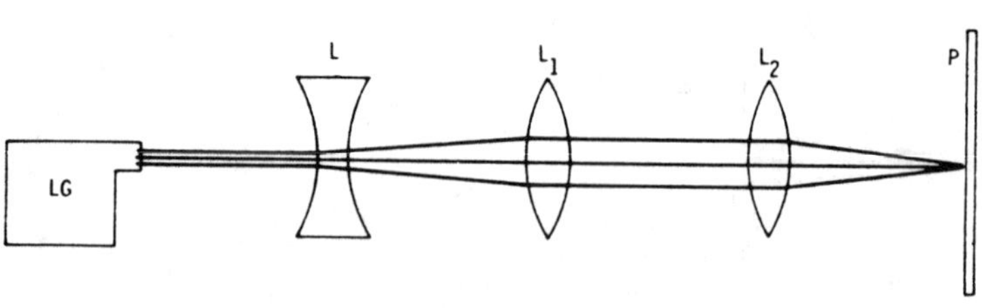

Figure 2-5 Focus formation of a collimated laser beam.

A collimated laser beam of pencil diameter or smaller is used in measuring distances, target ranging, surveying, aligning large structural equipment parts or building sections, and laying pipe in the ground. In aerial optical communication, a collimated laser beam will travel farther than a noncollimated laser beam and without losing significant intensity. Depending on weather conditions and the optical elements used, the laser beam can be projected as far as 50 miles in a line-of-sight situation. When the first collimated laser beam was projected to the moon, it diverged to an incidence area of 2 miles in diameter due to the scattering effect of the contaminated atmosphere. Thus, 100% parallelism through the atmosphere over long distances is not obtainable in practice.

Refraction of a Laser Beam

It was shown earlier that a laser beam travels in a straight line in a homogeneous, transparent medium regardless of the optical density of the medium, such as air, water, or glass. However, when a laser beam passes from a medium of one optical density to a medium of another optical density at some angle other than normal (perpendicular), the beam bends by an amount that depends on the optical density of the medium being traversed. For instance, if a laser beam L traveling in air is incident at an angle *i* with respect to the normal to the surface penetrated, such as water, the beam bends (refracts) in water toward the normal by slowing its speed of propagation in water, because of the higher optical density of water than that of air. On leaving the water and traversing the air of lower optical density, the beam bends away from the normal, as illustrated in Fig. 2-6.

The emergent laser beam L′ travels in air in parallel to the incident beam L. Thus, the change of direction of travel of the incident laser beam in water is due to the change of index of refraction, which is the term applied to the ratio of speeds of light rays

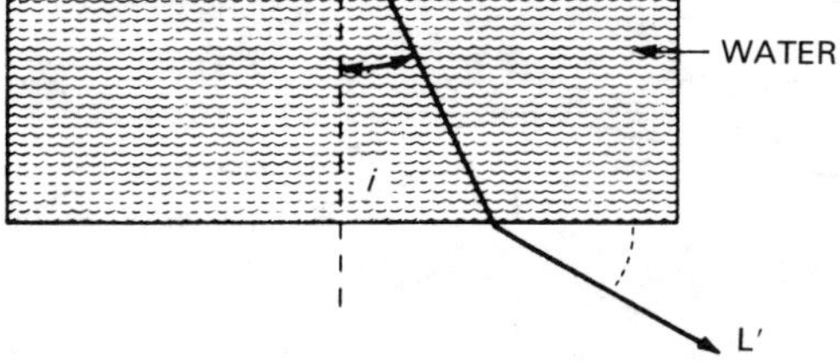

Figure 2-6 Refraction of a laser beam through water.

passing through media of different optical densities. Accordingly, a light beam going from a medium of lower density (air) to one of higher density (water) bends toward the normal, and a light beam going from a higher-density medium to a lower-density medium bends away from the normal. The angle at which the laser beam is incident on the surface with respect to the normal is called the angle of incidence. The angle at which the beam bends with respect to the normal in the second medium is called the angle of refraction.

There is a definite relationship between the angle of incidence and the angle of refraction. For instance, when the sine of the angle of incidence i is divided by the sine of the angle of refraction r, the resultant factor is a constant for the medium traversed by the beam. The relation may be given as

$$\frac{\sin i}{\sin r} = n \tag{2-12}$$

where the constant n is known as the index of refraction of the medium penetrated, such as water in the present case.

Since the speed of light is proportional to the optical density of the medium traversed, the above relation can be written

$$n = \frac{\sin i}{\sin r} = \frac{v_1}{v_2} \tag{2-13}$$

where v_1 and v_2 are the velocities of the incident and refracted laser beams, respectively.

Chromatic Aberration

Chromatic aberration is the dispersion of the laser light on the focused area due to the presence of different wavelengths in the projected polychromatic laser beam. Because the index of refraction of a lens varies with different wavelengths, the lens bends the multiline beam wavelengths by different amounts, resulting in loss of sharpness of the focus, as illustrated in Fig. 2-7.

In Fig. 2-7, only the wavelengths at the extreme spectral ranges are illustrated (exaggeratedly), with other spectral colors falling between these two wavelengths.

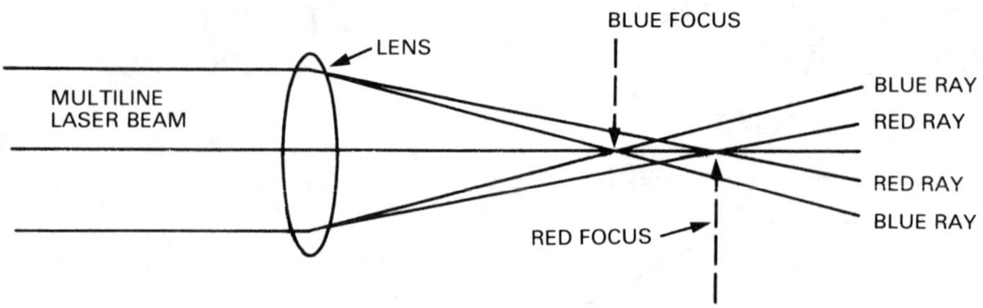

Figure 2-7 Diagram illustrating chromatic aberration.

Figure 2-8 Correction of chromatic aberration effect.

Although in laser work the focus of the beam usually does not suffer from this effect because of the monochromaticity of the laser beam, with ionic lasers, such as argon or krypton lasers, the beams are multiline, and with mixed-wavelength laser beams it may be necessary to correct the lens to eliminate the effect of chromatic aberration.

Chromatic aberration in a lens can be eliminated by combining a convex and a planoconcave lens; one lens annuls the dispersion of the other, as illustrated in Fig. 2.8a. Alternatively, two thin lenses may be combined to achieve this result, as shown in Fig. 2.8b.

An equation illustrating another method of correcting this optical anomaly may be given as

$$\frac{1}{f} = \Delta_1(n_1 - 1) + \Delta_2(n_2 - 1) - \Delta_1\Delta_2(n_1 - 1)(n_2 - 1)S \tag{2-14}$$

where f is the combined focal length of the two thin lenses, Δ_1 and Δ_2 denote the differences of the radii of curvatures of the lenses, n_1 and n_2 are the indices of refraction of the lenses, and S is the distance between the two lenses. Such a situation is seldom encountered with laser spatial-filtered beams, since the spatial filter removes most of the dispersed rays from the projected beam.

Absorption of a Laser Beam

In the subject treatment of this section the author does not refute the concept of population inversion as the basis of stimulated emission, but considers the terminology too abstruse for understanding by the uninitiated in the field of laser technology. For this reason, the author examines the intra-atomic lattice energy reactants in the light of the quantum concept of atomic excitation by photons. Thus, inference is made to atomic excitation as being brought about by inelastic collisions of electrons with atoms in a discharge tube, or photons encountering an ensemble of atoms absorbing energy in accordance with the well-known relation

$$I = I_0 e^{-\mu x} \tag{2-15}$$

where I is the intensity of the absorbed radiation, I_0 the intensity of the incident radiation, μ the coefficient of absorption, and x the thickness of the absorbing material. By measuring the radiation intensities I_0 and I and knowing the thickness of the material, one can calculate the coefficient of absorption of the lasing element, as illustrated in Fig. 2-9.

Cumulative excitation of atoms in a lasing element also proceeds exponentially

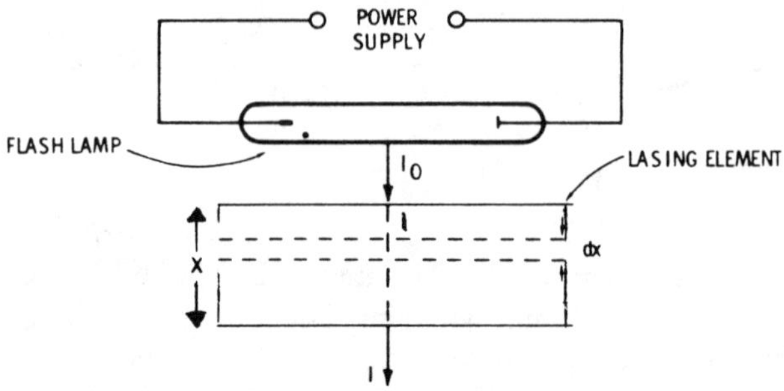

Figure 2-9 Interaction of photon energy with an atom of a lasing element.

and follows the laws of absorption of photonic energy. Taking the lasing cavity between two parallel mirrors, as shown in Fig. 2-10, the intensity of the radiation will grow in accordance with the relation

$$N = N_0 e^{ax} \tag{2-16}$$

where N is the total number of photons formed by the photonic oscillation process, N_0 the initial number of photons, a the coefficient of ionization (amplification) of the energy-absorbing atom, and x the space traversed by the initial photons in producing the stimulated radiation in the lasing element.

At absolute zero temperature, all atoms are at ground state. When their temperature is raised to the ambient temperature, the energies of some of the atoms will be raised to higher quantum states, so any external radiation whose frequency corresponds to the transition energy between the excited quantum states can be absorbed by those atoms. Thus, atoms that absorb monochromatic radiation from the resonant cavity will be excited, and they will ultimately return to the ground state with emission of spontaneous or stimulated radiation dependent on the irradiating energy and the coupling energy with which they are excited. At moderate to high-level excitation, the energy of the atoms will decay to the ground state with emission of laser radiation of frequencies corresponding to the energy absorption states.

It must be clear at this point that spontaneous emission of radiation from atoms is not stimulated radiation. For example, as shown in an earlier section, the emission

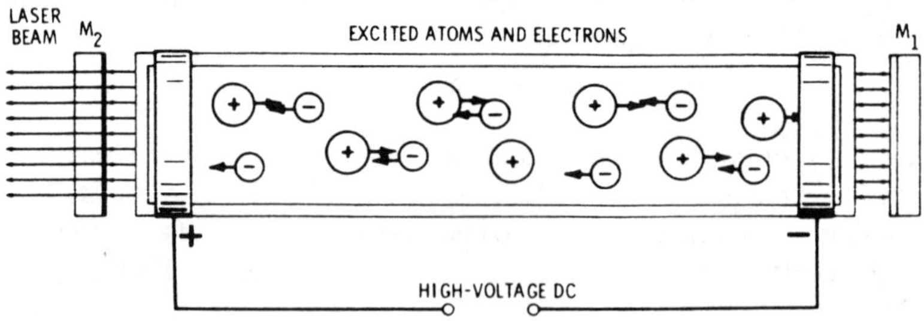

Figure 2-10 Laser emission by photon-atom interaction.

from a neon discharge tube is spontaneous emission, as is the fluorescence from phosphorescent minerals illuminated with ultraviolet light. The latter type of fluorescence can occur even in total darkness in cadmium sulfide or calcium sulfide that has absorbed light energy from sunlight or daylight. Their emission sometimes occurs for periods of seconds, as the ions decay to their ground states through a series of metastable states with emission between the metastable states.

In contrast to spontaneous emission, a stimulated transition takes place when negative frequency absorption by the atoms occurs. This condition can be attained when the frequencies of the stimulating photonic energies are in coherence with those of the absorbed energies of the stimulated atoms; in such a transition both the phase and the amplitude profiles are in coincidence, respectively, with those of the energy-transmitting photons. Stimulated emission occurs at a rate in direct proportion to the intensity of stimulation by the applied radiation, or the excess of stimulating radiation over the absorbed radiation energy. Thus, the resonant cavity functions as a feedback amplifier, producing a net closed-loop gain by the coherent oscillation between the end mirrors.

The increase of photons due to cumulative stimulation may be calculated to a close approximation. For uniform optical radiation of the atoms between the parallel mirrors, as shown in Fig. 2-10, assume that N_0 is the number of irradiation photons incident per second from a flashlamp with radiation wavelength compatible with the absorption characteristics of the lasing element, and N is the number of photons in the net stimulation per second. If we denote the number of photons produced by the lasing element per centimeter path length by a, then the number of photons produced by the irradiating photons in distance dx will be $a \times dx$. Thus, the increase in the number of photons within the path dx will be $Na \times dx$, or

$$dN = Na \times dx$$

and

$$\frac{dN}{N} = a \times dx \tag{2-17}$$

Integrating Eq. (2-17) from one plane of the lasing element to the opposite plane, we obtain

$$\frac{dN}{N} = ax\ dx + C$$

where C is a constant taken as the exponent of the Napierian logarithm base e and is equal to $\log_e$ of N_0.

When $x = 0$, $N = N_0$, and we can equate

$$\log_e N = ax + \log_e N_0$$

where

$$\log_e N - \log_e N_0 = \frac{\log N}{\log N_0}$$

hence

$$\log \frac{N}{N_0} = ax \qquad \text{and} \qquad \frac{N}{N_0} = e^{ax}$$

or

$$N = N_0 e^{ax} \tag{2-18}$$

where a is the coefficient of stimulation of the particular laser element.

Since the number of photons stimulated per second is directly proportional to the intensity of the emitted radiation thus formed, we have

$$I_0 \approx N_0 \quad \text{and} \quad I \approx N$$

therefore

$$I = I_0 e^{ax} \tag{2-19}$$

where I_0 is the intensity of the irradiating light and I is the total stimulated radiation from the lasing element.

Equations (2-18) and (2-19) indicate that by increasing the distance between the end mirrors the stimulated radiation will also be increased. However, there are limitations beyond which any increase in the resonant cavity length will produce a decrease in the laser output. This is due to a number of factors, some of the important ones arising from the erratic thermal stability of the system, intermittently occurring lens effects in the lasing element, and capricious heating of the lasing element beyond its thermal threshold.

From the preceding discussions, it is clear that optical amplification can result from the negative absorption of quantum frequencies within the domain of a positive feedback reaction provided by the end mirrors of the resonant cavity. The energy of the quantum state must be equal to or in excess of the threshold frequency of the quantum energy before a laser action occurs. The mechanism of optoatomic interaction occurs when energy absorption, excitation, radiation, oscillation, and amplification, in that order, are achieved within the optical cavity. Some of the energy from the optical source that does not contribute to the stimulation mechanism will dissipate thermally, elevating the temperature of the lasing element and thereby reducing the coherent emission; for this reason, certain types of laser systems require cooling. The quantum energy is absorbed by the processes of collision of electrons with atoms, such as those in the interaction of excited gaseous states. The interaction in solid-state lasing elements, such as ruby or neodymium-doped hosts, proceeds by transfer of quantum energy from the irradiating source to the excited atomic lattice.

The photonic oscillation in the loop embodying the lasing element and the end mirrors provides the amplification in accordance with the exponential Eq. (2-19), and the axial modes with discrete optical frequencies thus produced give rise to a monochromatic and unidirectional laser output because the feedback loop in this case is an integral number of wavelengths in the stimulated radiation, associated with a factor $C/2L$, in which C is the radiation speed and L is the distance between the mirrors.

Stimulation occurs at a rate proportional to the applied radiant energy and is further dependent on the energy profile of the latter as well as the degree of negative absorption by the radiating atoms. The frequencies that exhibit optical coupling with the excited atoms in the electric field, as in a gaseous discharge medium, or optical energy, as in a solid-state lasing material, give rise to stimulated transitions. Gaseous elements are more susceptible to this type of laser action than solid-state laser elements because the gaseous medium is more accessible for excitation by an electric field than the optically excited laser rods and dye lasers, where the transfer of energy is entirely dependent on the coupling compatibility between the light frequencies and the lasable atoms. In this respect, lasing elements in so-called four-level stimulation states, such as Nd-YAG, respond more readily to the interaction than those having three-level excitation states, such as ruby. This subject will be taken up in detail in a later chapter.

Diffraction of a Laser Beam

We have referred to the laser beam as being rectilinear in its propagation through space. A laser beam, like a light beam, bends while traversing an optical medium that contains obstructive elements in its path. The beam can bend over an obstruction it encounters as well as when it passes through a slit. This bending is known as diffraction. When the distance to be traveled by the laser beam is relatively great and the medium traversed is contaminated by dust particles, smoke, water vapor, and the like, the beam diffracts, and this may cause deviation of the beam from its course to the objective. We may classify diffraction, with reference to laser beams to be discussed in this chapter, as Fraunhofer diffraction and Fresnel diffraction. Either of these modes of diffraction may be applied in spectroscopic work for analysis of materials or for measuring wavelengths and intensities of radiation. Although these modes will be discussed more fully under pertinent subject headings, in this section we shall at least define the difference between the two modes.

Fraunhofer diffraction patterns are obtained when a collimated laser beam is incident on an obstacle or aperture and projected on a screen as parallel rays, which may be observed by means of a telescope. The term diffraction applies to the spreading of the laser beam into the region behind the obstacle, and it occurs in accordance with Huygens' principle. It is a form of interference phenomenon. Taking a slit as the obstacle, if the slit is broad and the beam falls on a screen at a distance from the slit, only a broad strip will be seen on the screen. However, if the slit is narrow, a diffraction fringe ensemble will appear on the screen. When a lens is placed in the path of the rays distant from the obstacle, dark and light lines are observed which determine the wavelengths characteristic of the element emitting the radiation.

A Fresnel diffraction pattern is observed when the laser beam or the screen is at a finite distance from the slit; the slit or the aperture is usually circular, and the pattern formed on the screen is circular, as one would observe in Newton's rings. By using a Fresnel biprism, interference fringes that are the same as Fraunhofer fringes can be obtained. A lens of Fresnel diffraction type may be employed in a Fabry-Perot cavity as the rear mirror when the second surface of the lens is 100% mirrored; a multiline laser spectrum can be separated into its component wavelengths and the separated wavelength amplified before leaving the exit aperture. Fresnel optical elements are important in laser work and will be discussed further under pertinent headings.

2-4 LASER BEAM COHERENCE

When two or more laser rays are moving in the same direction with the same amplitude and wavelength, they are said to be coherent. We will discuss two types of coherence—temporal coherence and lateral coherence. Figure 2-11 shows two waves moving in the same direction. The waves have the same wavelength and advance with the same speed, so the wave peaks occur at the same time and at the same respective points. These waves are said to be in phase and therefore coherent, in both time and undulations.

In Fig. 2-12, two laser waves are propagating parallel to each other, are equal in wavelength and amplitude, and are in phase with each other. Any corresponding points, *A* and *B*, on the two waves are moving in the same direction and are positioned at right angles to the propagation axis, executing undulatory motions in phase with each other. Because these arbitrarily chosen points on each wave at the corresponding positions are in phase and normal to the path of the propagation, the waves are said to be laterally coherent.

Although there is no 100% coherence, a laser beam is the closest thing to a coherent beam, in addition to being unidirectional. When any two laser beams are temporally coherent but a constant phase difference exists between them, a condition of interference

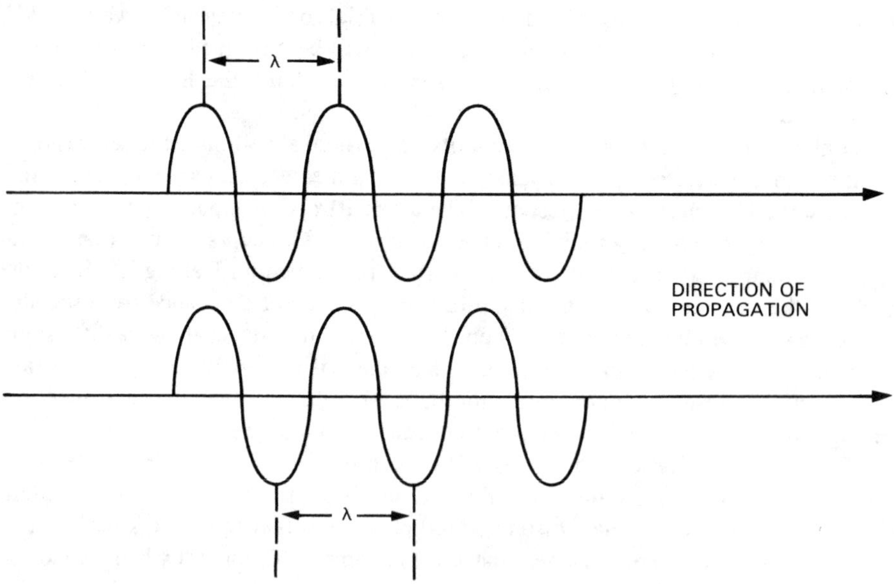

Figure 2-11 Temporal coherence of laser waves.

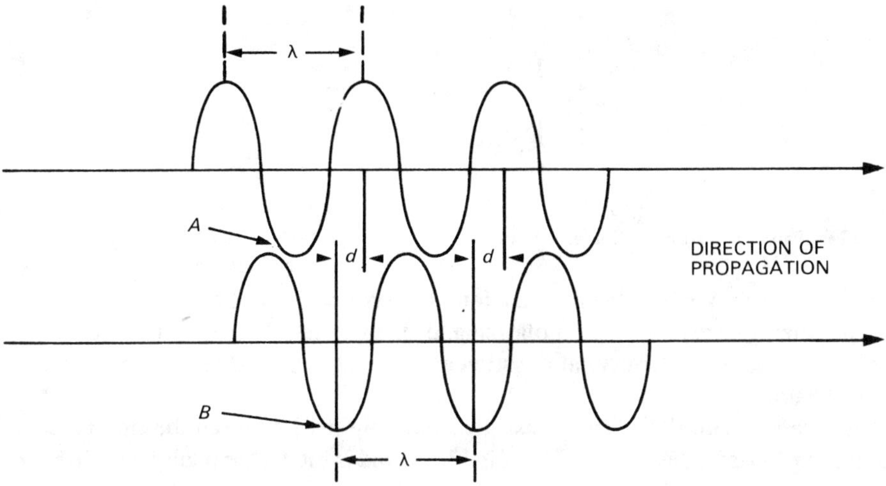

Figure 2-12 Lateral coherence of two laser beams.

results, the degree of interference depending on the amplitudes and their positions above or below the propagation axis. These phenomena will be utilized in our future discussions.

2-5 LASER BEAM POLARIZATION

We have already learned that a laser beam is a form of electromagnetic radiation. As the term implies, the radiation is composed of electric and magnetic field vectors executing vibrations at right angles to each other and to the propagation axis. We may consider the electric field as having an E vector and the magnetic field an H vector for purposes of discussion. Thus, if the E vector of the laser beam oscillates at a constant rate in a fixed plane during the propagation of the beam, the beam is known as plane polarized, or linearly polarized. A plane polarized laser beam can be obtained by use of a polarizing element, such as a film of Polaroid material. This effect is familiar to most people who wear Polaroid sunglasses on a sunny day at the beach. These glasses remove the extraneous light reflections from the view of the wearer to avoid eye strain.

To determine whether the beam from a laser system is partially polarized or unpolarized, the simple test setup shown in Fig. 2-13 may be employed. Since laser beams are coherent in general, one would expect to find some polarized effect in all laser beams as they emanate from the output mirror. In the setup in Fig. 2-13, a square white cardboard, C, is placed about 24 inches from the exit aperture A of a He-Ne laser source. When the laser beam is projected on the cardboard, a central bright area surrounded by a pink halo, which constitutes the fluorescent rays scattered from the laser beam, appears on the cardboard. If a polarizer disk P_1 is introduced between the aperture A and the cardboard C, a slight darkening of the projected laser beam occurs. As disk P_1 is rotated on its axis, the darkening effect becomes more dominant. When

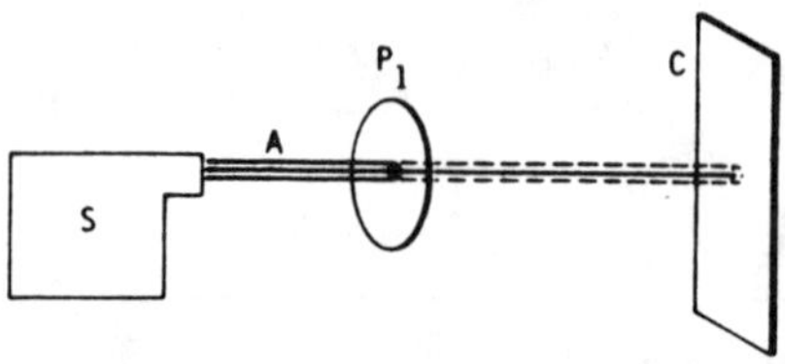

Figure 2-13 Partial polarization of a laser beam.

this effect is at a maximum, the polarization of the beam is at right angles to the planes of the polarizing crystals in the polarizing disk material. This effect is due to the orientation of the polarizer crystal constituents at right angles to the polarized rays of the laser beam.

If a second (neutral) polarizer disk P_2 is now inserted between the first polarizer P_1 and the cardboard C, as illustrated in Fig. 2-14, and is rotated gradually, an extinction of the laser beam will occur on the cardboard. The disappearance of the laser beam incidence indicates that the beam has been totally polarized. In this simple experiment, the second polarizing disk may be called an analyzer, and the transmitted beam intensity is a function of the rotary displacement of the analyzer. At any angle θ between the polarization planes of the two polarizers, the beam intensity transmitted through the analyzer may be given as

$$I = I_0 \cos^2 \theta \tag{2-20}$$

where I is the beam intensity transmitted through the analyzer, I_0 the beam intensity of the linearly polarized radiation incident on the polarizer, and θ the angle between the two polarization planes.

When the E vector polarization is parallel to the H vector polarization, or simply the analyzer orientation is parallel to the polarizer crystal orientation, we may write the relation as

$$A_r^2 = A^2\left(1 - \sin^2 2\phi \sin^2 \frac{d}{2}\right) \tag{2-21}$$

where A_r is the resultant amplitude of the beam, A the amplitude of the oncoming beam, ϕ the angle between the E and H vector polarizations, and d the phase difference of the emergent beam.

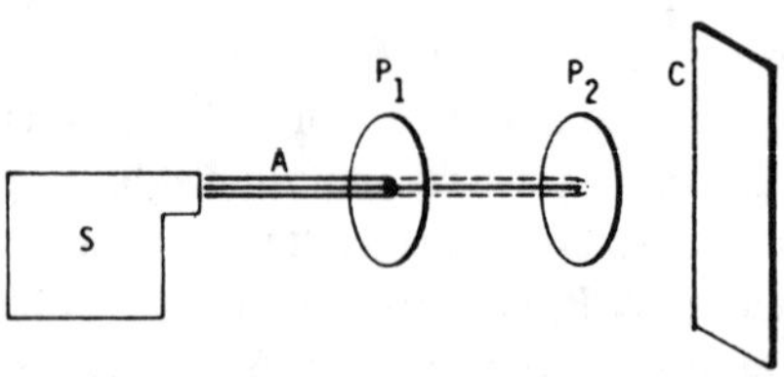

Figure 2-14 Total polarization of a laser beam.

In the event that a polarization retroaction occurs, a situation in which the polarizer and the analyzer are crossed at 90°, Eq. (2-21) reduces to

$$A_r^2 = A^2 \sin^2 2\phi \sin^2 \frac{d}{2} \tag{2-22}$$

In working with lasers, the occasion will arise that the system requires retardation plates, which are crystal plates used to induce phase differences between the E and H components of the beam. The path difference introduces a phase difference measured by

$$d = \frac{2\pi}{\lambda}(n_E - n_H)t \tag{2-23}$$

From this equation, one can calculate the thickness of the retardation plate required to produce the desired phase difference between the polarized beams. A half-wave plate then produces a phase difference of angle π, and a quarter-wave plate introduces a phase difference of $\pi/2$. Both of these plates are used extensively in laser work, as we shall have occasion to study in later chapters.

Some of the most common applications of polarized laser beams are in nondestructive testing of materials or parts that have been subjected to stress or strain. Defective honeycomb joints in aircraft wings, welded junctions, areas around bolted structures that have cracked due to stresses, and surface anomalies of metallic materials can be effectively examined by using polarized laser beams and analyzing the resultant effects with an analyzer. In some cases, vibrating the part by mechanical means or with ultrasonic waves will aid in accurate localization of the defective area. A system using this technique, which was devised by the author at Lockheed Aircraft Company, is shown in Fig. 2-15.

The system consists of a housing (1) with a coherent light source (He-Ne laser), a polarizer, and a biconvex lens with a cylindrical extension (4), which illuminates the surface of the workpiece (honeycomb wing, 5). The polarized laser beam projects divergently into a chamber (3), under which the honeycomb wing is located for inspection. An analyzer with a magnifying lens is located in a tube (2) through which the surface of the workpiece can be examined during the operation of the system. A camera (6) provided with a wide-angle lens is located on the roof of the housing to photograph the defect in the workpiece. In order to visualize the defective area, the workpiece is vibrated. This action is accomplished by means of an ultrasonic generating device (7) having a variable frequency that can be adjusted from 1000 to 100,000 hertz.

In operation, the workpiece is fed into the system by means of a continuously moving belt (not shown) and is vibrated at a certain frequency until the defective area becomes visible through the analyzer in tube 2. The frequency then is kept constant for the entire operation. Photographs of the defect are made while the workpiece is gradually moving under the housing. After one side has been inspected or photographed, the workpiece is passed under the housing a second time for examination of the other side.

It is important to note that when a laser beam is reflected from a target, as in range finding and missile guidance operations, or when the beam has traversed an optical

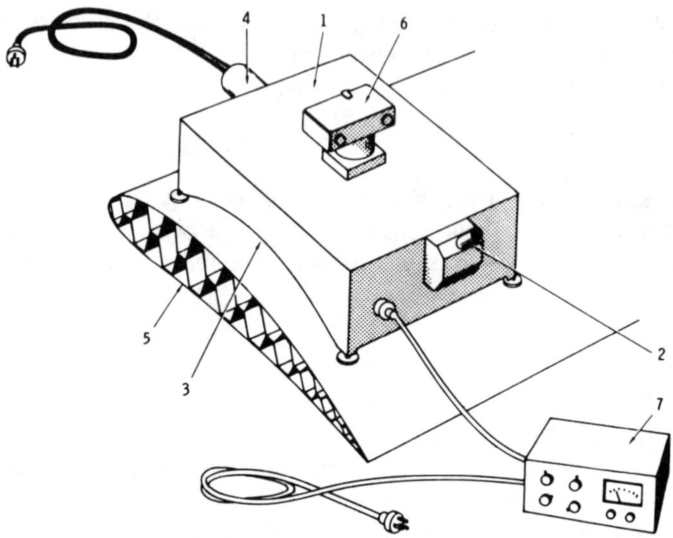

Figure 2-15 Simplified diagram of a system used for inspecting honeycomb structures.

system, the emergent beam is linearly polarized; this condition avoids fluctuations in the laser output power and enables better control of the beam intensity. It also permits the radiation to be used in modulation, as in communication and data transmission. Since the laser beam diverges to a slight extent as it emerges from the exit port of the laser system, provision should be made to reduce the divergence to a minimum to obtain as small an area as possible on the target. To achieve this condition without use of an optical collimating system, the beam emerging from the exit aperture must have a large waist diameter controllable by the design of the resonant cavity.

2-6 Q-SWITCHING A LASER BEAM

A counteraction exists between the laser optical pumping source and the laser stimulation. That is, the laser pumping tends to increase the excitation of the atoms and thus the laser output in the Fabry-Perot cavity, but the stimulated laser action tends to decrease this effect and thus restrict the power level of the resultant oscillation. The process of Q-switching (Q-spoiling) controls the resonance in such a manner that the counteraction between the optical pumping and the laser action is eliminated.

In Q-switching, when the atomic population inversion (stimulated radiation) is increasing cumulatively, the atomic excitation in the Fabry-Perot cavity is made nonresonant during the pumping of the laser element. This action increases the population inversion in excess of normal resonance and maximizes the inversion state. At this point, when the excited atomic system is suddenly made resonant by removing the Q-switching mechanism, the stored energy in the excited levels of the atoms (such as chromium ions in ruby rod) is suddenly stimulated to emission of laser radiation at a peak power. Because the conversion occurs in nanoseconds, the stored high energy bursts out of the laser element in a giant pulse with a laser power up to gigawatts per square centimeter in certain types of Q-switching. The giant-pulse technique can be employed for ruby, Nd-YAG, Nd-glass, alexandrite, and other solid-state lasing elements, with the exception of semiconductors that have pulse times too short for Q-switching.

Several methods for Q-switching exist: the Kerr cell technique, the rotating-prism technique, the liquid-bleaching technique, and the ultrasonic technique. Other techniques are also being developed, one of which utilizes the reflective property of semiconductor surfaces; the reflectivity of the surface increases with intense irradiation, and this property may be used to produce Q-switched giant pulses.

Figure 2-16 illustrates one of the most common Q-switching techniques, the Kerr cell method or electro-optical technique. In Fig. 2-16, the power supply section energizes the flashlamp F to illuminate the lasing element L. One end, M_1, of the lasing element is coated to make it partially reflective, and the other end, T, is uncoated. Mirror M_2 is positioned with its reflecting surface parallel to the face of the lasing element L as well as to the partial mirror M_1. Kerr cell K, containing the optically active liquid nitrobenzene, is positioned between the mirror M_2 and the lasing element terminal T and is connected to a direct-current source, DC, from which it receives a high voltage such as 10 kV. The cell is positioned so that when it is activated, the direct current field through it is at 45° with respect to the principal axis of the lasing element L. This condition produces a 90° rotation of the polarization plane of the laser beam as it traverses the cell.

In operation, the high voltage is applied to the Kerr cell to activate it; this action closes the optical switch. A fraction of a second later, the flashlamp F is triggered to pump the lasing element L. After about ¼ millisecond, the voltage to the Kerr cell is

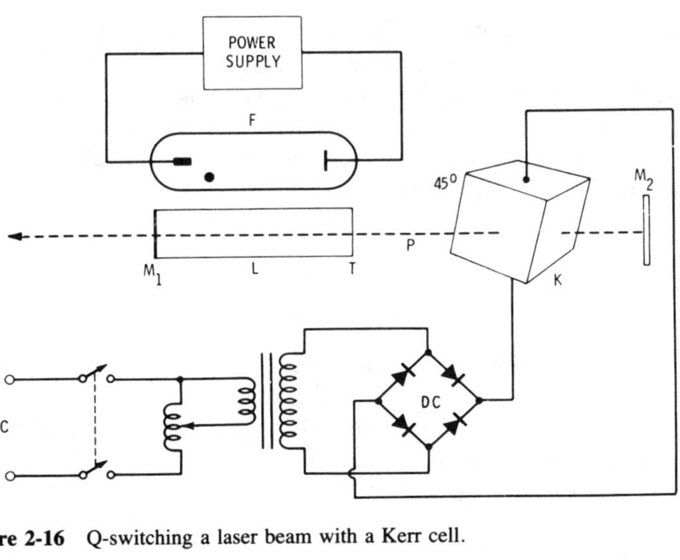

Figure 2-16 Q-switching a laser beam with a Kerr cell.

turned off and the optical switch (laser path between M_2 and M_1) is opened. An avalanche of photons in the optical cavity of the lasing element L begins to oscillate between M_2 and M_1. When the threshold excitation in the resonant cavity exceeds that which normally exists, a giant laser pulse bursts out through M_1 of the lasing element L. The radiation is then known as Q-switched.

Akin to the technique of Q-switching is cavity dumping, a process in which extremely short laser pulses can be obtained; the width of the pulses is determined by the length of the optical cavity. The technique consists of shutting off the cavity oscillation momentarily and then rapidly dumping the accumulated radiation by opening the optical shutter by means of an acousto-optic signal. This process is described in detail in Chapters 8 and 14.

2-7 TYPES OF LASER RADIATION

Since each of the types of laser radiation will be taken up in considerable detail in other chapters, we will give only an introductory discussion here to familiarize the reader with these types of radiation in a general way. The spectrum of laser radiation extends from x-ray waves at the shortest-wavelength end to the far-infrared region of the electromagnetic spectrum. Between these two extremes are the ultraviolet radiation near the x-ray end and the visible radiation between the ultraviolet and the infrared.

For practical purposes, we may arbitrarily categorize laser radiation into four general classifications: infrared, visible, ultraviolet, and x-ray laser radiations.

Infrared Radiation

Infrared radiation varies in wavelength from 7800 to 100,000 Å and slightly beyond. Special optical materials, such as quartz, sapphire, Pyrex, germanium, KRS-5, and lanthanate glass, are used for transmitting this radiation. The principal infrared lasing substances are carbon dioxide, Nd-YAG, carbon monoxide, nitrous oxide, YLF, Nd-glass, and GaAs and its many derivatives. The intensities of these materials vary from a harmless level to the most powerfully hazardous levels.

Visible Radiation

In this class, there are many lasing elements that produce visible radiation from 3900 to 7800 Å in wavelength. Some of the principal lasing materials in this category are ruby, He-Ne, Ar, Kr, Xe, He-Cd vapor, and others. The optical materials used in the transmission of this radiation are sodium glass, lead glass, Pyrex, silicon glass, Lucite, and many other visually transparent materials.

Ultraviolet Radiation

This radiation is also invisible and its wavelengths vary from 3900 Å down to long x-ray wavelengths such as 1100 Å. The lasing elements in this category include, among others, Ar, NH_3, He-Cd vapor, HF, ArF, Xe, KrF, Kr, and HeF. The optical elements that can transmit ultraviolet radiation are fluorite glass, Pyrex, quartz, sapphire, lanthanate glass, and others.

X-Ray Lasers

A variety of lasing materials have been tried to produce x-ray lasers. Among these, possibly Nd-YAG is the most popular active element. X-rays are produced by bombarding metals of high atomic weight, such as copper, with 0.5-J Nd-YAG radiation pulses to form a metal vapor plasma adjacent the surface of the metal. The x-ray laser power generated thus far is about 2 W, which is the maximum that can easily be produced by electron bombardment of the target. Although excimers can also be used to produce x-ray lasers, they are too weak to yield any useful output power. It is thought by the present experimenters that in time they will be able to create x-ray lasers with powers in excess of 100 watts by employing higher levels of laser pulse power, possibly at repetition rates of up to 10 Hz. The first application of current x-ray lasers is for lithography for photoresist exposures. Another characteristic of the x-ray laser is that it can be converged to a pencil beam, whereas radiographic or therapeutic-type x-rays cannot be converged. X-ray lasers are also being tried to preionize electric-discharge gas lasers in investigations of atomic fusion.

2-8 AMPLIFICATION OF LASER RADIATION

Ordinarily, laser rod materials provide the highest output of laser energy for solid-state-type lasers. The amount of laser energy produced depends on the coupling efficiency with which the laser rod is optically pumped. The optical pumping effectiveness is directly proportional to the surface area of the laser rod irradiated. Thus, the energy output is principally determined by the size, geometry, and type of the laser rod material. A physical factor to be overcome with such rod materials is the pump-radiation reflection loss at the laser rod–air interface, which becomes particularly significant with high-energy laser generators. The coherence of directionality of the emitted laser radiation is governed mainly by the geometry of the rod, and hence its cross-sectional area, as well as by increasing the optical pump energy. Although the concept at first appears very attractive, there are inherent limitations on the extent of increase that is possible:

1. The optical pump energy needed to produce metastable quantum states and to make the rod operable as a tubular body is effective only to a limited depth into the laser rod.

2. The core of the rod absorbs the heat from the optical pump energy; this heat tends to counteract the formation of excited metastable states.
3. The laser excitation at the core of the rod is reduced to nothing, and the thermal radiation, emitted radially from the center of the rod, further favors the reduction of emission.
4. On exceeding a certain rod length and the resultant emission gain, a condition of spontaneous avalanche occurs, which adversely affects the directivity of the radiation.
5. The formation of a nonuniform temperature condition within the rod creates distortion in the optical path, and this effect increases with heterogeneity in the rod material.
6. An attempt to obtain high laser power by increasing the volume and optical pump energy can ultimately lead to destruction of the laser rod.

These problems must be overcome before amplification of the laser output can be realized.

Several methods have been developed to overcome these adverse effects in order to achieve a high gain in laser emission level. One simple method has been to drill an axial bore centrally in the rod. This method alleviates the production and radial emission of heat from the core to the periphery of the rod, thus reducing the threshold of the metastable states and increasing the laser emission level. Another method takes advantage of the geometry of the rod by constructing it in a quadrangular or square cross-sectional dimension; this design increases the surface area exposed to the optical pump illumination and reduces the thermal effects of the core by more efficient radiation of the core heat. The difficulty in this method has been in obtaining optically flat surfaces and homogeneous rod materials, and progress has been slow and costly. Therefore, an ideal lasing element for obtaining amplified laser radiation would be one that does not require strictly homogeneous laser material, eliminates the thermal effects in the core of the lasing element, offers the greatest possible surface exposure to the optical pump radiation, eliminates the lens effect in the lasing element, and produces a uniform temperature distribution resulting in homogeneous metastable states throughout the laser rod.

General Electric Laser Amplifier

The suggested characteristics for the lasing element have been obtained by employing thin plates of the lasing element of quadrangular or oval configuration and arranging the plates in an axial array perpendicular to the axis of the array or at some suitable angle for eliminating surface reflection losses from the laser elements. At least two companies, General Electric Company and American Optical Corporation, have introduced this system arrangement, and later variations of the scheme have been successfully put into use. One example of the general design for such a scheme is illustrated in Fig. 2-17.

In Fig. 2-17, a rectangular housing H is provided with two flashlamps, L_1 and L_2,

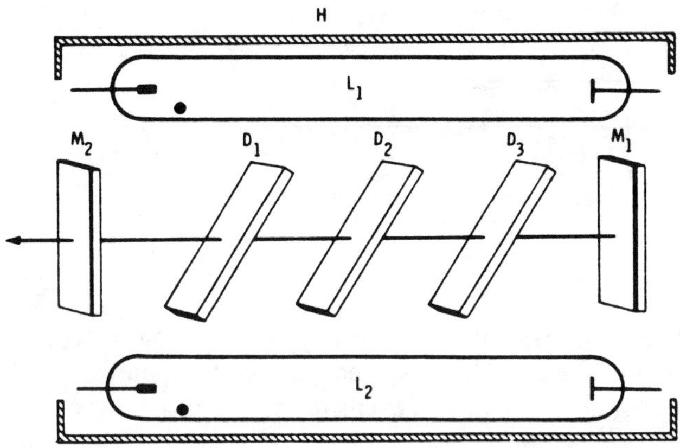

Figure 2-17 Amplifier disks arranged at Brewster angles.

for illuminating the laser disks, D_1, D_2, and D_3, which are arranged at Brewster angles and parallel to each other to eliminate plate-to-air interface reflection losses. For practical purposes, the tangent of the Brewster angle is equal to the refractive index of the laser disk. At each end of the principal axis of the plate arrangement are positioned the familiar mirrors M_1 and M_2. Mirror M_1 is heavily coated, and mirror M_2 is partially coated to permit the resultant laser beam to travel to the exterior of the resonant system. The laser disk is made of silica glass containing 3% neodymium. For a pulsing mode of operation xenon flashlamps are used, and for a continuous-wave mode electric flashlamps are used. The radiation from this laser system is collimated and highly coherent; it is emitted with a wavelength of 10,640 Å in the infrared spectrum. Such a system may be considered a module, and several such modules may be optically coupled in series to produce a high-power laser generator. The system is water-cooled; a stream of deionized distilled water circulates through it at a rate of 1½ to 2 gallons per minute per module. The operation of such a system is similar to that of a ruby or Nd-YAG laser generator.

ILS Laser Amplifier Systems

International Laser Systems, Inc. (ILS) manufactures systems with laser amplifiers cascaded in series with the main laser generator to yield a considerable gain. Since the laser systems are modular, several amplifiers can be added in series to a basic system in accordance with the power requirement of the user. The basis system, shown in Fig. 2-18, consists of a Nd-YAG laser rod 6.35 mm in diameter and 57 mm in length. The rod is immersed in a fluid contained in a close-coupled optical pump cavity provided with a xenon flashlamp for optical pumping of the laser rod.

The system operates with a frequency-doubled Nd-YAG lasing element emitting at 5320 Å. Two mirrors—a spherical mirror with 100% reflectivity used as the rear mirror and a flat mirror with 50% reflectivity positioned at the exit port—constitute the

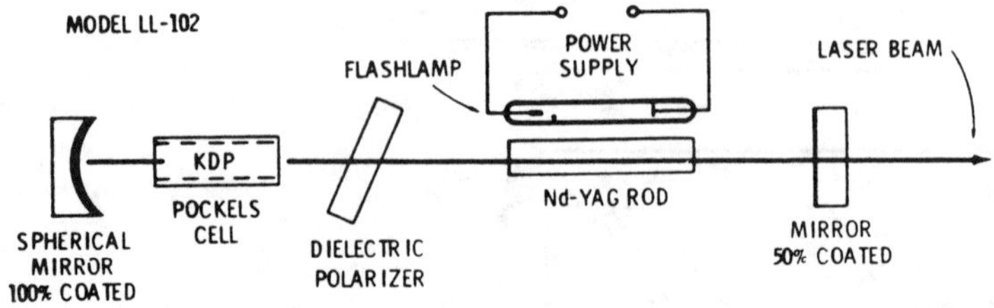

Figure 2-18 ILS basic laser oscillator system.

feedback resonant cavity. These components are movably mounted on an optical carriage. The system is powered with a 115-V, 15-A alternating-current source, which also furnishes electric power to the liquid-to-air heat exchangers used to cool the optical system.

The ILS basic laser system operates Q-switched and produces from 0.1 J at 20 pulses per second (pps) to 1.2 J at 20 pps. The system can be cascaded with laser rod amplifiers, shown in Fig. 2-19, to extend its capability to 2 J per pulse at 10 pps. Optical elements may be added to the system to produce outputs in the infrared, red, orange, green, and ultraviolet, from microjoules per pulse to several joules per pulse. It is also possible to obtain 200 mJ per pulse at TEM_{00}[†] operation at 10,640 Å with the use of a beam expander, or 60 mJ per pulse at TEM_{00} operation at 5240 Å (frequency doubled). The pulsewidth varies from picoseconds to milliseconds up to a repetition rate of 400 pps. The Pockels cell, provided with the system, is operable at high repetition rates and at low insertion loss. The extinction ratio is 100:1 at 10,640 Å, a power density of 300 mW/cm^2, and a nominal switching time of 15 nsec. Typical applications of the system are in high-speed photography, timing chemical reactions, holography, and work requiring precisely controlled coherent light pulses.

If the lasing elements in the basic system and the amplifier are ruby, uranyl glass may be used as a Q-switch in place of the Pockels cell. Uranyl glass is opaque to the ambient radiation. Thus, when a plate of this material is positioned between the rear spherical mirror and the ruby rod and both are simultaneously illuminated by a xenon flashlamp, the plate becomes transparent to the ruby radiation and an avalanche of laser radiation passes through it to the rear mirror. As a certain threshold is attained, an abrupt oscillation takes place in the optical cavity with the ensuing generation of a giant radiation pulse, as in a Q-switch.

[†] TEM_{00} signifies transverse electromagnetic mode. It is also known as Gaussian mode or fundamental mode.

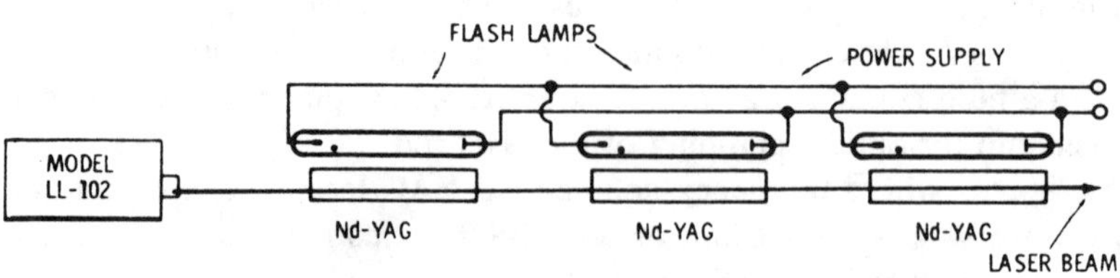

Figure 2-19 ILS laser amplifier system using three Nd-YAG laser rods.

CLC Laser Amplifier

Control Laser Corporation (CLC) has developed an oscillator amplifier that produces the highest possible laser power using a Nd-YAG laser rod generating 400 W of laser radiation at a wavelength of 10,640 Å. The system, Model 438, operates in a pulsed mode and produces 29 J/pulse at a 0.65 msec pulsewidth; its repetition rate is 100 pulses per second. The output beam diameter is 10 mm, with an exit port divergence of 10 mrad. A krypton arc flashlamp optically pumps each of the amplifier heads and has a lifetime of 10^6 pulses.

A schematic diagram of the Model 438 laser system is shown in Fig. 2-20, with its subsystems arranged in the optical axis and sequentially mounted on an optical rail forming the system unit. The system is provided with a resonator consisting of a Nd-YAG oscillator and a Nd-YAG amplifier. The oscillator has a mirror with 100% reflectivity forming the rear reflector and a front mirror of infrared-transparent glass with a partially coated surface to transmit the radiation to the amplifier through a shutter. Two krypton flashlamps irradiate the oscillator and the amplifier separately but in unison. Each of these laser heads is enclosed together with its respective flashlamp in an elliptical cavity, which is plated with nickel-gold for maximum reflectivity and durability.

The laser output beam of the oscillator is fed to the Nd-YAG amplifier, in which the radiation level is doubled to 400 watts in continuous mode. The amplified beam passes through a mechanical shutter, becomes expanded in the upcollimator, and then impinges on a dichroic mirror positioned at 45° with respect to the optical axis of the emergent beam. The reflected beam from the dichroic mirror is focused on the workpiece by means of a convex lens.

A stage illuminator is provided on the optical rail framework for illumination of the specimen, whose image is reflected through the dichroic mirror on a cube prism, which splits the image beam into two projections. One of the projected beams impinges on a binocular viewer and the outer beam falls on the detector of a television camera, which displays the image of the specimen on the television screen. The binocular viewer is used to localize the specimen under the laser beam.

The laser system receives its power from a 45-kVA, three-phase power source. It is provided with a high-voltage pulse-forming network charged through two silicon-controlled rectifiers (SCRs), which automatically regulate the line voltage. The optical cavity is cooled by deionized water circulating through the laser head at a rate of 15 gallons per minute (gpm) and maintained at a temperature not to exceed 80°F. The applications of this system include cutting, welding, drilling, and heat-treating metal alloys. Plastic and textile materials can also be cut or drilled with this equipment.

2-9 LASER BEAM–SHARING FACILITY

Research laboratories that use high-power laser beams tunable to various wavelengths are presently sharing the laser power generated at one central station. This is accomplished by distributing the radiation through optical channels to various research buildings for the use of individual laboratories. This scheme enables efficient utilization of

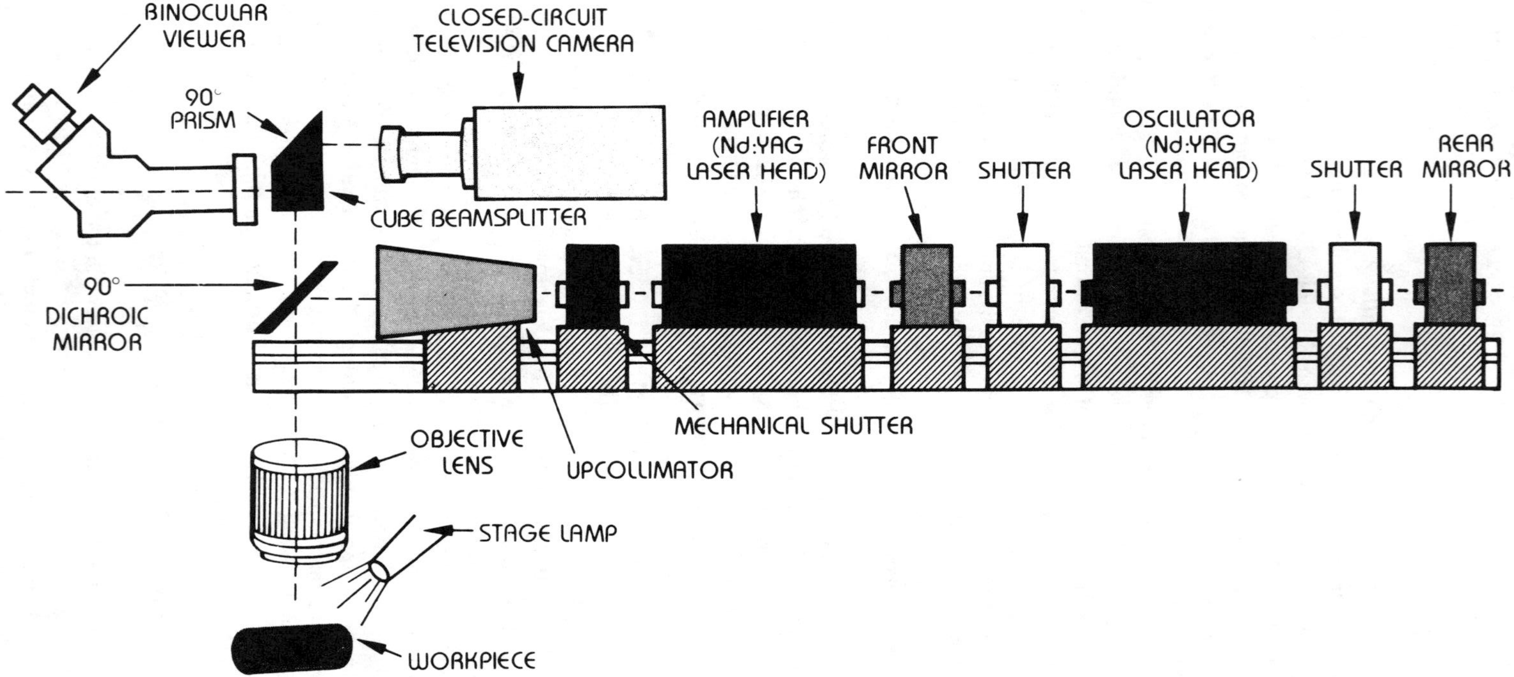

Figure 2-20 The CLC Model 438 laser system.

costly laser beams tunable for the various types of work in hand. The powerful laser beam generated at the central station may be transmitted directly or split by semitransparent mirrors into the desired number of channels. The beam can be tuned or frequency-multiplied, as desired, and varied to the full rating of the system, all operations being performed through the programmed use of a computer. Such laser beam sharing is currently being employed by Sandia and Livermore National Laboratories and others.

One possible laser-sharing system is depicted diagrammatically in Fig. 2-21. A laser beam A projects from the laser generating center LC and passes through solenoid-operated mirrors, which direct the beam by computer command to a particular laboratory in the facility. For instance, the operator in a certain laboratory, such as the first laboratory on the left in the figure, presses button B_1 on the instrument panel in that laboratory. This action energizes solenoid S_1, displacing mirror M_1 and opening the laser beam path to mirror M_4, from which the beam proceeds to M_5 and M_{13} and is reflected to laboratory 1. In the event that the total beam is being utilized by another laboratory, the red light signal L_1 comes on. In like manner, the operator at laboratory 4 may press button B_3, when the laser beam is available, to receive the beam through M_2, M_3, M_6, and M_7, the operation of the mirrors being under the computer control.

The facility radiation may also be arranged so that the laser beam is available at all times to each of the facility laboratories. In this case, mirrors M_1 and M_4 are made 50% transmissive and act as beamsplitters. Each laboratory then adjusts the power, wavelength, and frequency as desired for its particular use. If the laser beam is invisible, such as when it is radiating from a CO_2, Nd-YAG, or Nd-glass laser, provision is also

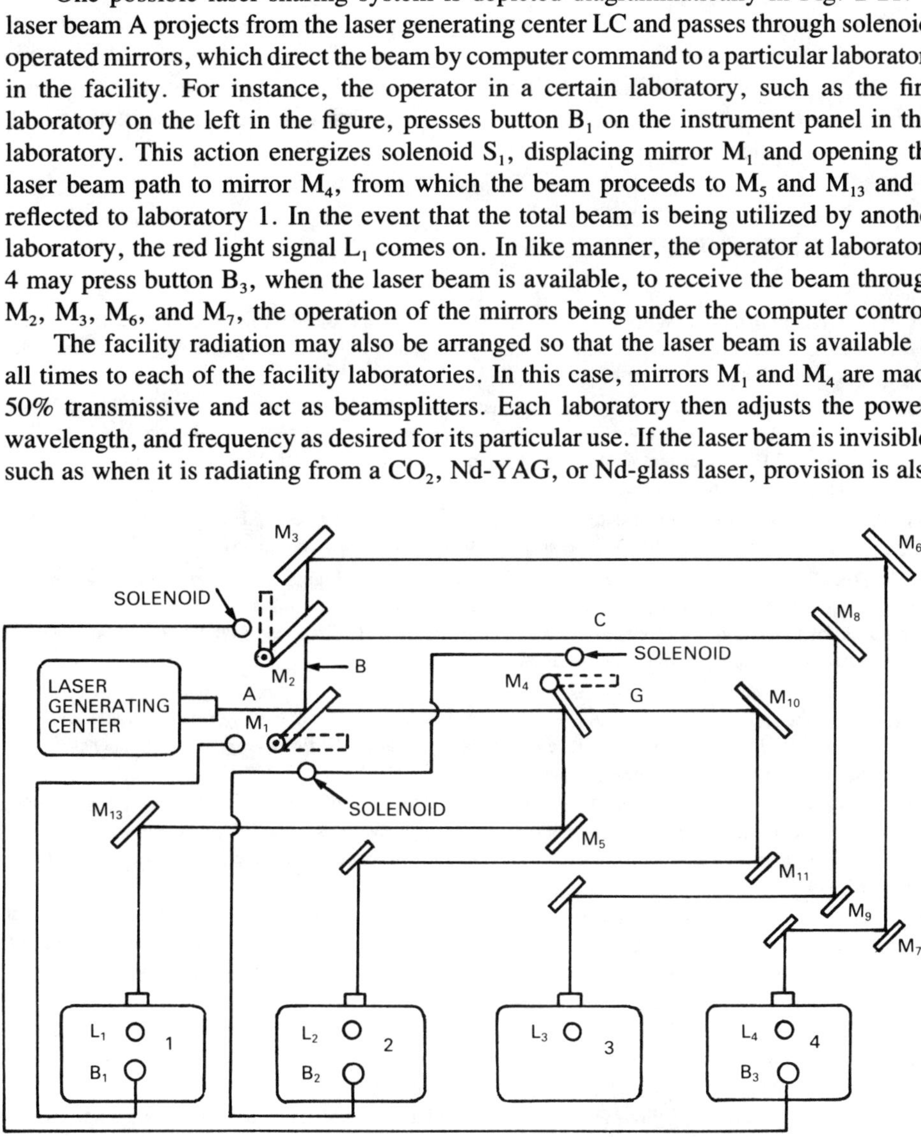

Figure 2-21 Schematic drawing of a laser beam-sharing facility.

made to mix the invisible beam with a He-Ne laser beam at the generating center and project the mixed beam to each laboratory. The beam mixing is illustrated in Fig. 8-16 of Chapter 8.

At Sandia National Laboratory, tunable laser beams are distributed to specific laboratories, which receive the beam continuously. A tunable dye laser and a frequency-multiplied Nd-YAG laser are located at the central station. The dye laser emits tunable radiation from 3900 to 7500 Å in 1.5 μsec, 5-J pulses with a repetition rate of 10 Hz. The Nd-YAG laser produces 1.5 μsec, 1-J pulses at 5320 Å at a repetition rate of 1 Hz. The applications of these lasers are confined to the diagnosis of a gas combustion environment in which various effluents are dominant. The research work at this laboratory also covers diagnosis of other pollutant emissions, Raman scattering measurements, fuel combustion efficiency measurements, and exploration of the use of substitute fuels for various applications.

QUESTIONS AND PROBLEMS

2-1 What is quantum theory? What is quantum mechanics? Who advanced the most accurate radiation theory that is now accepted? What is the energy in ergs of a photon if its wavelength is 4000 Å? What is the frequency of this photon wavelength?

2-2 Describe the structure of a typical atom. What are neutrons? What are protons? What are electrons? What are extranuclear electrons? What constitutes the weight of an atom? When is an atom charged? When is an atom at rest? How does temperature affect the quantum state of an atom? How does voltage affect the quantum state of an atom? What does electricity consist of?

2-3 If an electron accelerates toward the nucleus of an atom, what is the result? How many electrons produce 1 A of current? If a power of 10 kW operates for 10 seconds, how many joules of energy are produced?

2-4 Is an excited atom also ionized? Explain. Which effect requires higher voltage, excitation or ionization? Which of these two characteristics produces radiation when the atom regains its at-rest quantum state? What is threshold of ionization? In an electric discharge tube impressed with a very high voltage, would there be more excitation of atoms or ionization of atoms?

2-5 When an electron returns from an excited state of the atom to its original quantum state, what energy is evolved in the transition? Could the energy evolved be radiationless? If this energy is transferred by a voltage, how would you express the voltage energy of the electron? If this energy is 10^{14} frequencies, what will be the wavelength of the photon produced?

2-6 If an electron receives a voltage energy of 4 kW in an electric discharge tube, how would you express its energy? How would you express its kinetic energy? If this energy is completely converted into a photon, how would you express the photonic energy? How would you express the velocity of the electron in a tube impressed with 2 kW?

2-7 If electrical energy is expended at a rate of 2 kW per second and this energy is used for a period of 40 seconds, how many joules will be generated? If 20 J are used for 10 seconds, what will be power rate? If a laser beam is rated as 50 W/cm^2 and is used for 20 msec, how many joules of energy will be used?

2-8 What type of lens is used to converge a laser beam? What type of lens is used to diverge a laser beam? If a biconcave lens is placed about 20 cm from the exit port of the laser unit, and a convex lens is positioned after the biconcave lens at a point one focal length from the optical plane of the concave lens, what will the beam profile be after leaving the convex lens? What is this procedure called?

2-9 What is laser beam collimation? How would you collimate a laser beam? Discuss. How far apart are the concave and convex lens separated in order to obtain collimation? How is a collimated beam focused?

2-10 What is refraction of a laser beam? What is the relationship of the angle of incidence to the angle of refraction? What is refractive index of an optical material? How can one calculate the refractive index of a medium traversed by a laser beam? What relation does speed of light have to the angles of incidence and refraction? Discuss.

2-11 What is quantum concept of atomic excitation by photons? Explain the quantities given in the absorption equation. What is coefficient of ionization? Explain the difference between spontaneous emission and stimulated emission of laser radiation. Explain what negative absorption of frequency means.

2-12 Give an expression for the coefficient of stimulation for a particular lasing element. If the distance between the end mirrors is increased, will the stimulated radiation increase or decrease? Give an explanation for your response. What is the function of the feedback mirror in a laser discharge tube? Explain the mechanism involved in the interaction of quantum energy with excited atoms.

2-13 At what rate does stimulation occur in a gaseous discharge tube? Which lasing elements respond more readily to the interaction between the photon and lasing element? Explain this with respect to three-level or four-level stimulation states. What is meant by optical coupling? Can there be any optical coupling between gaseous elements and the photons? Explain.

2-14 What is diffraction of a laser beam? What important part does diffraction play in laser work? Is a beam traversing an optical medium affected by diffraction? What is Fraunhofer diffraction? What is Fresnel diffraction? What is the difference between the two types of diffraction? Explain.

2-15 What is meant by laser beam coherence? When is a laser beam coherent? What is temporal coherence? What is lateral coherence? What is the real difference between these two types of coherence? When there is a constant phase difference between two coherent laser beams, what phenomenon is the result? Explain.

2-16 What is polarization of a laser beam? How many components does a polarized laser beam have? Which component is more important from a practical standpoint? How does one determine whether a given beam is partially polarized or unpolarized? If a certain laser beam traversing an optical medium disappears from view, what becomes of that beam? What is an analyzer? Name a polarizer.

2-17 If a laser beam that is linearly polarized and is incident on a screen is manipulated by means of an analyzer, what beam intensity will be transmitted through the analyzer if the angular difference between the polarizer and the analyzer is 40°?

2-18 What is spatial filtering of a laser beam? How does one spatially filter a laser beam? Why should a laser beam be spatially filtered? What is the relationship between the focal lengths of the lenses that enter into spatial filtering of a laser beam? If an aluminum plate with a central hole was used in the spatial filtering, where would one focus the laser beam to filter it? Explain.

2-19 What is meant by the term Q-switching? Why should a laser beam be Q-switched? Does this action increase the laser power? Explain what it does. What is the functional relationship between optical pumping and Q-switching in a Fabry-Perot cavity? When is the atomic excitation in the Fabry-Perot cavity made nonresonant? When does the Q-switch open? What is then the result? What type of material is used in Q-switching a beam?

2-20 What are the types of laser radiation with respect to their spectral classification? Is x-ray radiation used for medical work a laser beam? How does it differ from an x-ray laser beam? Can an x-ray laser be focused to a point? Can an ordinary x-ray beam be focused to a point? Why? Explain. Give a general classification of laser beams with respect to their wavelengths.

2-21 How does one amplify a laser beam? In what type of lasers can amplification be produced? What is the one most important characteristic of a lasing material in amplifying a laser beam? Explain how the ILS amplifier operates, drawing diagrams to illustrate your explanation. Is it very common now to employ amplifiers in a laser system? For what application could an amplified laser beam be employed? Can Q-switching be employed in laser beam amplification? If your answer is yes, how is it utilized?

SELECTED SOURCES OF RELATED DATA

Basics of helium-neon lasers. 1984. *Lasers and Applications* May, pp. 77–89.

Brederlow, G., et al. 1983. *High-Power Iodine Laser*. New York: Springer-Verlag.

Dooley, D. 1987. Measuring CO_2 lasers with pyroelectric detectors. *Lasers and Applications* February, pp. 63–65.

Hagarty, J., et al. 1981. New probe of homogeneous linewidth. *Phys. Rev. Lett.* 49(13):930.
Haken, H., H. C. Wolf, and W. D. Brewer. 1984. *Atomic and Quantum Physics*. New York: Springer-Verlag.
Hammerling, P., et al. 1985. *Tunable Solid-State Lasers*, Vol. 47. New York: Springer-Verlag.
Holmes, L., exec. ed. 1987. Advances in commercial lasers 1986–1987. *Laser Focus* January, pp. 38–48.
Humphrey, L. M., et al. 1979. Doppler-Free Spectra from Dye Laser. Bell Laboratories, October.
Marathay, A. S. 1982. *Elements of Optical Coherence Theory*. New York: Wiley.
Moore, B. 1977. *Laser Photochemistry*. New York: Academic Press.
Oxford Lasers Ltd. 1984. Fluorine Lasers Produce 30 Millijoules, technical brochure. Oxford.
Pepper, D. M. 1982. Nonlinear optical phase conjugation. *Laser Focus* January, pp. 71–77.
Siegman, A. E. 1971. *An Introduction to Lasers and Masers*. New York: McGraw-Hill.
Simons, B. S. 1982. Laser Light Detection System. U.S. Patent No. 4,366,378.
Stark, H., ed. 1982. *Applications of Optical Fourier Transforms*. New York: Academic Press.
Stine, J. R. 1979. Two-laser photodissociation. *Los Alamos Scientific Laboratory Optical Communication J.* 312(Nov.):161.
Welch, M. 1986. Concept of polarization physics. *Lasers and Applications* January, pp. 67–71.
Wolfe, R. 1985. *Applied Solid State Science*. New York: Academic Press.

CHAPTER

THREE

TYPES OF LASER SYSTEMS

3-1 INTRODUCTION

The types of laser systems that are currently in commercial use are gas lasers, solid-state lasers, diode lasers, liquid dye lasers, excimers, metal vapor lasers, and combinations of such lasers to produce new optical characteristics. Such characteristics are polarization of the emitted beams, frequency multiplication of a laser beam by means of another laser source, beam expansion and collimation, beam modulation, and the like. By far the most widely used laser systems in industry are the ionizable gas-type lasers, which are employed in various applications including medical surgery, metrology, metal processing, semiconductor processing, military applications, and research and development work. Of the commonly used solid-state lasers, Nd-YAG, Nd-glass, and ruby play major parts in industry, surgery, and fusion work, although the use of ruby has been limited except in some types of military rangefinders.

Diode or semiconductor lasers find many applications in military field training, ranging, and certain communication systems. Liquid dye lasers are used where tunability of the laser beam is of prime concern, although other types of lasers are being developed that can be tuned to different wavelengths. Excimer lasers, which will be discussed under gas lasers, are particularly useful in the investigation of atomic fusion, although recent improvements in fusion work indicate that high-energy electron beams, x-ray lasers, and metal vapor lasers are also candidates for this type of investigation. For laser beam amplification, solid-state lasers, including alexandrite and beryllite, are very efficient emitters; these consist of ion-doped crystalline or glass rods and are pumped optically.

The laser radiation emitted from most types of lasing elements is polarized to some extent. The polarization may be linear, random, or circular, depending on the type of optical material the beam has traversed before emerging to the exterior of the emitting system. Furthermore, by use of etalons, Fresnel mirrors, optical filters, and neutral polarizers, the beam can be polarized external to the system in any way desired. Linearly polarized lasers are preferred when the beam is to be employed in precision measurements, holography, modulation of voice and data, rangefinding, and character recognition, since fluctuations are reduced to a minimum in a linearly polarized laser beam. For quality control work and alignment of structures, polarized beams are not of special concern. The spectrum of lasable elements covers the x-ray, ultraviolet, visible, and infrared regions.

3-2 GASEOUS LASER SYSTEMS

Gaseous laser systems include the lasing materials He-Ne, argon, krypton, xenon, carbon dioxide, nitrogen, ammonia, carbon monoxide, and combinations of these gases with other lasing elements. The inert gases, such as He, Ne, Ar, Kr, and Xe, combine with halogens in an electric field to form inert-gas halides known as excimers. Since the halides used with the inert gases are fluorine, chlorine, bromine, and in certain cases iodine, the resulting molecules, ArF, KrF, and the like, are highly toxic, and they may cause severe health problems when breathed accidentally. Even contact of these gases with human skin will cause injury because the gaseous material penetrates deep into the skin. Among gaseous lasers, the carbon dioxide laser provides the most powerful radiation, and it can operate in a continuous-wave mode or in pulsed mode at TEM_{00}; it can be employed in applications requiring tunability, such as spectroscopy, medical surgery, and materials processing, and in a destructive mode in military weapons.

In discussing the mechanism of laser radiation, we shall combine classical theory with quantum theory. According to the classical theory, interatomic energy transfer occurs by resonance of the colliding atoms and electrons. In the quantum theory, the radiation is emitted as a result of quantum energy (hf) transfer. Resonance excitation must occur between the excited atoms and the photons during their encounter in order to produce coherent, homogeneous laser radiation. To achieve this interaction, the discharge tube must have a reflector or mirror at each end, as shown in Fig. 2-10. During their travel across the tube, the emitted photons oscillate between the mirrors, and when they collide with excited atoms an emission of stimulated radiation occurs. As a result, net cumulative photonic energy of high gain is achieved.

Helium-Neon Lasers

Figure 2-10 shows a He-Ne laser system in which mirrors M_1 and M_2 provide quantum feedback for the stimulated laser emission. To conduct the amplified emission to the exterior of the resonant cavity (between the mirrors), mirror M_2 is an output mirror and is partially reflective and partially transmissive. The other mirror, M_1, has a thick coating that makes it 100% reflective. To cause the excitation and ionization of the atoms in the optical cavity, a high voltage of direct current is applied across the discharge tube; this unidirectional current reduces the recombination effect, which acts to diminish the dominance of excited atoms. The most prevalent wavelength in this emission is 6328 Å; two harmonic wavelengths are also generated, but they do not lend themselves to practical use in industry. The harmonic wavelengths are 11,520 Å and 33,910 Å.

The excitation of a He-Ne gas mixture occurs by transfer of energy from the excited helium to neon atoms at level 3s, shown in Fig. 3-1, during their transition to level 2p. Simultaneously, some of the atoms at 3s decay to level 3p, giving rise to a spontaneous emission. The stimulated red-colored laser beam is generally derived from transitions of neon atoms from the 2p level to the ground level, as illustrated.

The length of the tube varies from 40 to 100 cm, with a capillary bore 2 to 6 mm in diameter. The gases He and Ne are present at a ratio of 5–20 to 1, respectively, under relatively low pressure.

The essential part of a gaseous laser system is its plasma tube, which contains the excited and ionized atoms and electrons moving in opposite directions. The performance of the tube is directly influenced by the design and quality of its construction.

In operation, the inside wall of the plasma tube may become eroded in the course of time when ordinary Pyrex or quartz is used for the plasma bore tubing. This is especially true when the plasma consists of carbon dioxide. For this reason, recently

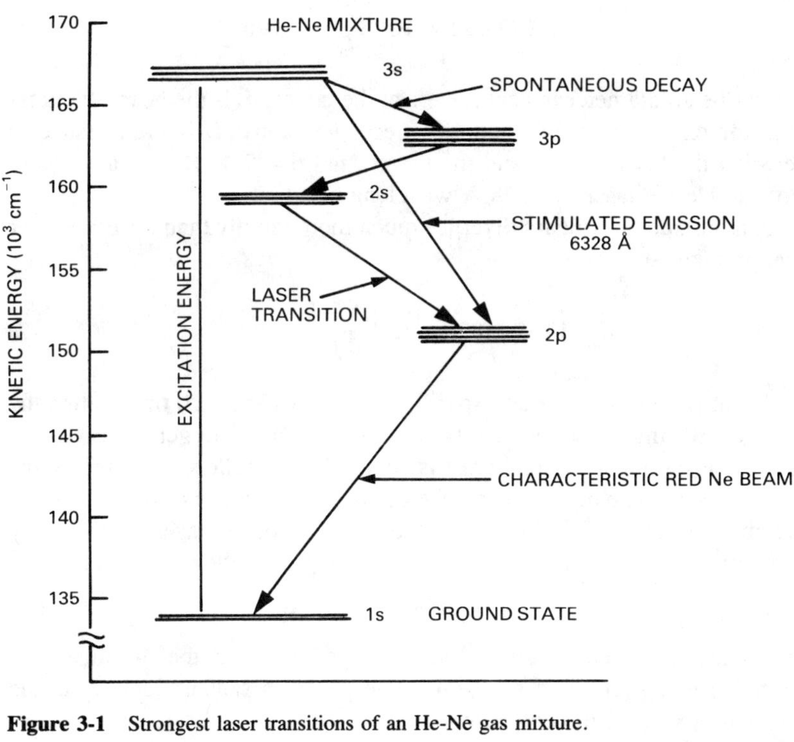

Figure 3-1 Strongest laser transitions of an He-Ne gas mixture.

developed gas tubes invariably employ beryllium oxide (BeO) for capillary bores in which the gas is in continuous motion. Beryllium oxide is rugged and possesses high durability and high thermal conductivity. This structure further permits the plasma tube to operate relatively cool to retain its laser emission efficiency. Sealed-in laser tubes need not be refilled for periods of 15,000 hours. Thus, for flowing-gas tubes, beryllium oxide has proved to be the ideal material. The plasma tube is usually mounted on an invar resonator framework because of its property of low thermal expansion. This structure offers reliable optical stability and low noise to within 0.2%. Accordingly, these lasers are capable of providing a diverse range of wavelengths at various combinations of power levels.

The next consideration in selecting a gas laser system is the power output density and the signal-to-noise ratio required in the system. This parameter requires some experimentation for the particular application. If the power requirement is not very high because of the distances within which the radiation beam is to be used, so that atmospheric attenuation is not of significance, then a lower-power laser may be employed with little or no concern regarding the signal-to-noise ratio. Manufacturers with extensive experience in problems of this type must be consulted. Many such manufacturers are listed in the appendix of this book.

A large laser beam waist diameter produces a gradually diverging beam usable to long distances. An equation that accounts for beam divergence with respect to target distance for He-Ne laser systems manufactured by Hughes Aircraft Company is

$$d = d_0\left(1 + 6.492 \times 10^{-9}\frac{D^2}{d_0^4}\right)^{1/2} \tag{3-1}$$

where d is the laser beam diameter in centimeters at the target, d_0 is the beam diameter in centimeters at the exit aperture of the optical resonant cavity, D is the distance in centimeters between the laser system and the target, and 6.492×10^{-9} is a constant characteristic of the He-Ne laser of 6328 Å wavelength.

Since a carbon dioxide laser beam diverges much more rapidly than a He-Ne laser beam, the equation takes the form

$$d = d_0\left(1 + 1.82 \times 10^{-6}\frac{D^2}{d_0^4}\right)^{1/2} \tag{3-2}$$

If the laser beam is collimated at an expansion of, for instance, 10 times, then the laser beam diverges 10 times less and the beam diameter at the target for distances beyond 60 feet will be significantly smaller. It is always best to follow the manufacturer's instructions precisely. The beam divergence of any laser system is generally given in the manufacturer's technical brochure, and the diameter of the laser beam at any distance from the exit aperture can be calculated by using the following equation:

$$d = S\phi + d_a \tag{3-3}$$

where d is the diameter in centimeters of the diverged beam at the distance S in centimeters from the exit port, ϕ is the beam divergence in radians, and d_a is the diameter of the beam exit port in centimeters.

Ionic Lasers

Among ionic lasers, Ar, Kr, Xe, and metal vapor lasers such as He-Cd and He-Se are in most common use. These lasers generally operate in TEM_{00} mode, pulsed or continuous-wave mode, and multiline or single mode. Argon ion lasers emit in multiline and yield the highest visible radiation in the spectral range of 4550 to 5145 Å, with output powers up to 40 W. The argon laser also emits in the ultraviolet at 3300 to 3600 Å in a doubly ionized transition.

Also noteworthy is the krypton ion laser, which emits laser radiation in the range from 3300 Å in the ultraviolet up to 7990 Å in the near-infrared spectrum in TEM_{00} mode. A mixture of krypton with argon radiates from the ultraviolet to 10,900 Å in the infrared. The emission may be in both TEM_{00} mode and multiline mode at an output laser power of 5 W. Although a krypton laser can produce more radiation in the infrared than a laser based on ruby, He-Ne, or similar gases, its resonant cavity length is only one-half that of a He-Ne tube. Ion lasers of this type generally are endowed with high amplitude and frequency stability at average powers and, therefore, are widely used.

In selecting an ion laser system for a particular application, several important factors must be considered. These factors are whether the radiation should be linearly polarized, what power intensity is to be projected on a given target, the degree of signal-to-noise ratio that is possible, and the focal spot size required to be incident on the target.

If the laser beam is to be employed for distance measurement, as for rangefinding, it should be linearly polarized to avoid output power fluctuations and to permit better control over the beam output intensity for various beam modulation purposes. Sufficient laser power must be available from the system in order that the retroreflected beam will contain sufficient intensity to be detected and recorded. Since the laser beam diverges slightly as it leaves the exit aperture, provision should be made to reduce the divergence to a minimum to achieve a focal spot as small as possible on the target. On the other hand, if the laser beam is to be used for applications such as structural alignment, linear polarization is not necessary and the benefit of cost saving can be obtained by selecting a randomly polarized laser system.

Among the many ion laser systems, Lexel Corporation's ion lasers deserve our study because of their advanced technological design. Two of the laser systems manufactured by Lexel are argon and krypton ion lasers that produce various wavelengths from the ultraviolet to the infrared at high power levels. The argon laser, shown in Fig. 3-2, emits in the range of blue and green wavelengths in multiline mode from 4579 to 5145 Å, the commonly used wavelengths being 4800 Å in the blue and 5145 Å in the green spectrum. When the mirrors in the Fabry-Perot cavity are changed to appropriate configurations, radiation of 10,900 Å wavelength may be obtained from the laser system. The argon in this system is doubly ionized when the biasing current is increased and produces ultraviolet rays in stimulated transitions.

Most ion lasers are inherently multiline emitters, since they contain in their plasma multiple wavelengths in various proportions, as depicted by the output beam content shown in Fig. 3-2. Monoline radiation can also be obtained, as illustrated in Fig. 3-3, by use of a rear prism in combination with an etalon (not shown). In multiline operation,

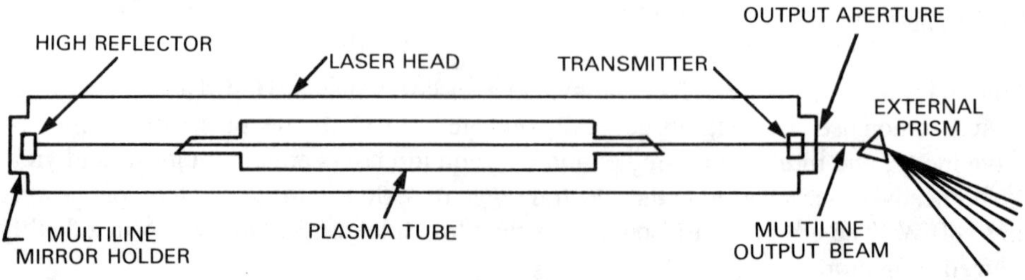

Figure 3-2 The Lexel multiline laser in the operative state.

an external prism is positioned in front of the output mirror; this prism disperses the plasma beam content into its component wavelengths. The feedback mirrors are aligned axially to the plasma tube, the rear mirror being 100% reflective and the output mirror, as usual, being partially reflective. The multiline unit emits six different discrete wavelengths, the 5145-Å wavelength being the most dominant.

The Lexel krypton ion lasers emit wavelengths through the entire visible spectrum. The red wavelengths, at 6471 and 6764 Å, are the most prominent and yield the best radiation performance. The power rating of this laser is based on 6471-Å output because the highest laser power of the unit is produced at that wavelength. This laser is also tunable to various wavelengths in its spectral content; the tuning can be performed either externally to the plasma tube or by means of a bilateral prism positioned between the rear mirror and the Brewster window. The tuned-wavelength photons are then amplified prior to leaving the tube. The radiation amplified in this manner would generally be of higher output than when the wavelength selection is made with a prism external to the output mirror.

The Lexel argon and krypton lasers both operate in the fundamental mode of TEM_{00} with a Gaussian power distribution. Operation in this mode has the smallest divergence characteristics and hence yields the smallest beam diameter. A Gaussian laser beam is measured at the $1/e^2$ power points corresponding to an aperture diameter stop that transmits 86.5% of the laser output from the tube. The beam diameters of Lexel lasers vary from 0.9 to 1.3 mm at the exit port, depending on the model type, with a beam divergence of 0.6 to 1.2 mrad. The polarization ratio in the vertical vector is 100:1. The laser units are energized by a 120-V AC, 50/60-Hz power supply, or by a 245-V AC, 50/60-Hz, three-phase source for larger units.

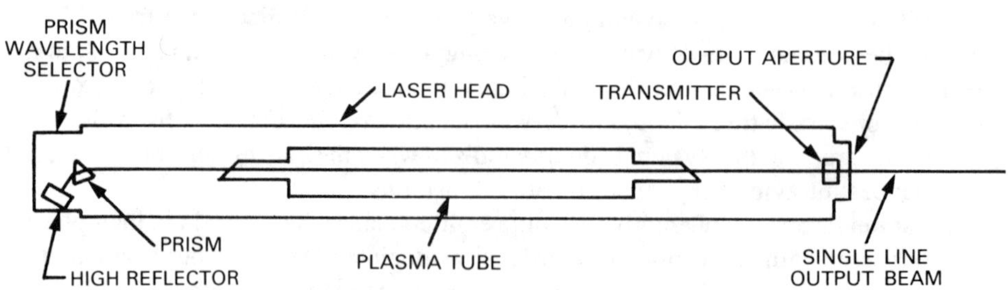

Figure 3-3 The Lexel monoline laser plasma unit.

One of the largest manufacturers of ion lasers in the United States is Spectra-Physics. This company manufactures high-performance scientific laser instruments in addition to He-Ne laser systems. An exemplary laser model taken up here for discussion is the Model 155 randomly polarized unit shown in Fig. 3-4. It is designed as an educational apparatus for demonstration of laser principles at schools, and in industry it is used as a low-power alignment laser. A neutral-density filter provided within the laser controls a constant 1.0-mW output limit. The laser emits radiation at 6328 Å at a beam diameter of 0.9 mm and a beam divergence of 1.0 mrad. Its noise characteristic at 1 to 100 kHz is 0.3%, with a longitudinal mode spacing of 550 MHz. The power requirement for the unit is 22 W from a 110-V AC, 60-Hz, single-phase power source.

Molecular Lasers

Molecular laser systems contain gaseous molecules for the active lasing media. The lasing medium consists of gases such as carbon dioxide (CO_2), carbon monoxide (CO), deuterium-tritium (DT_2), ammonia (NH_3), and excimers such as HeF, ArF, ArBr, and KrF. The most important of the molecular lasers in commercial use is the carbon dioxide laser. It consists of a mixture of 4.5 parts carbon dioxide, 13.5 parts nitrogen, and 82 parts helium. This mixture emits at 106,000 angstroms in the infrared spectrum and possesses the highest laser power among molecular lasers. Recently, CO_2 laser radiation has been investigated for atomic fusion work; however, high-power laser radiation and electron-beam plasmas appear to be the favored candidates for fusion. Some institutions have used an excimer laser for irradiation of the fusion pellet.

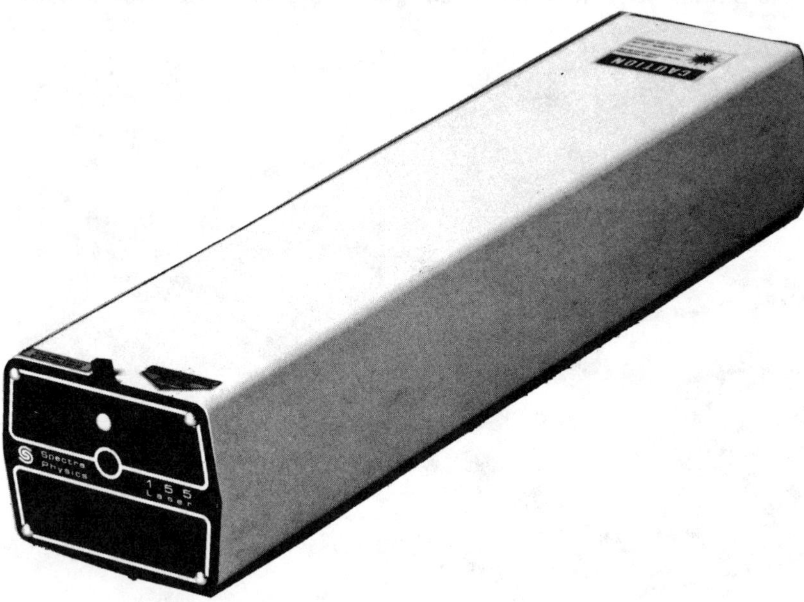

Figure 3-4 Spectra-Physics randomly polarized He-Ne OEM Model 155 laser.

Carbon dioxide lasers are manufactured in both sealed-in and continuously flowing gas forms. The sealed-in types are excited axial to the optical resonant cavity, and the continuously flowing types are excited transverse to the current flow in the discharge tube. Lasers of the latter type are known as TEA (transversely excited at atmospheric pressure) lasers. A TEA laser can be operated either continuously or in the pulsed mode. The system can produce kilojoules of laser energy per pulse and terawatts of peak power in the continuous mode. The TEA lasers are also employed as optical pumps to excite other molecular lasers. A sealed-in CO_2 laser is shown in Fig. 3-5.

Continuous tuning of the flowing-gas-type laser for more than 2000 different spectral wavelengths is possible, since the system operates at a pressure of 10 atmospheres or higher. The power produced by this type of laser is dependent on the length of the discharge tube. A power level of 10 kW per meter of the resonant cavity is easily achieved, and up to 15 kW per meter has been reported by AVCO Everrett Corporation. However, there is a tube length limitation, even when a folded cavity is employed.

The carbon dioxide laser shown in Fig. 3-5 is a versatile laser that can be used for applications requiring a single-line beam or tunable laser radiation. The resonant cavity is made of fused quartz with zinc selenide Brewster windows and feedback mirrors. It operates in single TEM_{00} mode, and the beam projected from it is vertically polarized at 106,400 Å. The laser power output is typically 3.5 W; however, up to 8 W have been obtained from units of this series with long-term stability of the radiation frequency. The beam diameter at the exit port varies from 4 to 5 mm with a beam divergence of 4 mrad, and less for high-power laser systems. The optical cavity length varies from 45 cm for the low-power lasers to 77 cm for the higher-power lasers. By tuning the Model C950 over a 150-MHz linewidth, the Doppler shift corresponding to speeds of targets moving at up to 0.5 km/sec can be measured. The system is cooled with deionized

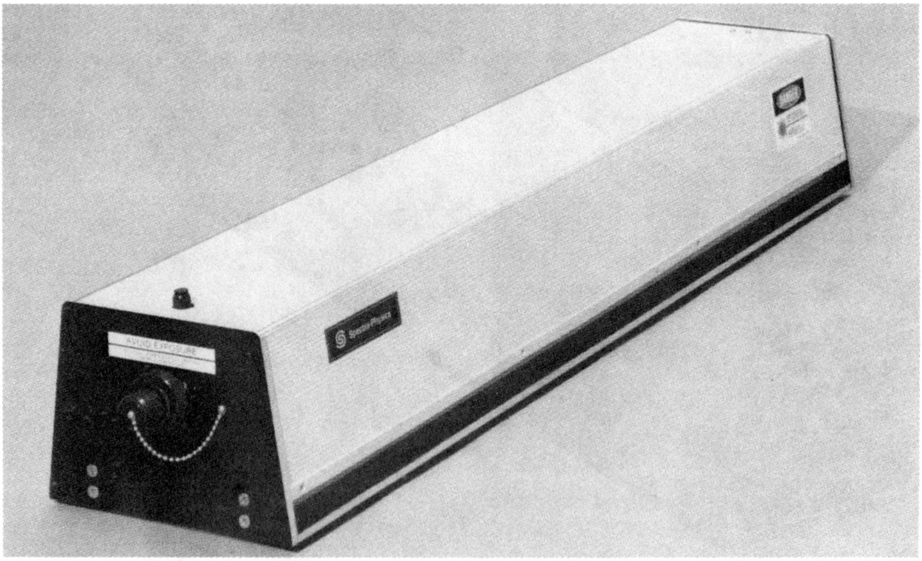

Figure 3-5 Spectra-Physics Model C950 tunable CO_2 laser.

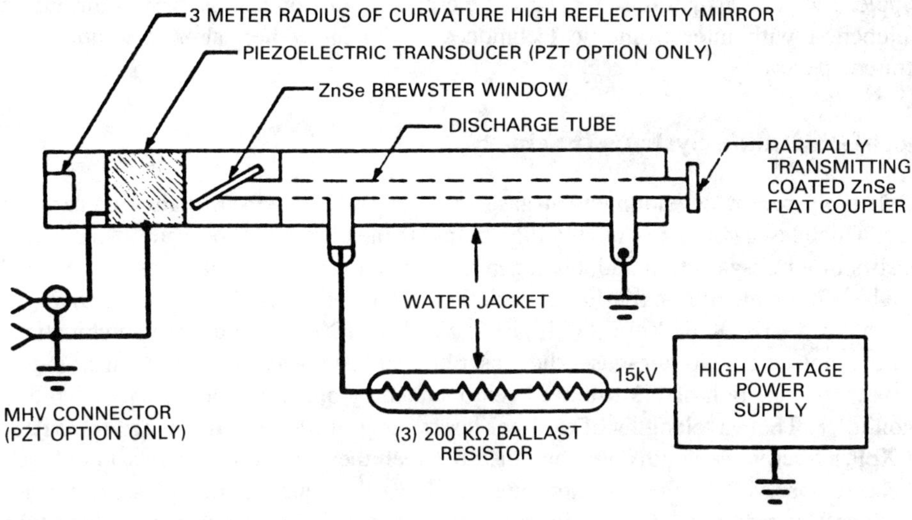

Figure 3-6 Block diagram of Spectra-Physics Model 948 carbon dioxide sealed-in laser system.

water flowing at a rate of 0.25 gallon per minute. The power supply for the system is a 115-V AC, 50/60-Hz, single-phase electric source. A block diagram of the Model C948 (same series) carbon dioxide laser system is shown in Fig. 3-6, and the vibrational energy levels for laser excitation of the Model 900 series are shown in Fig 3-7.

Because of their high frequency stability, Model 900 series units are used for pollution monitoring and detection, heterodyne communication and measurement,

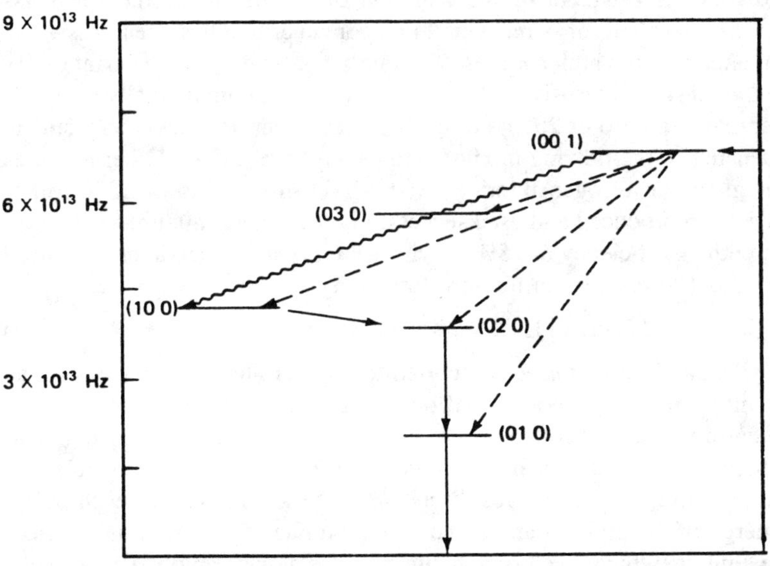

Figure 3-7 Vibrational energy levels for excitation of carbon dioxide laser.

Doppler radar investigations, testing for the presence of optical surface deformity in conjunction with interferometric techniques, and many other laboratory studies of infrared spectra.

Inert-Gas Halide Systems (Excimers)

A relatively recent development in gaseous lasers is the inert-gas halide or excimer laser, which lases with a reasonably high output in the ultraviolet spectrum. An excimer consists of a rare-gas atom and a halogen atom that are bound together in their excited state but dissociate after emission of radiation. Examples of excimers are the inert-gas monohalides ArF, KrF, XeF, ArCl, KrCl, XeCl, and XeBr and certain combinations of these molecules. No rare-gas iodide laser has yet been produced, although an iodine laser at a wavelength of 1.3 μm has been obtained by optical pumping with a 5145-Å argon laser. The wavelengths of excimer lasers range from 1930 Å for ArF to 3510 Å for XeF. Other wavelengths are contained between these two limits. Emission of NeF has been observed in the far ultraviolet at 1080 Å, which also represents longer-wavelength x-radiation. The excimer radiation occurs in an electron discharge tube with helium or argon gas used for the initial excitation of the lasing elements. Excimer lasers exhibit a high order of beam quality, pulse reproducibility, and tunability under laboratory conditions. Pulse durations up to 40 nsec and energy pulses up to 100 J have also been obtained with a laboratory-scale production unit. A photograph of an excimer unit is shown in Fig. 9-14.

The components of an excimer laser are mixed in the electron tube or just before the gases are introduced into the discharge cavity. As a high-potential current is applied to the tube, the inert gas–halide interaction takes place with excitation of the gas plasma followed by radiation. Subsequent to the emission of the radiation, the spent gas discharges into the atmosphere or is recovered in a container that has been passivated against chemical attack by the halide atoms. The discharge bore of the resonant cavity is also passivated against the corrosive action of the halides. A commercially available inert gas and fluorine at a ratio of 20:1 is diluted to 0.2% flourine content to facilitate storage and dispensing of the flowing mixture into the resonant cavity. Other mixtures of the dual inert gas with nitrogen fluoride (Ar-Kr-NF_3) in the ratio of 1300:130:1, respectively, have been produced and used successfully. The latter mixture has yielded a 1.5-J emission with an efficiency of 15%. Excimer lasers can operate in either pulsed or continuous mode. The reaction in the resonant cavity is

$$\text{(Excited) Kr}^*\text{F} \rightarrow \text{Kr} + \text{F} + hf\rightsquigarrow \tag{3-4}$$

The GTE Sylvania Corporation was reported to have produced 40 watts with an excimer laser emitting in the ultraviolet and operating at a pulse repetition rate of 1 kHz. The lasing gas flows at a rate of 15 psi gauge with an emission efficiency of 0.4%, which appears to be comparable to most commercial excimer laser efficiencies. A hydrogen thyratron is used to fire the gas discharge at the leading end of each pulse, achieving full energy of the pulse at any given repetition rate. The system is a sealed-off closed-cycle unit in which the krypton fluoride gas flows rapidly between the transversely located electrodes. (See Fig. 9-14.)

For a high-energy laser to become a commercially valuable tool it must operate in a constant-flow condition through the optical cavity. This condition requires that the excimer gases, after lasing, be expelled into the atmosphere or stored. This procedure makes the operation of an excimer costly. Consequently, several laser-producing concerns are in the process of developing traps for the flowing gas for reuse or recirculation. The conduits, including the optical cavity of the excimer system, are passivated by means of monel, Teflon, or the like. It is generally thought that excimer lasers will have many uses in industry as the development progresses and less costly practical systems are constructed.

3-3 SOLID-STATE LASER SYSTEMS

The solid-state lasers include radiation elements such as ruby, Nd-YAG, alexandrite, Nd-glass, and other synthetic crystalline rods doped with an active lasing element. These are ion-doped crystalline optically transparent rods and do not include the semiconductor diode lasers. The rods, which vary from 5 to 8 mm in diameter and 50 to 80 mm in length, are fabricated into solid cylinders whose ends are practically parallel and highly polished. The polished surfaces are coated with a reflective material such as silver, gold, rhodium, or magnesium fluoride (MgF_2). One end of the rod is 100% reflective and the other end is about 35 to 50% transmissive. All solid-state laser rods are optically pumped with a high-intensity flashlamp for excitation to radiation emission. Therefore, the energy transmission to the lasing material is through optical coupling of the photons from the flashlamp with the ions of the dopants in the crystalline rod. The dopant content varies from 1 to 3% in the host material, which may be sapphire, YAG, glass, calcium tungstate, or the like.

The flashlamp may consist of a quartz tube containing xenon gas or may be a krypton arc or tungsten-halogen type lamp. Certain of these lasers operate in the continuous mode and others operate in the pulsed mode. The flashlamp is either helical or linear in configuration, and its linear length is usually equal to or slightly greater than the length of the laser rod so that it entirely covers the rod, as illustrated in Fig. 1-20. A few turns of high-voltage triggering wire are sometimes looped around the flashlamp to start the electrical discharge in the tube. When the discharge occurs, the high voltage from the power supply sustains the potential across the tube. For pulsing-type lasers the flashlamp is pulsing, and for continuous-wave lasers the optical pump is a continuous type. The flashlamp usually starts with a higher voltage and settles down to a sustained lower voltage because the internal resistance of the tube stabilizes after the first discharge that occurs. The interelectrode resistance R across the tube is directly proportional to the electrical discharge length L and inversely proportional to the cross-sectional area A of the discharge column. Thus,

$$R \approx \frac{L}{A} \tag{3-5}$$

In a solid-state laser rod, the ionization that results from the flashlamp illumination occurs in the dopant rather than in the host material. In a ruby rod the dopant or the active element consists of chromium ions (CR^{3+}), which produce the laser light. In a Nd-YAG rod, the active element consists of the neodymium ions (Nd^{3+}), which triply ionize and give rise to radiation on returning to ground state. The flashlamp photons must have higher energy than the photons they produce from the rod in order to transfer energy to the dopant ions in an amount equal to or greater than that required for excitation of the atom, whether it is chromium or neodymium. After excitation, the dopant ions behave in the same manner as those in an argon discharge tube.

When they are optically pumped, the excited atoms are raised to a higher quantum state, and during their return to the ground state they emit photons with the energy of the excited quantum state rendered by the flashlamp photons; this action is called stimulated emission of radiation energy. When the energy transferred from the flashlamp photons to the dopant atoms is not sufficient to raise the atoms to an excited state, no radiation is emitted from the atoms and the energy is released in the form of heat. If the radiation is below the laser transition energy, spontaneous radiation may result, as illustrated in the energy transition curves in Figs. 3-8 and 3-9. When the energy is dissipated thermally, it may raise the temperature of the rod so that it loses its emission efficiency. Accordingly, the laser rod must be cooled by circulation of either air or pure deionized water, particularly in high-power laser generators.

Most of the extraneous heat produced by the flashlamp or by absorption of photons by the laser rod is concentrated in the central axis of the circular rod, where it acts as multiple lenses, decreasing the intensity of the radiation. For this reason, the author has tried a cylindrical laser rod with a central bore and found that the thermal stresses within the rod are reduced and cracking is prevented. Naturally, the resulting laser beam is tubular. However, for practical purposes, this configuration does not seem to have any significant effect on the focused density of the laser beam. The author has also used two flat-sided quadrangular bars cemented together. The cement acts as a reflector of heat from the bar. Both of these methods appear to reduce the thermal effect in a laser rod and retain the normal efficiency of the lasing element.

It should be easy to understand why a laser beam is said to be spatially coherent; that is, any two selected photonic projections describing a harmonic, unidirectional motion in the resonant cavity are of the same phase, frequency, amplitude, and wavelength. This is because the lasing photons are produced by equal-energy radiant photons. Since any increase or decrease in this energy transition will correspondingly increase or decrease the frequency and hence the photon energy hf, the resulting radiation will be of a different wavelength. When laser waves of equal frequency are not in phase, an interference phenomenon occurs in which the waves partially or wholly cancel each other, resulting in partial or entire extinction of the radiation. In an interferometer these wave cancellations appear as dark lines (fringes), which are of special interest in the spectrographic analysis of various metals and nonmetals.

The Ruby Laser

As mentioned earlier, the first solid-state laser system was developed and manufactured by Hughes Aircraft Company. This system employed a ruby rod as the laser generator. The system operated in the pulsed mode and emitted 1.4 J per pulse. Only two or three

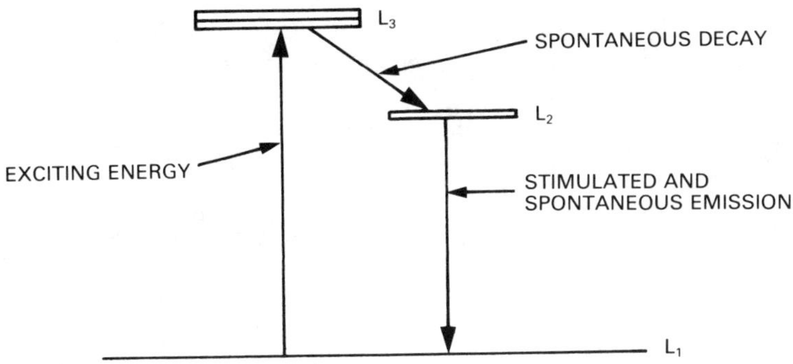

Figure 3-8 Simplified energy transition levels of Cr^{3+} ions in ruby rod.

pulses per minute were possible with the laser system, since it took about 20 seconds for the capacitors to become fully charged. This equipment was cooled by circulating air or nitrogen from a pressurized cylinder of gas adjusted to flow at a pressure of 2 atmospheres. Although this system, shown in Fig. 1-19, appears simple in design, the author used it to improvise a laser welding system with which thousands of welds were made with metallic materials from 1 to 3 mils thick. The energy transition states of the ruby laser are graphically represented in Fig. 3-8, which is a three-level transition diagram showing the energy absorbed by the ruby rod and the transitions corresponding to spontaneous and stimulated emissions of radiation from it.

In the diagram, the ground level is denoted by L_1; from the ground state the chromium ions are excited by the xenon flashlamp to the broad band L_3. The ions then return to the ground state by spontaneous decay in the first step to level L_2. As the energy state of the ions makes a transition from level L_2 to the ground state, both spontaneous and stimulated emissions are produced. The radiation emitted is red fluorescent laser light of a wavelength of 6943 Å. It should be noted that the excitation energy from the xenon flashlamp is greater than the radiation energy emitted from the ruby rod. In the excitation of ruby, half of the energy from the flashlamp produces no stimulated radiation. For this reason, more efficient emitters such as Nd-YAG or Nd-glass with four-level energy states would be necessary, as we well see in the next section.

Nd-YAG Laser System

In Nd-YAG lasers, the active medium is the neodymium ion (Nd^{3+}) and the host is yttrium-aluminum-garnet (YAG) of optically pure grade. This laser rod contains about 3% Nd^{3+}. The material has broad absorption bands and hence high emission efficiency because it offers effective optical coupling with the flashlamp radiation. Because Nd-YAG is a four-level solid-state laser, it conserves the energy that is wasted by a three-level solid-state laser. A four-level laser also meets the requirement of having a rapid spontaneous transition from level L_3 to level L_2, as shown in Fig. 3-9.

In Fig. 3-9, the quantum energy of Nd^{3+} ions is raised to level L_4 by means of a

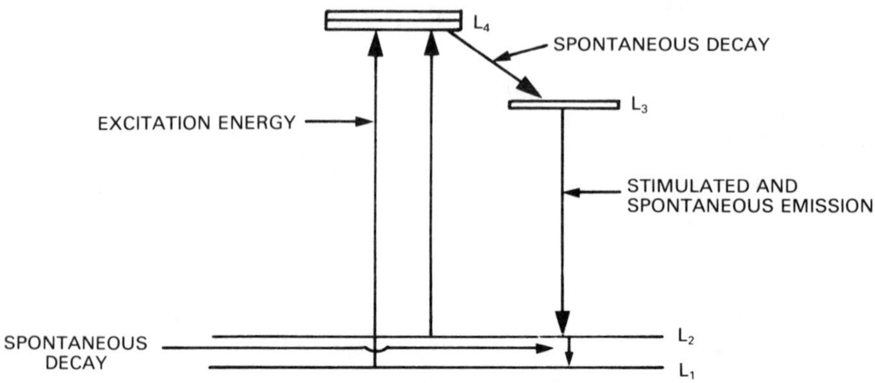

Figure 3-9 Simplified diagram of energy transition levels of Nd^{3+} ions.

flashlamp or a tungsten-halogen lamp. The excited atoms return from level L_4 to level L_3 by spontaneous decay and then drop to level L_2, giving off a stimulated emission of radiation. Finally, the atoms regain their ground-level state by further spontaneous decay. This type of laser, when pumped by a flashlamp, operates in pulsed mode; however, when a tungsten-halogen lamp is used as the optical pump, the device can be used in continuous-wave (CW) mode. An example of the laser system employing a tungsten-halogen optical pump is shown in Fig. 3-10.

The series TWO-10 Nd-YAG laser shown in Fig. 3-10 is a self-contained, single-mode unit. The tungsten-halogen lamp included in the system provides a long life and economical operation in TEM_{00} mode or multimode. The laser beam diameter at the exit port is 0.75 mm with a divergence of 3 mrad. The system operates unpolarized with a short-term stability of 6% RMS. It is water-cooled, and a thermostable resonator ensures maximum output stability regardless of the ambient temperature. The unit is powered by a 117-V 50/60-Hz power supply.

The dominant radiation from the General Photonics Nd-YAG rod is in the infrared at 10,640 Å. The power rating in this series varies from 1 to 18 W in the TEM_{00} mode. In pulsed-mode operation, the system can generate up to several hundred kilowatts per square centimeter at its peak. The laser beam from the Nd-YAG rod can be frequency-

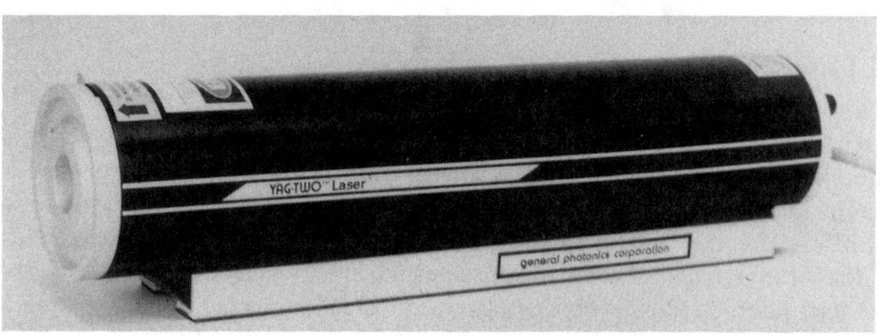

Figure 3-10 General Photonics TWO-10 series, general-purpose CW Nd-YAG laser.

multiplied to wavelengths of 5320, 3540, and 2660 Å in the ultraviolet. The Gaussian output from this laser gives the most uniform beam profile, with power limited by the size of the laser rod. Although Nd-glass operates with the same wavelengths as the Nd-YAG rod, it operates only in the pulsed mode but with better-quality radiation than is obtained with the Nd-YAG laser. A krypton or xenon arc lamp is used for optical pumping of the glass rod, which appears to be a good candidate for implosion of fusion pellets.

Another well-known commercial Nd-YAG laser system is the MY series pulsed Nd-YAG laser manufactured by Molecron Corporation. The system, shown diagrammatically in Fig. 3-11, has a unique design involving a polarization-coupled unstable resonator that delivers high laser energy; a near-Gaussian beam is produced at 10,640 Å with a beam divergence of 0.3 mrad. In the figure, the spherical mirrors M_1 and M_2 form the cavity feedback. In this unit, the Nd-YAG rod, an etalon, a quarter-wave plate, and a Q-switch complete the resonant-cavity elements. Between the Q-switch and the etalon is a horizontally polarizing element. During its oscillation between the mirrors, the beam passes through the quarter-wave plate and becomes elliptically polarized. After reflecting from mirror M_2 the beam again passes through the quarter-wave plate and becomes 75% vertically polarized and 25% horizontally polarized. The vertical component of the polarization is coupled out of the cavity by the polarizer, and the horizontally polarized component passes through the aperture to mirror M_1. The vertical component of the partially polarized beam passes through a second polarizer, which increases the polarization ratio of the beam. The resultant laser beam is amplified in the amplifier and subsequently enters the beam frequency multiplication section of the system, in which three harmonics of the beam become available to the exterior of the laser system when required.

The harmonic wavelengths are the frequency-doubled 5320-Å line, the third-harmonic 3550-Å line, and the original 10,640-Å line, any of which can be selectively obtained from the system. The pulse energy of this series varies from 200 mJ for the

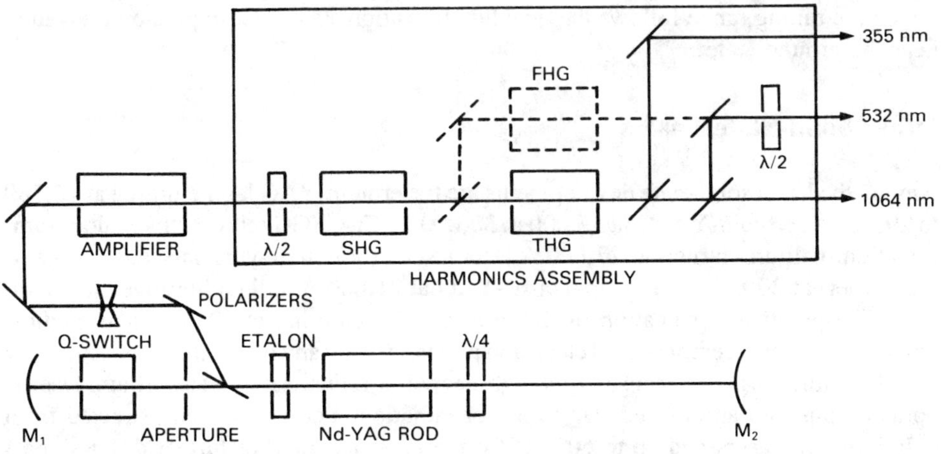

Figure 3-11 An optical schematic diagram of MY series pulsed Nd-YAG laser.

MY-32 model to 1000 mJ for the MY-35 model, at 10,640 Å. For the frequency-doubled beam, the output of the respective laser units varies from 75 to 350 mJ. For the frequency-tripled and -quadrupled harmonics, the power output correspondingly decreases. The system operates from a 208 ± 10 V, three-phase, Y-connected power supply.

Alexandrite Laser System

A new type of laser source that is akin to ruby has been developed and is classified as a solid-state vibronic laser. The lasing element is commercially known as alexandrite, is made from beryllium-doped aluminum oxide ($BeAl_2O_4$) and operates at room temperature, emitting at 7760 Å. The radiation from this laser material is due to vibrational quanta, known as phonons, and is intimately related to photonic emission. Alexandrite can operate like a three-level ruby laser or a four-level tunable vibronic laser at a low optical pumping threshold. The effective emission cross section of alexandrite increases as the temperature is increased, and the terminal-level population makes transition with the shortest-wavelength photons in the neighborhood of 6804 Å. The chromium content of this lasing element varies from 0.01 to 0.4%, where 0.1% of this dopant is equivalent to 3.51×10^{10} chromium ions per cubic centimeter of the crystal.

The alexandrite laser can operate either Q-switched or pulsed in the stimulated state. Operating in the continuous-wave mode, it behaves in the vibronic format. In the pulsed mode it is capable of operating as a three-level or a four-level laser, while in the continuous-wave mode the crystal operates as a four-level vibronic laser. The best performance is obtained when it is optically pumped with another laser. The radiation from alexandrite can be frequency-doubled, as in a Nd-YAG laser, with a tuned KD*P plate to operate in the ultraviolet spectrum between wavelengths of 3600 and 4100 Å. As the optical pumping power is increased, the emission increases because of the vibronic kinetics of the material. A photograph of the alexandrite laser system is shown in Fig. 3-12 and the energy transition levels are depicted graphically in Fig. 3-13, in which the tuning range follows the quantum transition state prior to phonon relaxation near the ground state.

Other Solid-State Lasers

Among the latest solid-state developments are the erbium-YAG laser emitting at 12,280 to 29,400 Å, erbium-YALO at 17,000 to 30,000 Å, Cr-GSGG (chromium–gadolinium-scandium-gallium-garnet) at 27,900 Å, Ho-YSGG (holmium–yttrium-scandium-gallium-garnet) at 20,800 Å, and Ho-Tm-Er-YAG at 21,000 Å at liquid-nitrogen temperature. The practical application of the erbium laser emitting at 12,200 is in medical work, where it has demonstrated clean and precise cutting and tissue ablation, especially in ophthalmic surgery because of its high degree of retinal safety. The erbium laser is optically pumped and pulsed. Up to 5 J of radiation energy has been extracted from holmium-type lasers and up to 600 mJ from erbium lasers. (For further discussion of erbium lasers, see Chapter 10.)

Figure 3-12 Photograph of an alexandrite vibronic laser manufactured by Apollo Lasers, Inc.

The MIT Laboratories have demonstrated the first tunable titanium-sapphire laser, emitting at 8000 Å in the near infrared with a tuning range of 3000 Å. Titanium lasers are generally optically pumped by a frequency-doubled Nd-YAG laser or an argon laser. Other research groups have also produced titanium lasers emitting in the range of 7000 to 10,000 Å. Titanium lasers can be produced in both pulsed and continuous-wave modes at efficiencies up to 40%. In pulsed mode a repetition rate up to 10 Hz at 1.6 W maximum output has been noted; in continuous-wave mode an output of 100 mJ has been obtained. These lasers, including the solid-state lasers mentioned earlier, can be pumped by means of diode lasers, such as AlGaAs or AlGaInP.

Extension of this work has given rise to the development of a titanium-sapphire laser pumped by a flashlamp instead of a laser beam. Several manufacturers have been

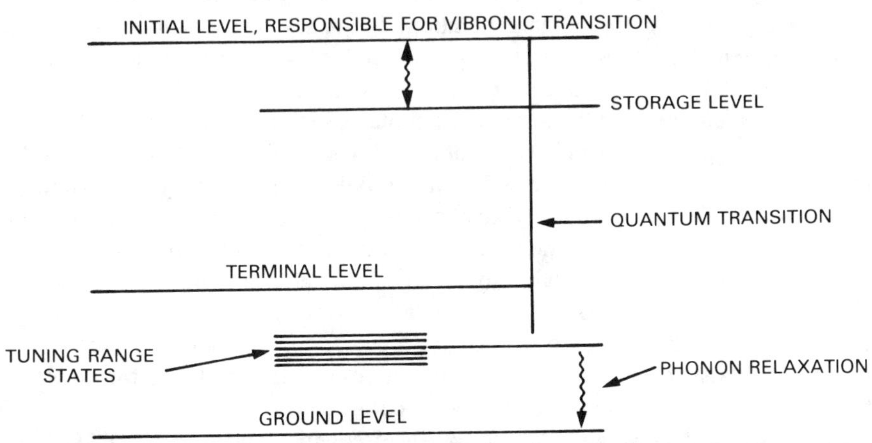

Figure 3-13 Quantum energy levels of chromium ions in an alexandrite laser crystal.

working on the concept, but progress has been hampered by excessive absorption of the output emission by the included cesium filter material. The problem has recently been overcome by using more Ti^{3+} ions than Ti^{4+} ions in the dopant, which was first suggested by Union Carbide and subsequently taken up by other manufacturers. The GTE Systems division has obtained 3 J per pulse at 8000 Å. This radiation is tunable from 6700 to 10,600 Å. Because the laser beam can be frequency-doubled, its spectrum is within the cesium-filter passband predominant in this system. This laser has been developed principally for satellite communications.

The new scheme provides ultrashort pulses, and the system can be pumped by other laser elements when desired. Its applications include laser radar transmission, imaging, medical instrumentation, and materials processing. Flashlamp pumping is less costly than laser beam pumping and gives reasonably high output efficiencies; the system has been reported to emit outputs up to 5.7 J per pulse at a repetition rate of one pulse per 50-μsec pulsewidth, using higher input power. Also, outputs up to 1.1 J a 40 Hz per 6-μsec pulsewidth have been reported by researchers. It is thought that with a longer flashlamp length and a longer laser rod, further increases in output can be expected with long lifetimes in the near future.

Radiation Frequency Multipliers

A frequency multiplier is a harmonic generator that increases the fundamental laser frequency and shortens its wavelength, thus increasing the photonic energy. For certain applications, both solid-state and liquid-dye lasers can be frequency-doubled or -tripled by employing nonlinear crystals, such as Kd*P (potassium-d-deuterium phosphate), KPR (potassium pentaborate), and similar crystals suitable to the type of application desired. For instance, a Nd-YAG laser of 10,640-Å wavelength can be frequency-doubled to 5320 Å or tripled to 3550 Å by use of a KD*P nonlinear crystal. It must be noted here that the first wavelength is in the infrared, the second is in the visible, and the third is in the ultraviolet region of the laser spectrum. Since the photonic energy increases in the process of harmonic generation, frequency multiplication is a direct function of the input laser power. The frequency-multiplied laser may also be mixed with the fundamental beam to increase the radiation energy.

One company employs a CD*A crystal doubler which is angle-tuned at 82° and produces 150 mJ of green laser radiation from a Nd-YAG laser operating in its fundamental wavelength mode; the laser emits 500 mJ per pulse. The company also manufacturers KBS ($KB_5O_8 \cdot 4H_2O$) for multiplying phase-matched frequencies at 1860 Å in the ultraviolet spectrum. The second- and third-harmonic lasers, emitting at 5320 and 2550 Å, respectively, are excellent optical pump sources for frequency-multiplying the dye laser. By using the fourth-harmonic laser of 2660 Å from Nd-YAG, copious amounts of ultraviolet radiation have been produced for medical applications. The frequency-multiplying crystals come in sizes of 14 mm square and 25 mm long; for emission stability, the crystals are housed in temperature-controlled ovens.

A lithium iodate crystal may also be used to frequency-double a Nd-YAG laser beam. The crystal is temperature-controlled and angle-tuned, similar to other crystal

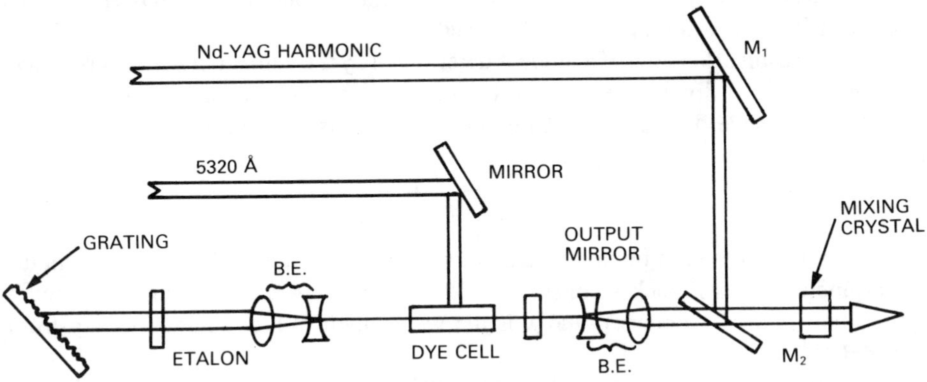

Figure 3-14 Nd-YAG harmonic beam mixing in a tunable dye laser.

multipliers. When the 10,640-Å beam from Nd-YAG is mixed with the 5600-Å beam emitted from a rhodamine 6G dye, an efficiency of up to 40% is achieved; the radiation from this system is 30 mJ. Use of the harmonic laser beam to increase the output beam energy is also possible. Figure 3-14 illustrates one method for mixing beams of multiple wavelengths.

3-4 SEMICONDUCTOR LASER SYSTEMS

Semiconductor laser systems are fabricated similarly to light-emitting diodes and have the same *n*-type and *p*-type active layers to form the *p-n* junction at which the light is produced. The difference between a semiconductor light-emitting diode and a laser diode is that the former operates below the threshold current of 0.10 A and the latter operates above the threshold current level. The most common semiconductor diode laser is the gallium arsenide laser. Other diode laser materials are gallium-aluminum-arsenide (GaAlAs), gallium-aluminum-phosphide (GaAlP), gallium-indium-aluminum-arsenide (GaInAlAs), and other derivatives of gallium arsenide. Certain of these diodes operate in the continuous-wave mode and emit between 7800 and 8500 Å; others operate in the pulsing mode with wavelengths between 8000 and 9050 Å.

A semiconductor diode laser has two active energy bands—a valence band and a conduction band, shown in Fig. 9-8—and the radiation occurs between these two bands. The emission wavelengths correspond to the energy difference between the valence bands and the conduction bands. The valence bands are replete with electrons and the conduction bands are depleted of electrons. Valence bands that are above the normal electron reserve bands and can accept electrons from the neighboring bands are called acceptors and are classified as *p*-type active layers. The active layers that are just below the conduction bands are called donors and are known as *n*-type active layers, since they have electrons in excess. Accordingly, the current flow in a diode laser is from the *n*-type active layer to the *p*-type active layer. The current across the *p-n* junction increases with an increase in forward voltage, and the output laser power from the junction increases as the current is increased.

The emission efficiency of a diode is determined by the number of photons produced at the *p-n* junction divided by the number of electrons (current) that have passed the junction to produce the photons. An expression for this relation is

$$N_q = \frac{P_t}{hc/\lambda} \tag{3-5}$$

where N_q is the number of photons (quanta) produced, P_t is the total laser power emitted at the junction, h is Planck's constant (6.625×10^{-27} erg-second), c is the speed of light (about 3×10^{10} cm/sec), and λ is the wavelength of the radiation in angstroms ($\text{cm} \times 10^{-8}$).

The electrical input power to the device and the emission output power of the laser may be expressed as

$$P_{\text{eff}} = (I \times V_i) - P_q \tag{3-6}$$

where P_{eff} is the diode power efficiency, I is the current passing across the *p-n* junction, and V_i is the voltage impressed on the diode. Rewriting Eq. (3-6) in terms of the wavelength λ and the rate of photon emission $\Delta I/e$ per second, we obtain

$$P_{\text{eff}} = \frac{\Delta I \; hc/e\lambda}{I \times V_i} \tag{3-7}$$

A schematic diagram of a typical diode laser cross section is shown in Fig. 3-15. It will be noted from the figure that the peak emission wavelength of the diode radiation can be controlled by varying the proportion of aluminum with respect to gallium arsenide in the active layer. Because diode lasers can be produced in commercial quantities by conventional methods at low cost, these devices are very attractive for use in many applications, such as in military rangefinders, pollution detection and monitoring systems, aerial and fiberoptic communication systems, burglar detection and alarm systems, target illuminators, marksmanship training systems, and the like.

3-5 DYE LASER SYSTEMS

In a dye laser, the active, radiant medium is an organic dye in solution. Dye lasers are excited to radiation by optically pumping the dye with a flashlamp, Nd-YAG laser, nitrogen laser, krypton laser, excimer laser, or radiation from a harmonic beam of a

Figure 3-15 Schematic diagram of a GaAs semiconductor diode laser.

solid-state laser. The radiation may be pulsed or in continuous-wave mode. In the pulsed mode of operation, the dye laser is optically pumped by a nitrogen laser, excimer laser, Nd-YAG laser, or coaxial flashlamp. In the continuous-wave mode of operation, an argon or krypton laser is used to pump the dye to emission of radiation.

A broad range of wavelengths can be obtained from pulsed lasers by using optical elements such as etalons, prisms, or birefringent filters; their output can be frequency-multiplied by Nd-YAG harmonics. The continuous-wave dye lasers, on the other hand, are commercially attractive because of their narrower linewidths and higher stability than pulsed laser types. When additional narrowing of the spectral bandwidth is desired, a diffraction grating of proper grazing angle may be used at the output end of the optical resonant cavity. The combination of a grating and an etalon provides a further improvement in narrowing the output bandwidth down to about 0.025 angstrom.

Continuous Dye Laser System

For continuous optical pumping, argon or krypton laser radiation in the spectral range of 3500 to 8000 Å is used. A high-power krypton laser, as well as an argon laser, has been used to optically pump the dye from the far ultraviolet through the visible to the infrared, employing different dye solutions. In a sealed-off dye laser, the dye solution flows continuously through the capillary bore of the dye cell by means of a mechanical pump, and the irradiating beam projects at right angles to the axial flow of the dye solution. In such a system, the dye solution is replaced periodically with a fresh solution, since the dye loses its laser emission strength with constant exposure to the pump radiation.

Figure 3-16 shows a diagram of a typical continuous-wave dye laser in which the dye solution is optically pumped by an argon flashlamp (F) energized from a high-voltage power supply (H.V.). The dye solution flows continuously through the capillary tube T by means of a mechanical pump P. One end of the capillary tube, marked by M_1, is 100% reflective, and the other end is provided with a Brewster window B. The output mirror M_2 is about 40 to 50% transmissive, and an etalon E between the Brewster window and mirror M_2 serves to tune the output laser beam within the spectral range of the dye. It is readily seen that the dye laser is self-cooled by the flowing dye solution

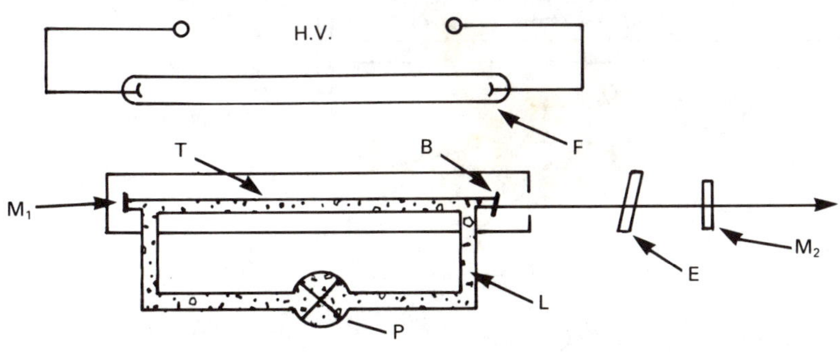

Figure 3-16 Simplified diagram of a continuous-wave dye laser system.

for short-period applications. For continuous use, the laser system may be cooled by circulating water around the capillary tube enclosed in a sealed jacket. The continuous-wave dye laser beam may be employed in Raman spectroscopy, since it can be tuned to isolate any discrete line from the beam.

Commercial CW dye lasers are constructed differently from that shown in Fig. 3-16, since they are improved production units, and each manufacturer has a proprietary system configuration. Figure 3-17 shows a typical commercial dye laser design in which the dye solution flows continuously through a flat cell C positioned at the foci of two spherical mirrors M_1 and M_2. The optical pump mirror, M_3, has a short focal length. The fully coated mirrors M_1 and M_2 produce oscillation of the emitted photons, and mirror M_2 further reflects the photons to output mirror M_4, which is about 50% transmissive. The laser output beam is tuned by means of the birefringent filter or an etalon, designated by E. The optical pump illumination is provided by the laser source L.

Pulsed Dye Laser Systems

In pulsed operation, the duration of the pulse varies from a fraction of a microsecond to several microseconds, depending on the power capability of the dye laser system. For short pulses, Q-switching may be used. Pulsed dye lasers have much higher peak output powers than CW dye lasers; however, the average power is much less. The power output from a dye laser provided with a coaxial flashlamp is greater per pulse (up to 20 J) at a low repetition rate than that of a dye laser pumped by a linear flashlamp, although the output per second from both types remains about the same. Also, pulsed dye lasers provide a broader range of wavelengths, and because their output power is sufficiently high they can be used for frequency doubling and for multiline tuning. Additional narrowing of the spectral bandwidth is accomplished by inserting a diffraction grating of proper grazing angle at the output end of the optical resonance cavity. The laser can then be tuned merely by rotating the grating.

A schematic diagram of a pulsed dye laser system containing all the optical components discussed in the previous paragraph is shown in Fig. 3-18. The laser dye

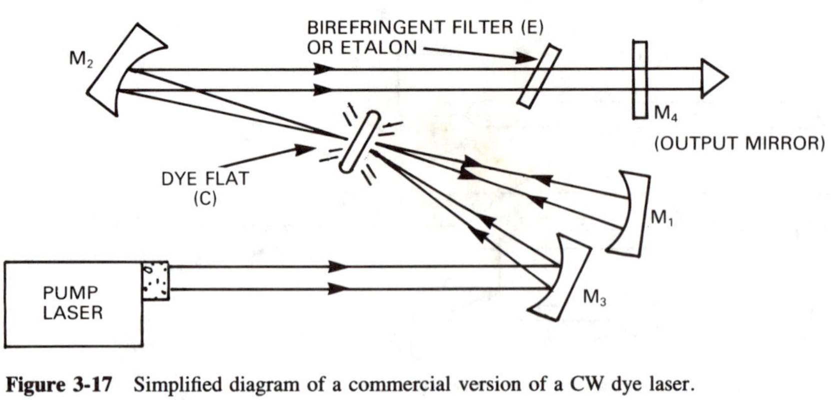

Figure 3-17 Simplified diagram of a commercial version of a CW dye laser.

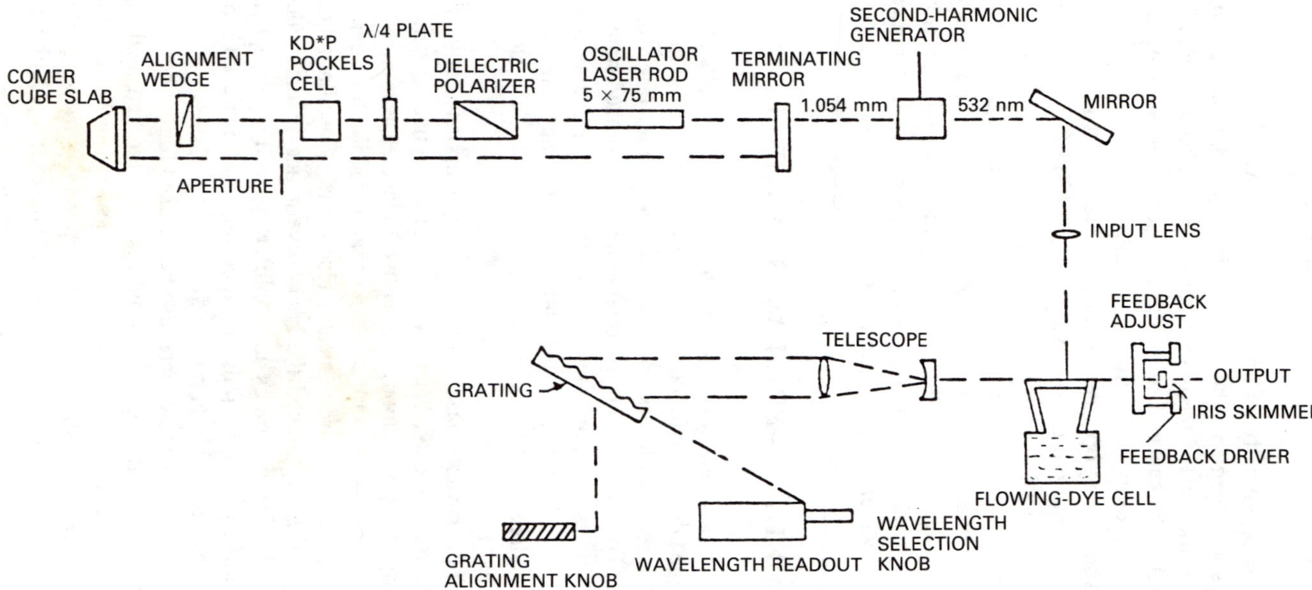

Figure 3-18 Schematic diagram of a pulsed dye laser.

solution is optically pumped with a Nd-YAG crystal harmonic wavelength at 5320 Å. In this arrangement, the output mirror (terminating mirror) serves two purposes—as the rear mirror and the output mirror. The reflected beam from the "rear mirror" passes through a Pockels cell and a quarter-wave retardation plate back to the Nd-YAG crystal and to the terminating mirror, as shown.

The system operates as follows: the dielectric polarizer mounted between the Nd-YAG rod and the quarter-wave plate transmits a polarized beam to the Pockels cell, which is biased to polarize the beam circularly. When the circularly polarized beam is reflected from the rear mirror and passed through the Pockels cell a second time, it becomes linearly polarized and oriented 90° from the original direction. The laser beam then passes through the KD*P second-harmonic generator to shorten the original 10,640-Å wavelength to a 5320-Å wavelength, which optically pumps the flowing dye. Further wavelength selection is made by means of the grating.

3-6 METAL VAPOR LASER SYSTEMS

As laser materials development progresses, new types of lasing materials are emerging from the laboratory for commercial use. Among these materials are several recent vapor phase rare-earth lasers. A potentially important laser for fusion work is the terbium vapor (Tb^{3+}) laser. This material in its triply ionized state emits at 5450 Å in the visible spectrum, with an optimum gain of 0.5 per centimeter in the active cavity. Rare-earth lasers yield high pulse rates with high efficiency, and the $TbCl_3$-$AlCl_3$ vapor complex is considered to be one of the best candidates for atomic fusion work by Argonne National Laboratory.

Copper vapor lasers also have high optical conversion efficiency for optically pumping tunable dye lasers. Workers at Lawrence Livermore National Laboratory have developed copper vapor lasers with efficiencies as high as 40%. Although conventional copper vapor lasers with stable resonators exhibit high beam divergence, the Livermore investigators have used unstable resonators by oscillating the emission in the transverse mode and thus reduced the beam divergence several orders of magnitude. The laser system has been used as an amplifier, as well as for optically pumping a dye such as rhodamine 6G in the pulsing mode, and has achieved the highest optical conversion efficiency attainable at present.

Copper vapor lasers are also being used for the detection of latent fingerprints, and it is thought that with additional research the age of the print could be determined. Comparative tests of an argon laser, now in current use by forensic specialists, and a copper vapor laser showed that the copper vapor laser is more efficient. The fingerprint is treated with a laser dye solution and examined under radiation from the copper vapor laser. The dye fluoresces and readily makes the print distinguishable from the background. The print can then be photographed for future observation as desired. Also, it was found that pulsed excitation of the dye fluorescence to isolate the print from the background is preferable to CW radiation from the copper vapor laser.

Xerox Corporation in Pasadena, California, manufactures He-Cd lasers for use with the company's high-speed printers. It is reported that this system uses less power per emission rate than a He-Ne laser used for the same purpose. The He-Cd laser requires 4 or 5 minutes to go from start to full power, whereas it takes only 1 minute for an argon laser to attain full power; also, the argon laser operates with a higher discharge current and lower voltage than a He-Cd laser. These and other significant advantages of the He-Cd laser render it greater importance in applications to reprographics and medical instrumentation. This laser's other possible applications are in the semiconductor, audiodisk, and videodisk industries and as a substitution for the argon laser in other uses.

Helium-cadmium lasers manufactured by Liconix Corporation emit a single blue line at 4420 Å. These are low-current, high-voltage vapor-discharge lasers using currents from 50 to 100 mA at up to 1500 V. The power outputs range from 5 to 50 mW for an expected life of 5000 hours, the lifetime depending on the ratio of helium to cadmium vapor in the discharge tube. Since He-Cd lasers are temperature-sensitive, they must be operated below 40°C and above room temperature. At higher temperatures the cadmium-to-helium ratio increases, spoiling the gain, and below ambient room temperature (about 20°C) the cadmium vapor condenses and its ratio with respect to helium decreases. Furthermore, He-Cd lasers must operate at only one power level to avoid instability in the resonance cavity.

3-7 F-CENTER LASERS

An F-center laser emission occurs between quantum levels at crystal defect site. The F-center laser is a powerful instrument for high-resolution infrared spectroscopy. The laser operates in the continuous-wave mode and is continuously tunable from 23,000 to 33,000 Å with a narrow linewidth. It has relatively high output power and low beam divergence. A 1-W optical pump power can produce 15 mW of laser power in the Burleigh FCL-100 F-center laser shown in Fig. 3-19.

The United States Air Force employs an F-center laser to monitor the drift of an orbiting satellite by determining its deviation from a geosynchronous orbit. The procedure is a one-way ranging technique in which infrared photodetectors on the satellite acquire the beam from the ground-located F-center laser emitting at 29,000 Å. The Burleigh F-center laser produces a 40-mW laser beam optically pumped by a 1-watt krypton ion laser manufactured by Spectra-Physics. This is a distinct improvement over the previously used hydrogen fluoride laser, which has disadvantages including hazardous fumes and extensive maintenance problems. The F-center laser station is located on the island of Maui, and the beam projected from the laser system appears as a bright point to the infrared detectors aboard the satellite. When the satellite deviates from its assigned location, the bright point alters its position and compels repositioning of the satellite in its normal orbit.

Recently, German physicists have produced an F-center laser that they employ frequency-doubled at 13,000 Å in the infrared spectrum. Earlier, Burleigh researchers had produced a frequency-doubled F-center laser beam with a peak power of 200 picowatts at 13,000 Å, using an angle-tuned lithium iodate crystal. Comparative tests showed that the frequency-doubled F-center laser and the fourth harmonic line of a

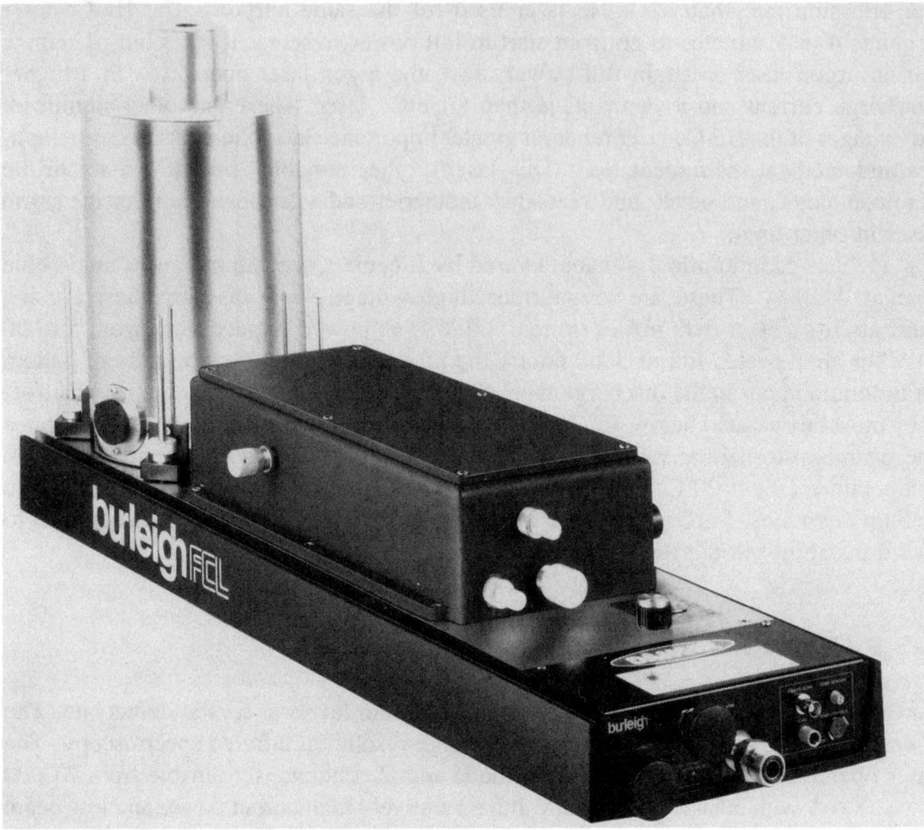

Figure 3-19 The Burleigh FCL-100 F-center laser.

carbon dioxide laser, as well as the line from a neodymium pentaphosphate laser, appeared to produce radiation with identical properties and with possible applications in satellite monitoring.

3-8 DIODE-PUMPED LASER SYSTEMS

The diode-pumped solid-state laser system was developed by the author for a typewriter eraser in 1968 and patented in the early 1970s. Since then, nearly a dozen manufacturers in the United States and abroad have made various versions of the first device. Laser diode pumping of solid-state laser elements can be accomplished in several ways: (1) end pumping, in which a laser rod is illuminated axially from one end of the rod; (2) peripheral pumping, in which the rod is excited to emission of radiation by the illumination from an array of laser diodes arranged peripherally to the rod; and (3) fiber-coupled end pumping, in which the laser diode or a group of diodes are located remotely from the lasing rod and their illumination is conducted through a fiberoptic conduit to one end of the rod to end-pump it to emission.

Peripheral pumping achieves the highest laser laser output power because any number of laser diodes can be arranged on the periphery of the rod; the pump and the laser rod must be cooled, preferably by a thermoelectric method. The fiber-coupled end-pumping design has the smallest laser emission head, since the laser diode and the lasing rod are located remotely from the laser head and are in a separate container. The end-pumped design offers the highest efficiency but low laser output power.

In this type of laser system, the laser-pumping diode is AlGaAs or its derivative and the output laser source is generally Nd-YAG, Er-YAG, or alexandrite. The advantages of diode-pumped solid-state laser systems include high efficiency and reliability of operation; in addition, they can be produced in a small compact package and are easily handled, so the device head can be moved to almost any area in a metal-processing laboratory or to any site on the human body for surgery or therapy. The output from such a system can be tuned, frequency-multiplied, Q-switched, or cavity-dumped. The usual laser beam from Nd-YAG is in the infrared at 10,640 Å and, when frequency-doubled, is in the visible spectral region at 5320 Å. The output power of the laser is variable from 100 to several hundred milliwatts in the continuous mode and up to 20 W in pulsed mode. When the laser is Q-switched, the power can increase to several kilowatts. There are many variations of the design, frequency, and output power, which is increasing almost every day, as new diode elements and rod materials are developed and used with diode pumping. The pulse duration also varies from 50 μsec to several hundred microseconds, depending on the mode of operation. The available pulse energy can be increased by substituting yttrium-lithium-fluoride (YLF) for YAG, using a neodymium lasant; the YLF can develop higher pulse energy. A laser diode-pumped device used in processing thin metals and in urgent medical surgery is shown in Fig. 3-20.

The technology of semiconductor diode lasers is moving toward achieving high-power infrared lasers with narrower linewidths at shorter infrared wavelengths. The preferred laser would be one emitting at visible wavelengths. A frequency-doubled infrared diode laser recently developed in Japan is transmitted through a lithium niobate waveguide. A U.S. company has achieved an AlGaInP diode emitting at 6210 Å that operates at room temperature. By using monolithic diode lasers in a phase array, very

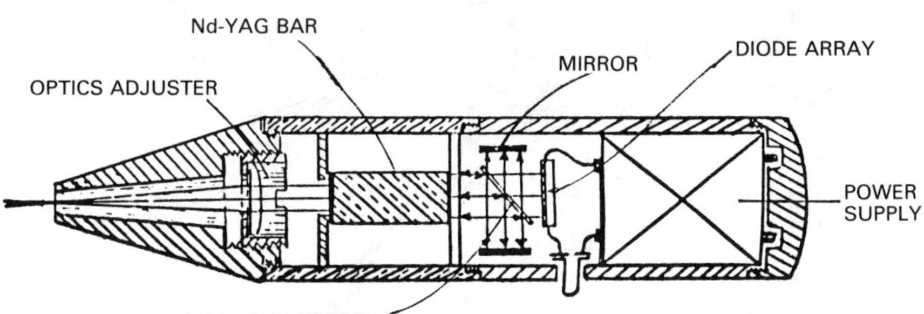

Figure 3-20 A laser diode-pumped laser device developed by the author (simplified).

high power, such as 10 W in 200-μsec pulses at repetition rates of up to 100 Hz, can also be achieved. Linear diode arrays have also been produced with up to 1 kW/cm^2 of continuous-wave output.

Although semiconductor laser-pumping diodes are efficient emitters, nearly 50% of the energy applied to them is wasted as heat. Accordingly, thermoelectric coolers are employed, which have helped to some extent in reducing this dissipation. The Lightwave Company, using Q-switched diode lasers, has produced repetition rates up to 10 kHz in Nd-YAG at powers up to 400 mW. Although the laser energy produced in laser-pumping diodes is in microjoules, their place in a variety of commercial and medical operations outweighs their low-energy emission. Another version of the diode-pumped laser system is shown in Fig. 3-21.

Figure 3-21 illustrates another version of the system with two component units optically connected by means of a fiberoptic cable. One of the units contains the laser-generating component and the second is the dispensing unit or stylus. The laser-generating unit receives a 110-V alternating current from an external source and converts the energy into a low-voltage current to supply the diode pumps as well as the related switching parts. The laser radiation emitted from the diode-pumped source passes through a frequency doubling module and into the fiber-optic cable, which is surrounded by a peripheral tubular space through which a gaseous element can be transmitted to and from a surgical site during a medical operation. The dispensing unit receives the laser beam from the fiber-optic cable and permits it to pass through a Q-switching section and a beam-collimating optical unit, from which the beam impinges on a focusing lens, which projects it on an object to be processed.

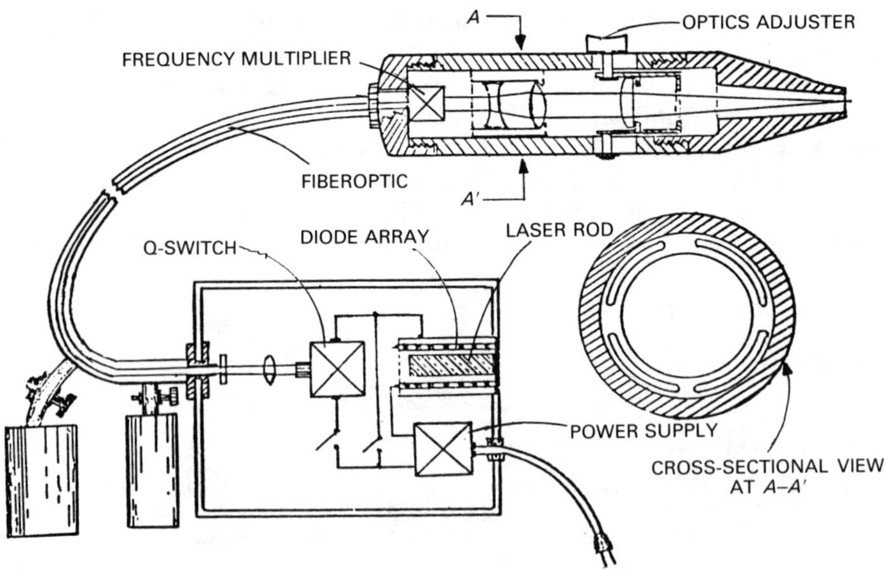

Figure 3-21 Another version of the diode-pumped laser system with a fiber-optic waveguide and dispensing stylus developed by the author.

Surrounding the laser-generating rod and the pump diodes is a thermoelectric heat dissipator, which is also used to modulate the wavelengths of the laser beam by controlling the temperature of the laser generator. When the laser rod is operating in the continuous-wave format, Q-switching provides shorter and more intense pulses than when the rod is pulsed by usual methods. The Q-switched laser beam is collimated in the dispensing unit and then projected on an output lens, which shapes the laser beam focus. The focusing lens is mounted in a cage that is movable to and from the point of the stylus to change the focus area of the lens. In medical work, this function of the stylus is very important because the surgeon can alter the focus size by moving the output lens with his thumb.

In microelectronic lead-bonding operations, the position-alterability of the lens is again important because the technician can change the bonding beam intensity of the laser by moving the lens up or down the tubular stylus. This changes the focus size and hence the beam intensity, since certain board thicknesses require higher intensity and others require lower intensity of the beam for welding, drilling, and scribing.

QUESTIONS AND PROBLEMS

3-1 Name the different gas laser systems. What are ionic lasers? At what mode do ionic lasers operate? What is the difference between a multiline and a monoline laser? At what wavelength does an argon laser emit? At what wavelength does a krypton laser emit? What are the relative cavity sizes of a krypton laser and a He-Ne laser?

3-2 What is a metastable transition state? How is photonic amplification achieved? Explain the difference between classical theory and quantum theory. How is He-Ne optically pumped? What is the recombination effect in a resonant cavity? Why is a beryllium oxide capillary used in a plasma discharge tube? Why is an invar resonator framework used for supporting a plasma tube? Discuss.

3-3 Why should an ion laser be linearly polarized? Should a laser beam be linearly polarized when used in a rangefinder? Explain why. In what application does signal-to-noise ratio become important? If a laser beam is collimated at an expansion of 10 times, how much does the beam diverge at the exit port? When do argon ions doubly ionize?

3-4 What will be the diameter of a laser beam incident on a target 80 m from the exit port, if the beam has a divergence of 0.1 mrad and the exit aperture is 8 mm?

3-5 Are most ion lasers inherently multiline emitters? How can one obtain a single line from a multiline laser emitter? Can such a laser emitter be tuned? If an external prism is used in a laser head, what function will the prism have? Explain. At what fundamental mode does argon laser operate? What is a Gaussian power distribution? At what power points is a Gaussian laser beam measured?

3-6 How does a randomly polarized laser differ from a linearly polarized laser? What are molecular lasers? How do they differ from ionic lasers? What is the most important molecular laser used in industry? What lasing elements enter into the composition of a carbon dioxide laser? In what ratio do these elements exist? In what two forms are carbon dioxide lasers manufactured? What are TEA lasers?

3-7 What other lasers do TEA lasers optically pump? Which CO_2 laser is more powerful, a sealed-in laser or a continuously flowing CO_2 laser? Explain why. Can continuous tuning be achieved with a flowing-gas carbon dioxide laser? How many wavelengths can be obtained from it? How much laser power per meter can be obtained from a CO_2 cavity?

3-8 If the carbon dioxide laser is tuned over a 150-MHz linewidth and is directed at a target in air, to what speed of the target can Doppler shift be measured? For what purposes are Spectra-Physics Model C900 series employed in their applications? Draw the vibrational energy levels for excitation of a CO_2 laser.

3-9 What are excimers? Are excimers stable outside the resonant cavity? Explain why. Where does interaction between an inert gas and a halide take place? Why is the discharge bore of an excimer laser passivated? Can excimer lasers operate either pulsed or in continuous mode? Why is a hydrogen thyratron used in conjunction with an excimer laser? How are the electrodes located in an excimer?

3-10 What are solid-state lasers? Is alexandrite a solid-state laser? In what general sizes do solid-state laser rods come? What is a dopant? Describe a finished solid-state laser rod. How are solid-state lasing elements optically pumped? What are some of the optical pumps that are used for pumping the lasing rod? What is the relation of the interelectrode resistance to the electric discharge length and to its cross-sectional area? How can interelectrode resistance be decreased? What is the dopant in a ruby rod? In Nd-YAG?

3-11 What is the difference between spontaneous radiation and stimulated radiation? How is each type produced? What are phonons? How are thermal stresses in a laser rod reduced? Why is a laser beam spatially coherent? When laser beams of equal frequencies are not in phase, what phenomenon occurs? If the laser beams of equal frequencies are out of phase, what phenomena can occur? Explain.

3-12 How many energy transition levels do ruby lasing ions have? Between what energy levels is spontaneous light emitted? Between what levels does stimulated emission occur? At what wavelength does a ruby laser emit? What is the dopant in a ruby laser?

3-13 How many energy transition levels does a Nd-YAG laser have? Which transition levels include the stimulated emission? What type of optical pump is used for a Nd-YAG laser? At what mode does a Nd-YAG laser operate? At what wavelength does Nd-YAG emit? What is the divergence range of Nd-YAG? What is the function of a quarter-wave plate when located in the resonant cavity? Can Nd-YAG operate in the ultraviolet spectrum? Explain how Nd-YAG is made to emit in the ultraviolet.

3-14 Explain how the polarization ratio of a Nd-YAG laser beam is increased in the system. State the harmonic wavelengths of Nd-YAG. How is a second harmonic of Nd-YAG produced? In what spectral range is the second harmonic of Nd-YAG?

3-15 What is an alexandrite laser? Explain. What are the phonons in an alexandrite laser transition? At what wavelengths does alexandrite emit? In what energy state does alexandrite operate in the continuous mode? What is vibronic format? Is this laser tunable? How many energy transition levels does it have? How is alexandrite frequency-multiplied? Explain.

3-16 Name three frequency multipliers. What is a nonlinear crystal? Why is a frequency-multiplied laser mixed with a fundamental beam of another laser? Draw a diagram for mixing a Nd-YAG harmonic beam with a tunable dye laser beam. Name three diode lasers. Do they operate in the pulsed mode or in continuous-wave mode? What are donors and acceptors? How does one determine the emission efficiency of a diode? State the equation.

3-17 How does one express the rate of photon emission per second in a laser diode? How can the peak emission of a laser diode be controlled? What are some of the applications of a laser diode? What is the radiant medium in a dye laser? How are dye lasers excited to radiation? Are dye lasers tunable? In what range? How can narrowing in a dye laser be accomplished? Which dye lasers have narrower linewidths? Explain why.

3-18 Draw the diagrams of continuous-wave and pulsed dye lasers. What are the differences between the pulsed and CW dye laser generators? Explain. Why can dye lasers be used for frequency doubling and for multiline runability? If narrowing of the spectral bandwidth of a dye laser is required, how can it be accomplished? What is the function of a Pockels cell in a dye laser? Explain.

3-19 What are metal vapor lasers? Give examples. Which metal vapor laser is projected for fusion work? Why? For what industry is the He-Cd laser used? With what type of resonator can beam divergence be reduced? What single line does an He-Cd laser emit? How do high temperatures affect He-Cd laser emission? What are F-center lasers? How are they produced and at what wavelengths?

SELECTED SOURCES OF RELATED DATA

Blauvelt, H., and A. Yariv. 1982. Higher pulsed power from AlGaAs lasers. *Appl. Phys. Lett.* 40(12):1029.

Brau, C. A. 1988. Free-electron lasers are ready to prove themselves. *Laser Focus/Electro-Optics* April, pp. 20–26.

Brederlow, G. 1983. *High-Power Iodine Laser*. New York: Springer-Verlag.

Cunningham, R. 1985. F-center laser emits nearly one watt at 1.6 μm. *Lasers and Applications* September, p. 32.

Diode pump stabilizes Nd-YAG laser output. 1985. *Lasers and Applications* May, pp. 50–52.

Forrest, G. T. 1988. Diode-pumped solid-state lasers. *Laser Focus/Electro-Optics* June, pp. 57–74.

Giuliano, C. R., et al. 1983. Can phase-conjugate resonators enhance laser performance? *Laser Focus* February, pp. 55–64.

Hilbig, R., and R. Wallenstein. 1982. Mixing in Xe and Kr produces tunable VUV. *Appl. Opt*. 21(5):913.

Hohla, K. L. 1982. Excimer-pumped dye lasers. *Laser Focus* June, pp. 67–74.

Holmes, L. 1986. Tunable dye lasers. *Laser Focus* February, pp. 70–82.

L&A Staff. 1984. Erbium solid-state laser. *Lasers and Applications* May, p. 50.

Letokhov, V. S. 1983. *Nonlinear Laser Chemistry*. New York: Springer-Verlag.

MIT's phototransistor InGaAsP as alternative to avalanche photodiode. 1980. *Laser Focus* February, pp. 81–92.

Muncheryan, H. M. 1989. Diode-Pumped Laser Instrumentation System. U.S. Patent No. 4,808,789.

Phelps, F. M. 1982. *MIT Wavelength Tables,* Vol. 2. Cambridge, Mass.: Massachusetts Institute of Technology.

Rudermann, W. 1979. Solid-state tunable laser sources. *Electro-Optic Systems Design* September.

Tsang, W. T. 1985. *Semiconductor Injection Diodes*. Orlando, Fla.: Academic Press.

Ungar, J., et al. 1985. High-power GaAlAs window lasers. *Lasers and Applications* September, pp. 111–114.

CHAPTER

FOUR

HOLOGRAPHY AND INTERFEROMETRY

4-1 INTRODUCTION

Holography is the laser-imaging process in which a three-dimensional recording of an object is made by use of two split laser beams. Each of the split beams is expanded sufficiently to cover both the object and the photographic plate on which the object is imaged. The hologram thus produced consists of wave patterns that possess both the amplitude and the phase variations of the wave front. An ordinary photographic plate can sense only a two-dimensional view—the width and the length or height—and the amplitude of the light received from the object. Human eyes can receive both amplitude (intensity) and phase (direction) variations of light reflecting from the object; therefore, a sensation of depth as well as width and height is perceived by the eyes.

In holography, the two split laser beams take the place of the eyes. One beam directly illuminates the photographic plate as a reference wave front, while the other beam, after having been dispersed from the surface of the object, impinges on the photographic plate. The developed photographic plate contains a latent image pattern comprising a mixture of phase fronts and laser light intensity. The image is diffused, formed by the interference of the two split laser beams. In order to view the image of the object, the photographic plate (hologram) is illuminated with a laser beam that is expanded and projected on the plate from its back side and at the same angle of incidence as the original laser beam used to construct the image. The process of forming an image on a photographic plate is illustrated in Fig. 4-1, and the process of image reconstruction is shown in Fig. 4-2.

It should be noted that photography utilizes only the intensity and the wavelength of the light from the object, while holography uses the intensity, wavelength, and phase

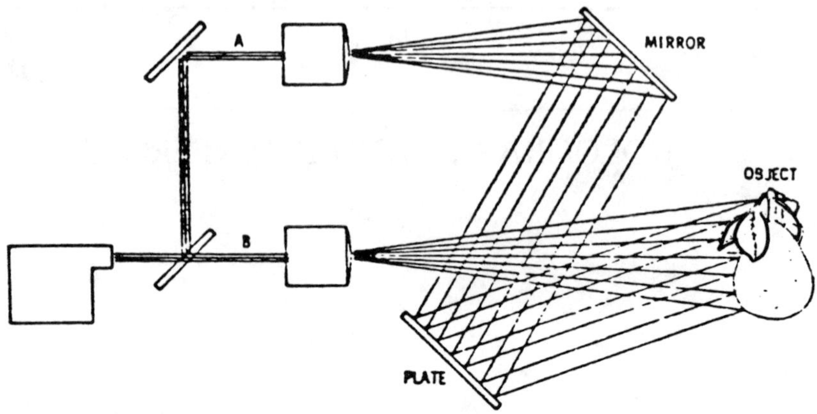

Figure 4-1 Holographic image formation by two laser beams.

content of the laser light; the two laser beams replace the optical lenses of a binocular camera. The hologram is an image of the interactions of these three attributes in a pattern of interference. This pattern is formed by the two interference fringes of the laser waves advancing from two different directions with two different time relations. Accordingly, the wave fronts incident on the photographic plate are reflected with different time and space relations from different angles of the holographic subject.

In Fig. 4-1, one of the split laser beams, such as A, is the reference laser beam and the other, B, is the beam dispersed from the illuminated object. The wave fronts of the two beams become superimposed on the photographic plate to impress on it a visually incoherent latent image. That is, the image appears to be a conglomeration of wavy lines and clusters of spots with no "picture" when viewed with transmitted light. However, when the hologram is held against a spatially filtered and expanded laser beam at an angle equal to the original laser illumination angle and is observed at right angles to the plane of the plate, as shown in Fig. 4-2, a three-dimensional image of the

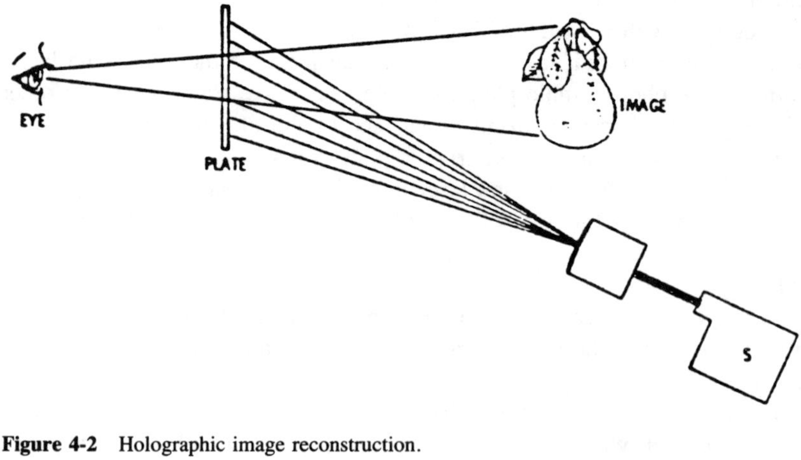

Figure 4-2 Holographic image reconstruction.

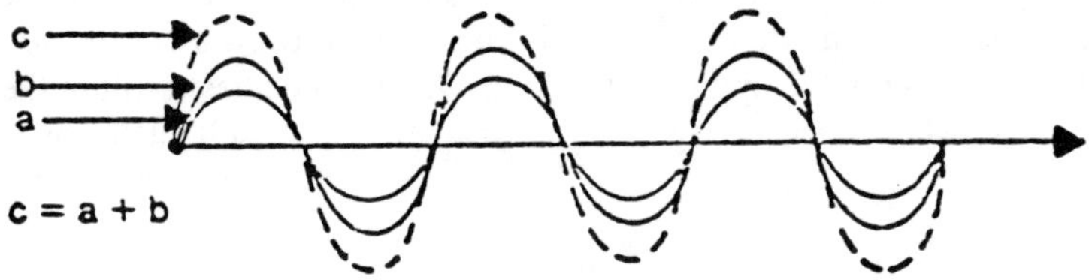

Figure 4-3 Illustration of constructive interference.

object can be observed. As the viewer moves his eyes sideways, he can see the sides of the object as if it were suspended in air.

An interesting effect is that if the hologram is cut into, for instance, four sections, each section will still be capable of displaying the entire hologram or image of the object, with some sacrifice in resolution. Furthermore, either a photographic negative plate or a positive plate can be used to make the hologram. Of course, in either case, the viewing angle should correspond to the angle of the holographic beam setup.

Akin to holography is interferometry, which is the measurement of wavelengths or small distances by means of an interferometer, in which two wave trains are superimposed to interfere with each other. There are two types of interference: constructive interference, in which the resultant intensity is greater than the intensity of the separate waves, and destructive interference, in which the resultant intensity is smaller than either of the two beam intensities and may be zero. An interferometer is an instrument that uses at least two interfering light waves in the measurement of wavelengths of light and relatively small distances. One of the most popular devices is the Michelson interferometer, which will be studied presently.

A laser beam travels through space by transverse vibrations characterized by amplitude and wavelength. Let us assume that two laser beams, as illustrated in Fig. 4-3, are propagating in the same direction with equal amplitudes and phase formats. Since the peaks and troughs of the two beams occur at the same time, the two waves are said to be in phase and temporally coherent, and their amplitudes will add; this will be constructive interference. When the amplitudes of the two waves are of different magnitudes and opposite in direction, as illustrated in Fig. 4-4, the resultant amplitude will be smaller—destructive interference.

In Fig. 4-3, beam b has a larger amplitude than beam a, and since they are superimposed, their magnitudes add and the resultant beam c is larger in amplitude. The optical effect of this phenomenon will be one of magnified beam intensity, as depicted by the dashed lines c.

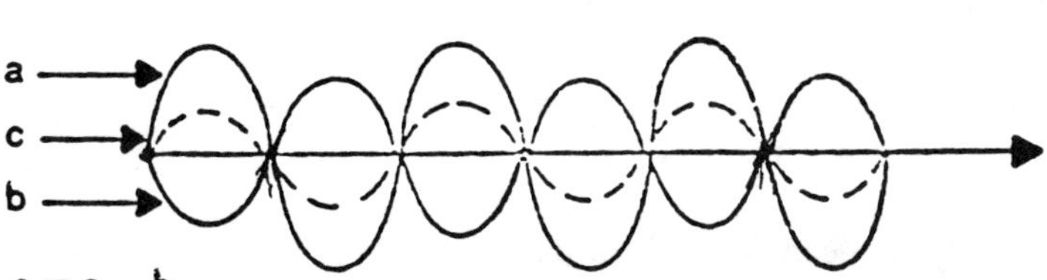

Figure 4-4 Illustration of destructive interference.

In Fig. 4-4, two rays of different amplitudes and 180° out of phase are propagating in the same direction, the ray a amplitude moving in the positive (upward) direction and the ray b amplitude moving in the opposite direction. Their intensities result in the third ray, c, which is the difference between rays a and b. The resultant beam is dimmed.

4-2 HOLOGRAPHIC PRINCIPLES

Since all laser beams emit several different frequencies, and hence wavelengths, the coherence length of the beam should be known, so one may determine how much tolerance the laser system possesses—that is, how large an object can be holographed with the laser beam. It was already mentioned that a He-Ne laser emits three wavelengths, the 6328-Å wavelength being the most dominant, and is used for holographic work. The frequency difference between the several beams can be calculated to determine the coherence length or the temporal coherence of the beam. Thus, in order that the two coherent beams used in holography interfere, the distances that they travel before becoming incident on the photographic film must be equal to optimize the contrast of the latent image in the film. The relationship between the beam frequency and the cavity length (distance between the two mirrors) may be given as

$$\Delta f = \frac{C}{2L} \tag{4-1}$$

where Δf is the frequency difference between the adjacent modes, C is the velocity of laser light, and L is the distance between the two mirrors or the optical cavity length.

The principle involved in this discussion, including Eq. (4-1), may be proved by a simple experiment. Using a beamsplitter, two split laser beams are reflected by means of mirrors onto the photographic plate, with one impinging on the plate directly from one of the mirrors and the other impinging on the plate after it is reflected from the object. The two distances can be measured by means of a metric ruler. If the distances are made equal, the contrast of the hologram in the plate will be maximum. If the distance of one of the mirrors is changed slightly, such as by several centimeters, the hologram will lose contrast, indicating that the two beams are slightly out of coincidence. When the distance is increased still further, the hologram may completely disappear or become indistinguishable.

Accordingly, in making a hologram the operator must observe the following simple rules to achieve the optimum results:

1. The table on which the laser system and the photographic plate are set up must be completely stabilized so that no movement can occur during exposure of the photographic plate.
2. An inexpensive laser system can be used for the work.
3. To shorten exposure time, either a fast photographic plate must be employed or the laser intensity must be increased to a suitable power level.

4. If a professionally constructed holographic table is used, the time of exposure can be increased to produce high-contrast diffraction fringes in the latent image; this procedure results in very sharp holographic detail.
5. Use of an optical filter may improve the contrast.
6. For safety reasons, the operator should never look directly in the direction of the oncoming beam, either in making the hologram or in viewing the reconstructed hologram by transmitted laser light.

4-3 HOLOGRAPHIC APPLICATIONS

The holographic technique has numerous industrial and medical uses. In industry, defects in structural parts or materials, porosity in castings, and unbonded areas in bonded honeycomb wing sections of aircraft can be visualized by the holographic technique. (See the latter application in Chapter 2.)

A unique application of holography is found in the medical field (see Chapter 10). Using this method, three-dimensional views of malignant growths can be obtained. The diseased structures of tissues in the human skull, heart, lungs, liver, and other body cavities can be visualized. External to the body, a hologram can show incipient breast cancer, dermal afflictions, ocular dislocations, superficial stress areas, and so forth.

Another possible application of holography is in making three-dimensional television image sets by employing blue, green, and red laser beams—preferably from dye lasers, since they can be tuned to different spectral colors. Work in this direction is reported to be under way by several laser companies in the United States and Japan.

Hadron Holocamera

A versatile holographic system, formerly manufactured by the KORAD division of Hadron, is shown in Fig. 4-5. This system is a completely self-contained unit weighing about 200 pounds. It can be used for reflective holography in nondestructive testing of materials and for transmission holography to produce three-dimensional profiles of stationary parts, moving particles, liquids or gases for particle sizing, flow visualization, and so forth. Because of the Q-switched operation at 20 nsec exposure time, it "freezes" the moving particles on a permanent hologram. A conventional photographic camera can be mounted on it to photograph the holograms. The unit operates from a 115-V, 15 to 20 A, single-phase power source and produces a single-mode pulse of energy ranging from 0.025 to 10 J, with a pulse repetition rate of 1 to 60 ppm (pulses per minute). A hologram up to 5 inches square can be obtained with this system.

In Fig. 4-6 the Hadron Holocamera laser system is shown with its cover removed to display the various optical and electronic assemblies it contains and their positional relationship to each other. It should be noted that this system provides the smallest packaging construction possible in order to obtain the maximum flexibility of use and versatility of the equipment. It is a self-contained and self-sufficient instrument of highest utility.

Figure 4-5 Hadron holographic laser system Model KHCl. (Courtesy KORAD Division, Hadron, Inc.)

IBM Document Scanner

A document scanner that consists of a rotating ring of individual phase-volume holograms and uses a laser beam to illuminate the hologram from the inside of the ring has been developed by IBM researchers. The interferometric laser beam is captured by a detector, and the data thus obtained are translated by a computer-aided system. By

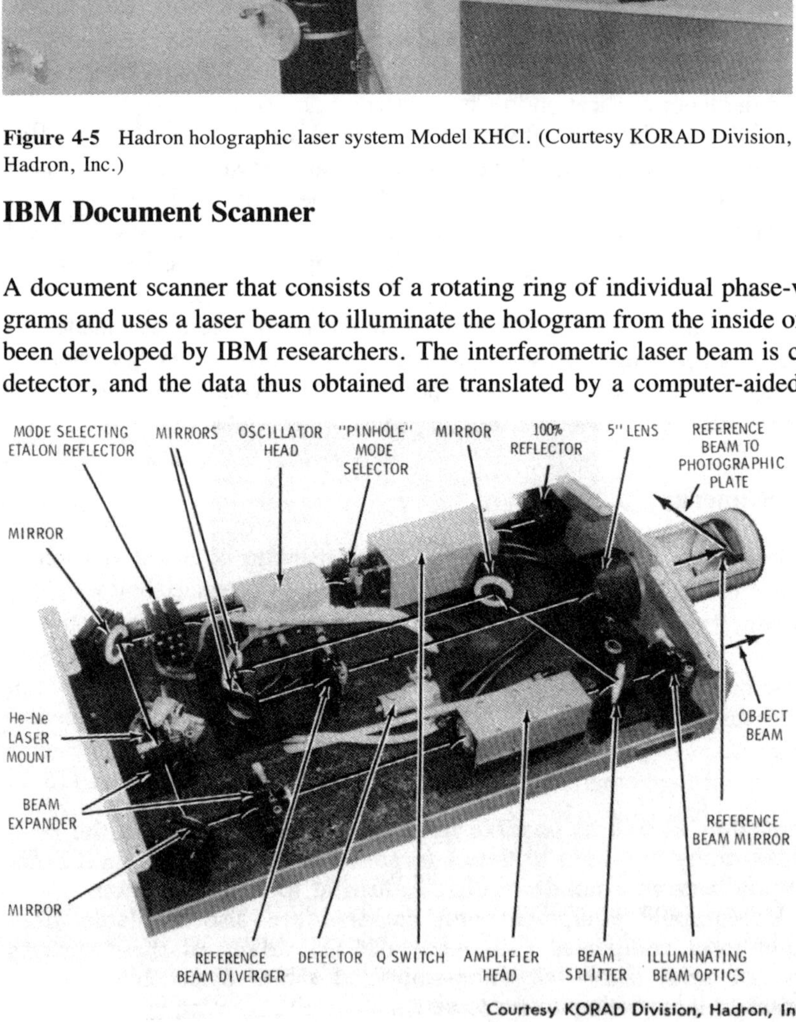

Figure 4-6 KORAD Holocamera laser system with cover removed to display internal features.

rotating the ring of holograms, the document is scanned across its width with a resolution of 60 μm on the finished product and a light-collection efficiency of 0.5%.

Although the system employs a He-Ne laser beam for economy, it is said to be capable of scanning and recording documents in color by using various laser beam wavelengths corresponding to the fundamental colors. The different-wavelength colors are sensed by corresponding detectors, and the combined wavelength effects form the final color copy of the document. It is claimed by the developers of the process that the system eliminates geometric problems that are ordinarily encountered with conventional rotating-mirror scanners.

IBM Two-Photon Holographic Technique

A two-photon holographic technique developed by IBM employs both ultraviolet and infrared radiations. The hologram consists of a self-developing polymer, such as polycyanoacrylate, in which the organic molecule biacetyl is dissolved. The ultraviolet radiation from a mercury vapor lamp raises the biacetyl molecules to an excited singlet state, from which they decay by a nonradiative process to a triplet state. Following this, the infrared radiation from a krypton ion laser beam excites the biacetyl molecules to a higher-energy triplet state, which produces a latent image that is irreversible. The hologram formation is due to the density change upon radiation and hence the change in refractive index of the plastic material, with an efficiency of more than 10%.

Apollo Lasers Holographic Camera

A holographic camera that uses a thermoplastic film material for photographing has been developed by Apollo Lasers, Inc. The camera, Model HSB-100, incorporates a nonchemical development process feature within its housing. The film is electrostatically sensitized immediately before exposure. On exposure of the film, a static charge forms at the exposed areas because of the photoconductivity in the process. The static charge distribution is identical to the exposure light distribution pattern on the sensitized film. The exposed film is developed within 0.15 second by warming it; the hologram is a relief image that is consistent with the charge distribution pattern of the film. It can be viewed 3 seconds after the development. The image may also be recorded on a videocassette for viewing on a television screen.

Single-Beam Holography

A hologram can also be made with a single laser beam projecting from the laser source. The beam is divergent as it leaves the exit aperture of the source, as shown in Fig. 4-7. The divergent beam, after reflection from a mirror, falls on the photographic plate or film supported on a fixture erected on a table. The object is placed between the film and the table top, as shown in the figure. The divergently projected laser beam is the reference beam, and after it passes through the film it impinges on the object. The

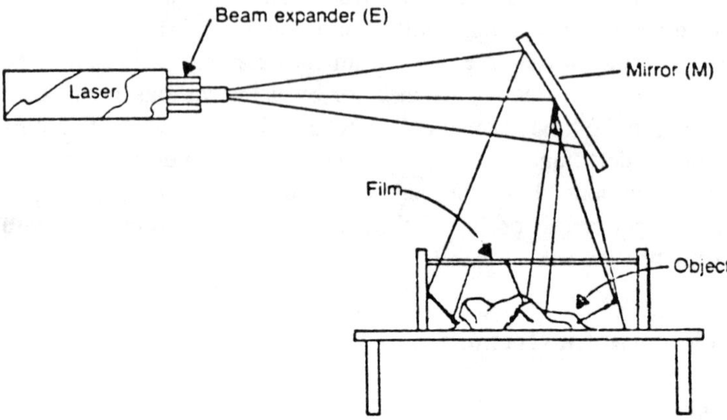

Figure 4-7 A single-beam retroreflected-beam holographic technique.

retroreflected laser beam from the object falls on the bottom side of the film, superimposed on the reference beam incident on the upper side of the film. The reference and retroreflected beams interfere with each other on the film and form the hologram. After development of the film, the hologram is viewed by the usual technique, a transmitted laser beam.

The field of holography is growing rapidly as new techniques are devised for its application and instruments are developed to implement these techniques. In industry, besides its routine use for nondestructive tests with specialized equipment, holography is invaluable in the design, development, and testing of high-performance fuel nozzles for gas turbines and carburetors and fuel injectors for combustion chambers and solid- and liquid-fuel rocket engines. Wind-tunnel flow visualization tests of projectiles passing through air can be made conveniently with a mobile holographic system having a mounted camera, such as the system shown in Fig. 4-5. This equipment is easily positioned at any angle with respect to the projectile. Also, holograms of objects of art can be made in three-dimensional views for authentication and restoration purposes.

4-4 INTERFEROMETRIC CONSIDERATIONS

We have already learned that when two coherent beams are executing sinusoidal motion and propagating in the same direction, their amplitudes either add or subtract, so that the beams interfere with each other and form interference fringes. We have also learned that these fringes may be employed to measure wavelengths or small distances. The instrument with which such measurements are made is known as an interferometer. The most original and well-known interferometer is the Michelson interferometer. Other types of interferometers are either modifications or derivatives of this instrument. An interferometer can also measure the quality of lenses, prisms, and optical mirrors, as well as the index of refraction of optically transparent materials. A diagrammatic illustration of the Michelson interferometer is shown in Fig. 4-8.

The Michelson interferometer consists principally of a coherent light source, two reflecting mirrors M_1 and M_2, a beamsplitter B, and an eyepiece E. One of the mirrors, such as M_1, is adjustable so that its distance can be altered with respect to the beamsplitter. When a beam of coherent light is projected from the light source on the beamsplitter B, the beam is partially reflected to mirror M_1 and partially transmitted to mirror M_2. After reflection from mirrors M_1 and M_2, the beam returns to the eyepiece E. If the distances of the two mirrors from the beamsplitter are equal, there is an integral number of full wavelengths in each distance. The beam will interfere constructively and a bright field will be observed at the eyepiece. If the distance between the beamsplitter and, for instance, M_1 is increased, for example, by a quarter wavelength ($\frac{1}{4}\lambda$), the optical wavelength of beam 1 will increase by $\frac{1}{2}\lambda$, destructive interference will occur, and a dark field will be observed at the eyepiece. These bright and dark fields observed at the eyepiece are known as optical fringes, and they aid in the measurement of length or distance with the instrument. For a given distance measurement, a definite number of fringes is counted and translated into a numerical quantity in terms of the wavelength of the beam. An equation giving this relation is

$$\lambda = \frac{2D}{N} \tag{4-2}$$

and

$$D = \frac{N\lambda}{2} \tag{4-3}$$

or

$$N = \frac{2D}{\lambda} \tag{4-4}$$

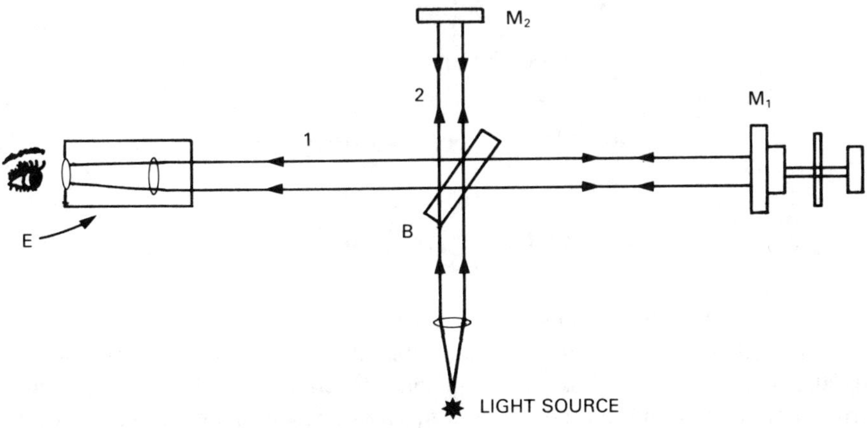

Figure 4-8 The Michelson interferometer, simplified.

where D is the distance mirror M_1 is displaced, N is the number of successive dark fringes counted, λ is the wavelength of the light beam, and the numeral 2 stands for the two passages of the beam through the beamsplitter.

Example The distance between two points has been measured with a Michelson interferometer employing a coherent beam from a 6328-Å He-Ne laser. If 600 fringes have been counted as the mirror M_1 is displaced, calculate the distance between the two points.

Solution

$$\begin{aligned} D &= \frac{N\lambda}{2} \\ &= \frac{600 \times 6328 \times 10^{-8}}{2} \\ &= 3 \times 6328 \times 10^{-6} \\ &= 0.018984 \text{ cm} \end{aligned}$$

Example Mirror M_1 in a Michelson interferometer was moved 0.004 cm and 200 fringes were counted. What is the wavelength of the radiation beam in this measurement?

Solution

$$\begin{aligned} \lambda &= \frac{2D}{N} \\ &= \frac{2 \times 0.004 \times 10^{8}}{200} \\ &= 0.004 \times 10^{-6} \text{ cm or 4000 angstroms} \end{aligned}$$

4-5 FOURIER TRANSFORM HOLOGRAPHY

Jean Fourier, a French mathematician, advanced the theorem that any periodic function can be represented by a trigonometric series. In this function, the frequencies of waves are integrally associated with those of the original function. Therefore, any periodic waveform, such as a simple harmonic motion, can be resolved into amplitudes and phases. Thus, when any two or more simple harmonic waves having equal frequencies

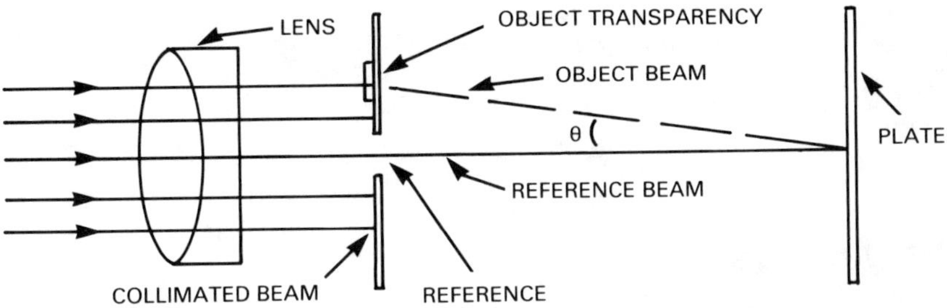

Figure 4-9 Simplified diagram for producing a Fourier transform hologram.

at varying amplitudes and phases are superimposed, the resultant wave contains implicitly the same frequency and may be represented by a simple sinusoidal function. When such an element cannot be represented by a simple periodic function, it is indicative of complex waves, which would be outside the scope of this study and unrelated to our concept of holographic representation of optical waves.

Fourier's theorem, however, can be applied to producing three-dimensional holograms, using the Fourier optical transform concept. Such a hologram is a photographic record of the interference patterns intermediate to the object amplitude and a Fourier transform, taken against a coherent background created from a reference point. To elucidate this, assume a condition of propagating waves with equal frequencies at different amplitudes and phases, as illustrated in Fig. 4-9, which depicts lensless Fourier transform holography. This hologram can be reconstructed when the far-field condition has been satisfied. This is accomplished by uniformly illuminating the hologram at its zero-order phase relation. Both real and virtual images can be reconstructed by projecting a collimated laser beam through the transform. The resultant image can also be photographed as desired.

4-6 USES OF HOLOGRAPHY IN INDUSTRY

A high-performance holographic recording system for nondestructive testing of materials and parts is available from Newport Corporation, Fountain Valley, California. The system is capable of measuring thermal effects, vibrational conditions, and strain and stress points on the parts measured. Such parts may be computer disk drives, automobile frames, and various other materials. The system is widely used in engineering design applications. The system, shown in Fig. 4-10, records test measurements on a thermographic medium, on which the hologram is processed electronically within 10 minutes. The record can be electronically erased and reused for hundreds of other holographic measurements. The system employs a 7-mW He-Ne laser for holographic measurement of objects up to 1.5 feet in size. For larger structures, an argon laser beam is used with the same apparatus, shown in Fig. 4-11.

With this system, stresses, minute abnormal structures on surfaces, deformations in manufactured parts, and other flaws can easily be examined and measured. The holographic test results from this system can be correlated with topographic stress

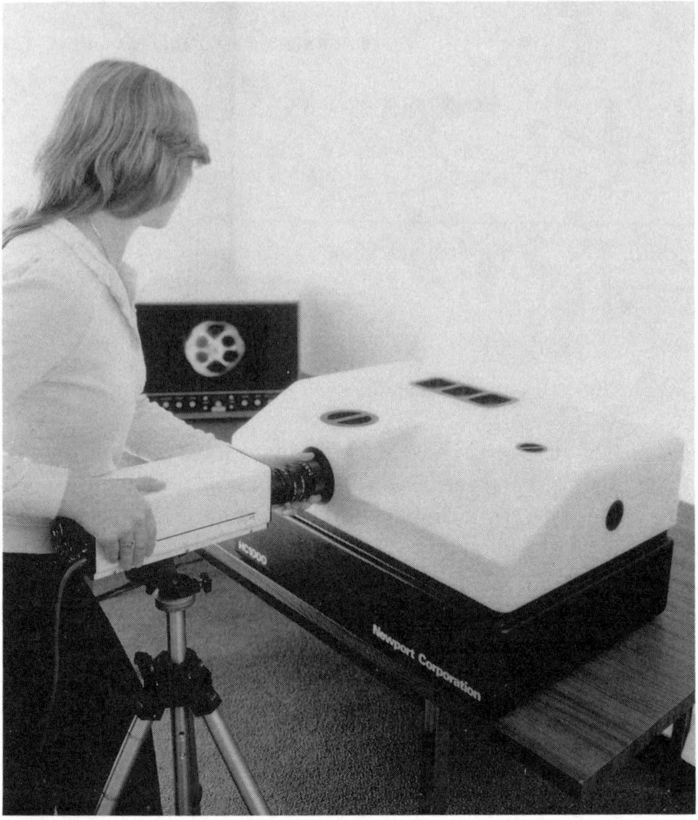

Figure 4-10 The Newport Corporation's Holocamera.

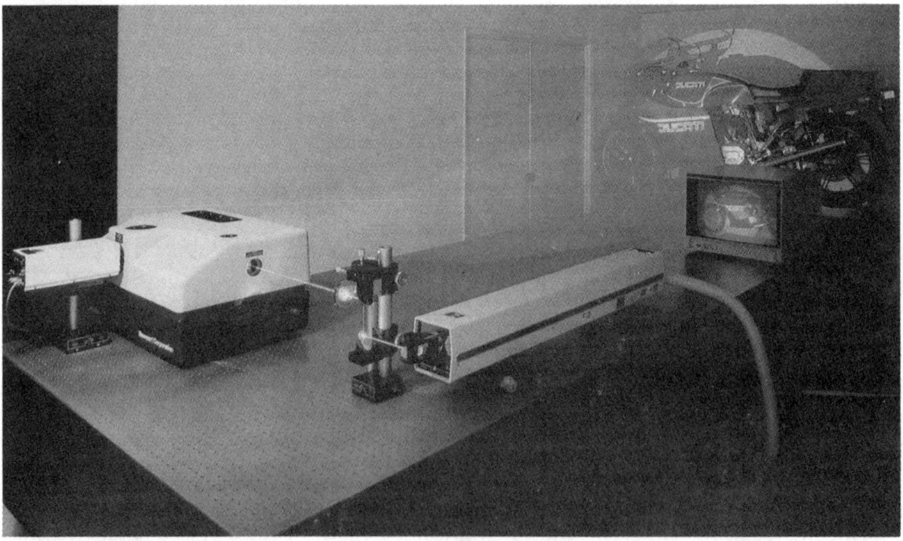

Figure 4-11 Newport Corporation's Model HC-1000 holographic camera, showing the test setup for measuring the mechanical resonance behavior of a motorcycle fairing.

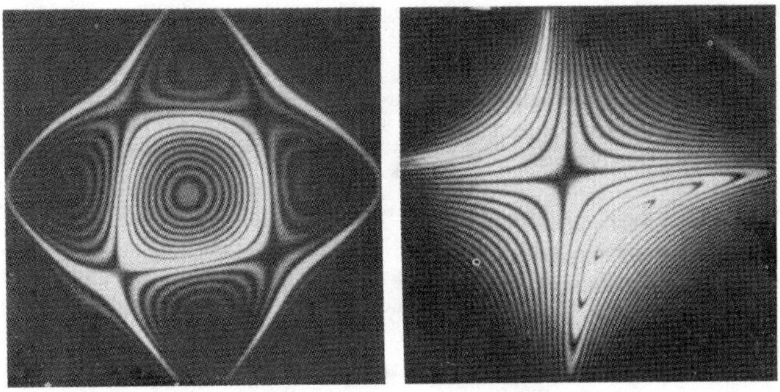

Figure 4-12 Resonant modes of a vibrating plate. Courtesy of Newport Corporation.

analysis known as hybrid structure design analysis to aid in the design of engineering parts to precision tolerances. Resonant modes of a vibrating plate made with the system are shown in Fig. 4-12.

QUESTIONS AND PROBLEMS

4-1 What is a hologram? What do the recordings in a hologram consist of? How is a hologram made? Explain. How is a hologram viewed? What is an interference fringe? Can you destroy a hologram by cutting it into four equal parts? What is interferometry? What is constructive interference? What is destructive interference?

4-2 If two laser beams are traveling in the same direction with different amplitudes but temporally in phase, what will be the resultant interference phenomenon? If two rays of different amplitudes and 180° out of phase are moving in the same direction, what will be the resultant interference phenomenon?

4-3 Why should the coherence length of a laser beam be known? What is temporal coherence? What relation do two split beams forming a hologram have to each other? In Eq. (4-1), calculate the frequency difference between the adjacent modes if the cavity length is 40 centimeters. When will the contrast in a hologram be maximum? Under what conditions will the contrast be poor? State six rules for producing acceptable holograms.

4-4 What are some of the applications of holography? What are the component parts of a holographic camera? How does holography work in the inspection of honeycomb wing sections of an aircraft? Why is an ultrasonic device used in honeycomb inspection? How would you determine a defect in a honeycomb section of a wing?

4-5 What is meant by reflective holography? What is meant by transmission holography? Why is Q-switching used for holographic work? What unique application is found in the medical field? How does it operate? Can holography be used in three-dimensional imaging of television pictures? How is a rotating-ring-type holographic setup operated? How does an IBM holographic system work? How good is the resolution that can be obtained from the IBM system?

4-6 How does an IBM holographic system record documents in color? What is the advantage of the IBM system over other holographic systems? How does the Apollo Lasers camera using a thermoplastic film material operate? How rapidly can a holograph be produced with the Apollo Lasers camera? What is a two-photon holographic technique?

4-7 How does the IBM holographic technique compare with that of the Apollo Lasers camera? How are biacetyl molecules excited to produce the holographic image? What is the mechanism of image formation in the IBM holographic technique? Discuss.

4-8 How is holography used in aircraft structural parts? What one factor aids the design of these structures? When is a holographic image considered to be real? When is it considered to be virtual? What is the importance of making a holograph cylindrical? Discuss a technique for making a hologram in a wide-mouthed glass jar.

4-9 What is a Michelson interferometer? What is its function? Can distances up to 100 cm be measured with it? Explain how. What does a Michelson interferometer consist of? Referring to Fig. 4-8, if the distance between the beamsplitter and mirror M_1 is increased a quarter-wavelength, what becomes of the optical wavelength of the laser beam? What is the resultant type of interference called? What would then be observed at the eyepiece?

4-10 What are the optical fringes that are seen in a Michelson interferometer? What are these fringes used for determining? What is the relation of the laser wavelength with respect to the position of mirror M_1 and to the number of successive fringes that pass across the interferometer field?

4-11 The distance between two points has been measured with a Michelson interferometer employing a coherent beam of 4000-Å wavelength. If 200 fringes have been counted as mirror M_1 (Fig. 4-8) is displaced, calculate the distance in millimeters between the two points.

4-12 To determine the wavelength of a laser beam, the Michelson interferometer is used, and mirror M_1 is displaced 0.008 cm. If 400 fringes were counted, what was the wavelength of the beam used? Convert the wavelength and state it in angstroms.

SELECTED SOURCES OF RELATED DATA

Abramson, N. 1982. *Making and Evaluation of Holograms*. New York: Academic Press.

Apollo Lasers, Inc. 1984. Technical brochure on holography. Chatsworth, Calif.

Beiser, L. 1983. Holographic scanning. *Electro-Optics Laser Technol.* February, pp. 20–21.

Caulfield, H. J. 1970. *Applications of Holography*. New York: Wiley-Interscience.

Caulfield, H. J. 1983. Computer generated holography. *Lasers and Applications* May, pp. 59–64.

Holometry for mining problems. 1985. *Photonics Spectra* January, p. 68.

Iscoff, R. 1988. X-ray holograms created at Livermore. *Lasers and Optronics* January, p. 28.

Iuchi, T., and F. Tanaka. 1987. Interferometric Measurement of Ultrasonic Displacements. U.S. Patent No. 4,619,529.

Kelly, E. 1985. Holographic inspection procedures. *Lasers and Applications* June, pp. 85–86.

L&A Staff. 1984. Holograms catch public eye. *Lasers and Applications* May, pp. 38–42.

Lamping, S. 1986. Holograms open new window of opportunity. *Lasers and Applications* January, pp. 18–22.

Letokhov, V. S. 1983. *Nonlinear Laser Chemistry*. New York: Springer-Verlag.

Levanson, M. D. 1982. *Introduction to Nonlinear Spectroscopy*. New York: Academic Press.

McKeller, A. R. W., et al. 1982. *Laser Spectroscopy V*. New York: Springer-Verlag.

Mikes, T. 1983. Applications of holographic elements. *Photonic Spectra* November, pp. 40–44.

Mishima, T., and K. C. Kao. 1982. Laser interferometry for high spatial resolution (thin films). *Appl. Opt.* 21(6):1101.

Multiple-diode-laser gas-detection spectrometer. 1988. Jet Propulsion Laboratory, NASA Technol. Briefs, vol. 12, no. 4 (April).

Pepper, D. M. 1982. Nonlinear optical phase conjugation. *Laser Focus* January, pp. 71–77.

Pinnow, D. A. 1984. Holographic nondestructive testing: Aiding mechanical design. *Laser Focus* July, pp. 60–66.

Riech, S. M. 1980. Resonant Laser-Beam Choppers, technical brochure. Woodside, N.Y.: Bulova Watch Company.

Smith, H. M. 1975. *Principles of Holography,* 2d ed. New York: Wiley-Interscience.
Solimar, L., and D. J. Cooke. 1981. *Volume Holography and Volume Gratings*. New York: Academic Press.
Stark, H., ed. 1982. *Applications of Optical Fourier Transforms*. New York: Academic Press.
Tolansky, S. 1973. *An Introduction to Interferometry*. New York: Wiley-Interscience.
Valdmanis, J. A., and G. Mourou. 1986. Picosecond opto-electronics. *Laser Focus* February, pp. 84–96.
Van Bally, G. 1979. *Holography in Medicine and Biology*. New York: Springer-Verlag.

CHAPTER

FIVE

LASER METROLOGY

5-1 GENERAL REVIEW

One of the most flourishing businesses in technology since 1960 has been the metrological application of laser radiation in industry. Equipment using gas laser generators has been used in laser gyroscopes, aligning, surveying, rangefinding, target designating, and gauging of manufactured parts. Since the laser beam is monochromatic, high-intensity, and unidirectional, measurements once considered to be difficult or impossible have become feasible using a laser beam. Precision instruments for metrological applications have been constructed and are in use at the present time. They perform various metrological operations not only accurately but rapidly and effectively, requiring fewer operators and more reasonable costs than are possible with conventional methods.

Some of the most common and important metrological laser instruments are those that are employed in tunnel construction, pipelaying, machine tool aligning, land surveying, rangefinding, and gauging of manufactured parts and cast hardware. Laser instruments are also used for guidance of earth-moving and trenching equipment.

5-2 LASER ALIGNMENT SYSTEMS

A laser alignment system consists of a continuous-wave laser beam generator, such as a He-Ne or a GaAs laser, mounted on a platform or on a sturdy tripod; a telescope to view the location of the laser beam target; and a remote detector or screen on which the laser beam is projected. The function of such a system is to determine straightness, levelness, and accuracy of the part or structure measured.

The laser beam employed in metrological work is spatially filtered and then expanded, when necessary, to about 10 diameters or more after it leaves the output mirror of the unit. The expanded beam is collimated to a perfectly symmetrical cross-sectional configuration, free from ghost fringes or birefringent effects. Ordinarily, a 0.5-mW He-Ne laser beam is sufficient to accomplish the required operation. It is relatively safe for the operator to use the beam without wearing safety goggles (although it is recommended to use them at all times). Thus, an excellent compromise between high visibility and operating safety can be achieved.

In operation, the collimated laser beam is projected from the laser unit, supported at a starting point, to a target, which may be a specially prepared screen, a photodetector, or a retroreflector. In pipelaying, the hub of the pipe may be marked on the screen and the laser beam projected on this mark as a reference. This procedure ensures that all structures to be aligned or jigs to be erected for building the system are aligned accurately with respect to the laser beam. Since the laser beam does not sag, as a string or a wire reference line does, all parts built around the setting can be aligned to an accuracy of less than 1 mm in a structural length of 80 to 100 feet. Although the He-Ne laser beam has low divergence, such as 0.2 arc second, for long distances up to 1000 feet the beam diverges slightly. However, the incident beam on the target element is still sharp for easy determination of the center point where the mark is located.

The advantage of employing a photodetector is that the *X-Y* axis can be accurately localized by an indicator provided on the laser head. The indicator is then electrically connected to the photodetector through an elongated conductor. Alternatively, the indicator may be mounted on a unit containing the photodetector with an *X-Y* axis. An illustration of a pipe alignment system in operation is shown in Fig. 5-1. In this case, the laser beam is projected over the pipes for alignment.

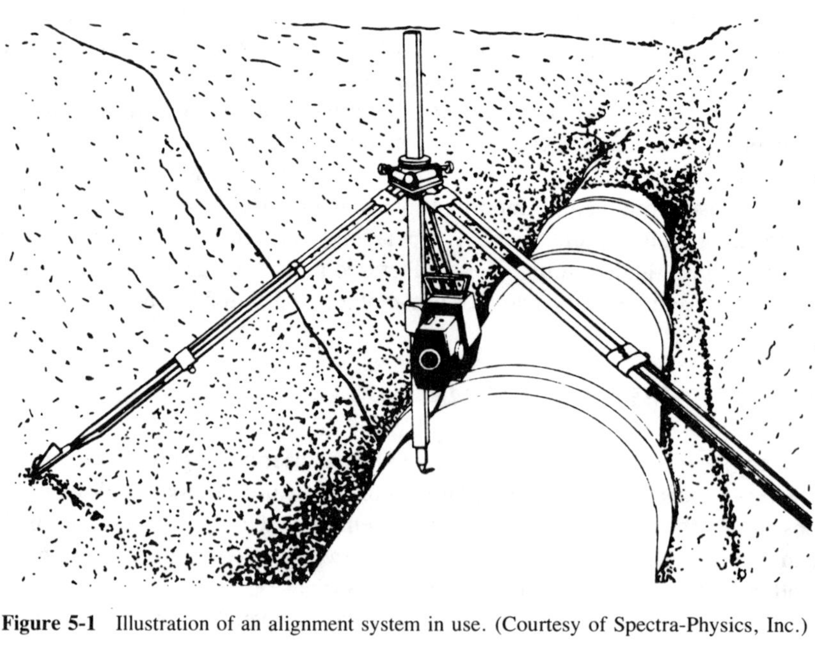

Figure 5-1 Illustration of an alignment system in use. (Courtesy of Spectra-Physics, Inc.)

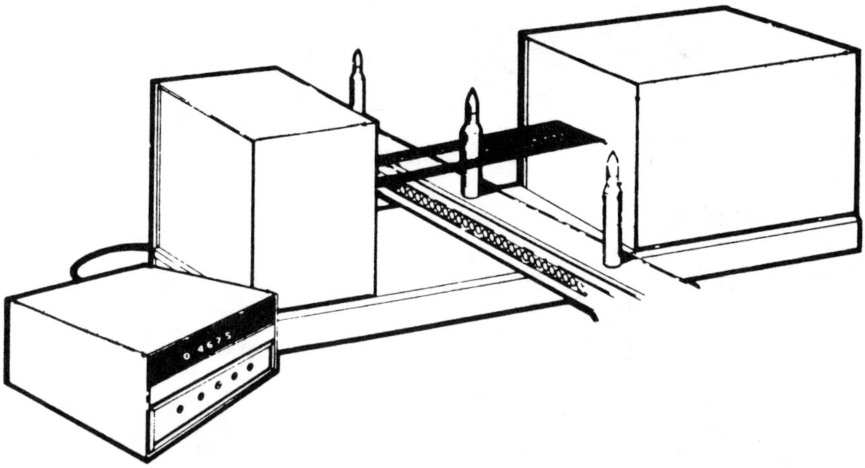

Figure 5-2 A typical laser gauge in operation.

5-3 LASER GAUGING INSTRUMENT

A laser gauging instrument for rapidly measuring the dimensions of parts without physically containing them is shown in Fig. 5-2. The device consists of a 1-mW He-Ne laser generator whose beam is collimated into a ribbon configuration. Parts to be measured pass through the beam, and their dimensional profiles are numerically displayed on a digital readout panel at the rate of 10 readings per second. The range of the gauging system is 0.015 to 1.75 inches, and it has an accuracy of ±0.15%, as reported. A large number of quality assurance measurements can be made with it very quickly by setting up the instrument to the particular tolerance requirement.

5-4 LASER GAUGING AND SORTING SYSTEM

An automatic laser gauging and sorting system improvised by the author is schematically represented in Fig. 5-3. An expanded and collimated He-Ne laser beam from the laser source (L) is projected through the beam collimator (BC) onto a mirror, which bends the beam 90° and reflects it through a 30-mm-long slit (SL) onto a beamsplitting mirror (M_1). Mirror M_1 is exactly 50% transmissive and 50% reflective, so half of the beam intensity passes through mirror M_1 and the other half is reflected to front-surface mirror M_2. The beam then reflects from M_2 to front-surface mirror M_3, which deflects the beam 90° and projects it to photodetector PD_1. The beam transmitted through the beamsplitter M_1 impinges on photodetector PD_2. Thus, both of the photodetectors receive laser radiation of the same intensity. Each of the photodetectors, PD_1 and PD_2, forms a leg of a resistance bridge circuit (BR), and the other two legs of the bridge are formed by resistors R_1 and R_2. Bridge BR is energized by a rectified current generated at REC; this current is impressed on the bridge by a voltage from 6 to 10 V. The output of bridge BR is connected to an amplifier (A), whose amplified output is fed to a

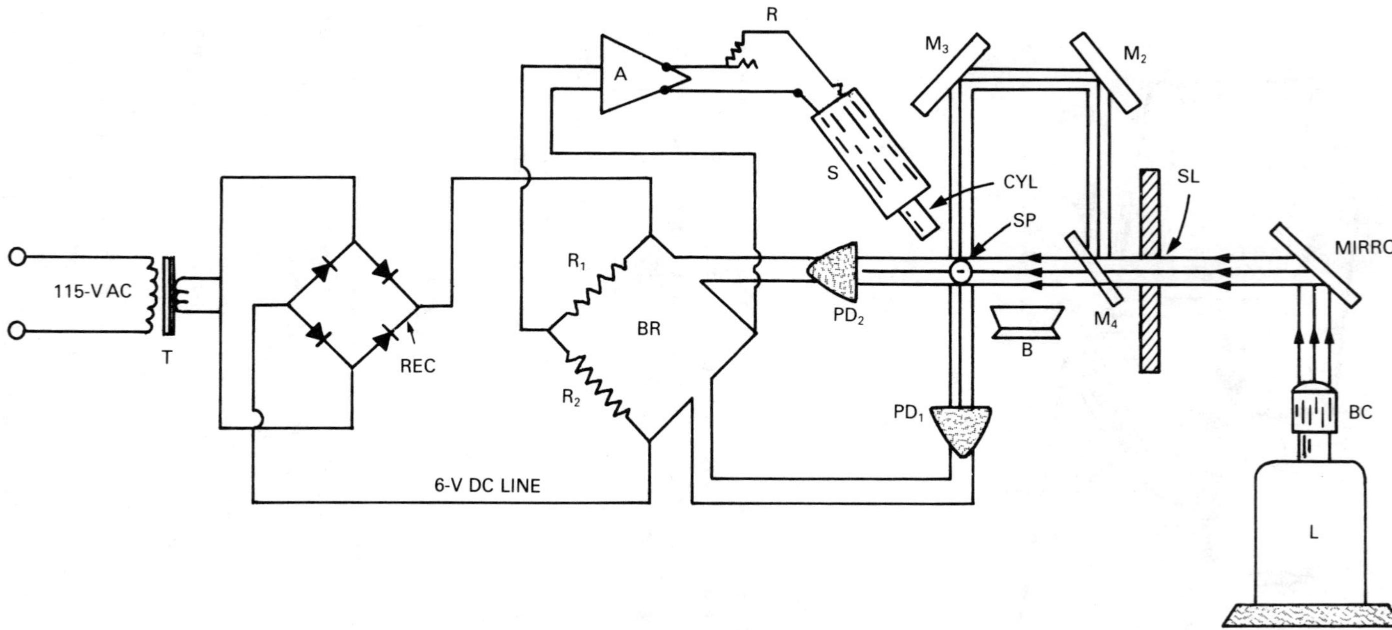

Figure 5-3 Schematic diagram of an automatic laser gauging and sorting system.

solenoid (S) through a potentiometer or a variable resistor (R), by which the sensitivity of the solenoid pull is adjusted.

The specimen to be gauged is positioned by means of a continuously moving belt, in production testing, at the crossing of the two laser beams, as shown at SP. The two photodiodes are matched for their optical sensitivities, so that their dark-current resistances are equal; since resistances R_1 and R_2 are also equal in the bridge setup, the bridge is balanced when no specimen is present at the crossing point of the laser beams.

In order to calibrate the apparatus shown in Fig. 5-3, a standard specimen with accurately gauged dimensions is positioned at the center of the beam crossing. The laser radiation is turned on and the potentiometer or the variable resistor (R) is adjusted until the solenoid cylinder (CYL) is just retracted into the housing of the solenoid (S), so that a slight change, by part dimension change, in the photocurrent of detector PD_1 or PD_2 upsets the balance of the bridge (BR) and a current flows into the solenoid (S) through the amplifier (A). The solenoid then ejects the cylinder (CYL) outward to hit the specimen (SP) and discard it into the bin (B). The change in the photocurrent, of course, is due to the dimensional deviation of the specimen from that of the standard specimen, which means that it does not meet the dimensional tolerances set forth for the standard specimen.

5-5 QUANTITATIVE HARDNESS TESTER

The laser beam is also employed in hardness testing of metals and comparing irregularities, such as pits, scratches, voids, and nicks, in the surface profile of the inspected parts with a previously inspected sound part. The parts are conveyed by means of a moving belt in the path of the laser beam, which scans the surface of the part. A logic circuit incorporated in the inspection laser system decodes the information received from the part and rejects any part that does not meet the reference logic requirement set by the scanned sound specimen. The system can be employed in metal fabrication industries that require accuracies within a few mils of deviation from the reference (standard) specimen.

5-6 LASER GYROSCOPE

A laser gyroscope with no moving parts has been developed independently by several manufacturers. The system principally utilizes a three-axis inertial guidance subsystem to provide a missile midcourse guidance and autopilot function to control the course of a missile. The gyroscope uses two coherent laser beams traveling in opposite directions in a triangular ring that has a plane perpendicular to the input axis under measurement. The error signal, or difference between the clockwise and counterclockwise frequencies of the two beams, is a measure of the change of the missile attitude about the input axis. This signal is fed to the guidance control system to control the vehicle and to interface with the terminal-seeker head of the missile.

5-7 LASER VELOCIMETER

A laser-aided velocimeter developed by Singer-General Precision, Inc., measures the speed and length of moving parts on a belt with an accuracy of 0.1% of the velocity of the moving part. The system comprises a sensor head, a velocity indicator, and signal-processing electronic circuitry. In operation, a laser beam is projected on the moving object to illuminate its surface and then is reflected to the sensor head, which provides an output signal proportional to the speed of the object. This signal is processed in the electronic circuitry and converted to direct speed and length readings. Applications of this device, in addition to measuring length and speed, include monitoring the automated processing of textiles, rubber, metals, and plastics.

5-8 POLLUTANT DETECTION SYSTEMS

A tunable laser system for identifying and measuring the quantity of pollutant in the atmosphere has been developed by Stanford University researchers. The system consists of a laser beam transmitter in which a lithium niobate crystal emits a narrow laser beam that is projected to pollutant particles in the atmosphere and reflected from them to a 16-inch telescope. The received beam is transmitted to a computer-aided analyzer, which analyzes and measures the pollutant content, as well as indicating the source of the pollutant located up to 8 miles away from the system. The laser beam can be tuned to different frequencies characteristic of the different pollutant molecules, so that a mixture of pollutant particles—for instance, carbon dioxide, sulfur dioxide, methane, nitrous oxide, Freon, and ethylene, some of which are constituents of automobile exhausts—can be separated and quantitatively analyzed by the computer.

The system is mounted on a mobile vehicle and used by the Environmental Protection Agency to monitor effluents from manufacturing plants located up to 8 miles away. The EPA may then instruct the plant owners to correct any pollutant emission in excess of that allowed by EPA regulations. Also, the system can monitor cancer-causing agents such as vinyl chloride emission from plastic fabricating plants, methane emission from oil field deposits, and leaks from refinery pipelines. A further application of the system is in remote sensing of acetic anhydride, which is given off by plants engaged in illegal manufacture of heroin. A scheme of pollution detection is shown in Fig. 5-4.

The technology of remote sensing of pollutants is also being investigated by NASA's Atmospheric Science Division. This investigation consists of measuring atmospheric moisture condensation, hydrochloric acid by-products, aerosols, and other airborne molecular species. The procedure involves monitoring the effluents from industrial smokestacks, gas leaks from pipelines, and other atmospheric pollutants by means of a laser system located on a satellite or other space vehicle. The technique, known as light detection and ranging (lidar), uses a carbon dioxide laser capable of heterodyned sensing of the light backscattered from pollutant particles that fluoresce, altering the wavelengths; these are detected as Raman fluorescence in which Doppler action is also sensed. The concentration of the pollutant particles is determined by

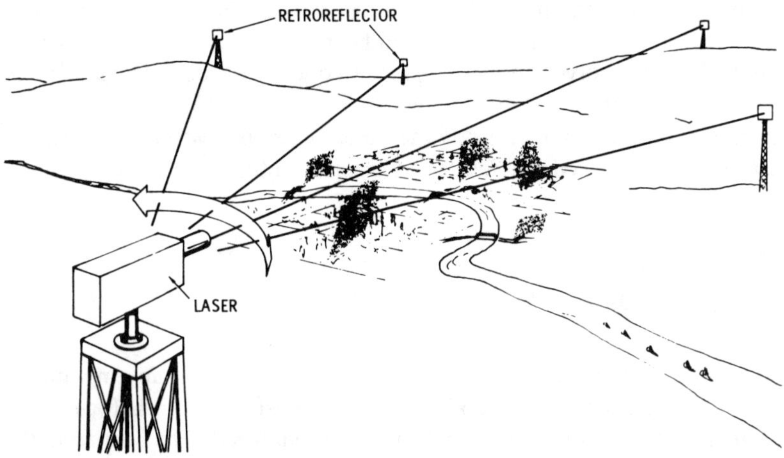

Figure 5-4 A scheme for monitoring air pollution.

tuning on and off the projected beam frequency and comparing frequency differences of the projected and backscattered signal strengths. The system is capable of sensing as few as 10^6 hydroxyl radicals per cubic centimeter and is found to be much more sensitive than any other conventional technique. It can detect and identify the backscattered molecular signals at distances up to 20 kilometers.

The researchers employ two types of laser systems. One is ground-based and uses a ruby laser emitting at 6940 Å and the other is located on an airplane and uses a combination of ruby radiation and a Nd-YAG beam at 10,600 Å; the second set can scan the atmosphere in both elevation and azimuth. These lidars are capable of remote detection of acid rain as well as effluents from nuclear plant accident emissions. Using the same procedure, Jet Propulsion Laboratory monitors the hydrazine by-product fuel used in launching space shuttles, since the gas is hazardous to humans as well as explosive in concentrations of a few percent in the atmosphere.

This investigation is being continued in search of more sensitive and continuously tunable laser materials, such as alexandrite, which is tunable from 7000 to 8000 Å, and the newly discovered cobalt-magnesium fluoride, which is tunable from 15,000 to 23,000 Å. The work is also being carried out by scientists in Germany, Sweden, Canada, and Japan. Japanese investigation is generally concentrated on the use of optical fiber sensors to monitor particularly high-temperature, contaminated, and dangerous media at rather limited distances.

5-9 LASER ANEMOMETER

A gas flow anemometer using optical frequency shifting of laser radiation passing through a radial diffraction grating has been developed by the Atomic Energy Research Establishment at Harwell, England. The system determines both the magnitude and direction of the gas flow velocity with accuracy unattainable with conventional flowmeters; it can be used to measure fluid velocity in the range of 60 to 260 m/sec. The constant frequency between the control and shifted beams defines the volume passed, and the direction of flow is resolved by the Doppler frequency of the dispersed signal relative to the frequency shift. This development makes it possible to optimize the design of industrial gas burners by measuring the velocity profiles within the furnace, permitting operation with increased efficiency and precise control over the temperature profile of the flames.

5-10 LASER SPECTROCHEMICAL ANALYSIS

The system for this type of chemical analysis is a laser-aided spectrograph in which a single 1- to 2-J Q-switched laser pulse, usually of 10^{-6}-second duration, is focused on a minute area of the material whose constituents are to be analyzed. The heat from the laser energy produces a partially ionized plasma, which is then completely ionized by passing a spark gap discharge through it. The ionized gas emits a radiation with wavelengths characteristic of the material constituents. The flash of the ionized gas is sampled by means of an optical system that projects the sampled radiation through the standard slit of the spectrograph onto a strip of photographic film. The various wavelength fringes are recorded on the film and are compared with a standard film recording of a known sample to determine the constituents.

Laser-aided analysis has several advantages: (1) the vaporization and subsequent ionization of the material under study are accomplished rapidly without resort to the conventional method of first making the sample electrically conductive; (2) a sample as small as 20 μm in diameter or less can be analyzed without destroying any significant part of the parent material; (3) the method eliminates any direct contact with the material under study; (4) the optical system used to perform these functions could be constructed in a relatively simple configuration.

5-11 LASER FLAW FINDER IN SEMICONDUCTOR WAFERS

A laser scanner for detecting flaws in semiconductor wafers has been developed by Aeronca Electronics, Inc. The system consists of a 5-milliwatt He-Ne laser, a rotary multifaceted scanning mirror, and two light-collecting fiberoptic bundles. Wafers up to 12.5 cm in diameter are conveyed by a moving belt under the laser beam at a speed of 3.5 m/sec and a rate of up to 600 wafers per hour. The wafers are scanned at a beam incidence angle of 15° with respect to the optical normal. One of the fiberoptic elements collects the scattered light from the wafer and the other receives the specularly reflected light. The light collected by the fiberoptic elements is transmitted to the respective photomultipliers, and an error signal is generated between the signal differences. This signal is then transmitted to a microprocessor, which determines the flaw character. Up to 12 types of flaws can be determined with the system, as reported by the developers. The laser inspection system is shown in Fig. 5-5.

Figure 5-5 The Aeronca semiconductor wafer inspection laser system.

5-12 LASER-INDUCED CHEMICAL ACTION

There is growing activity in the use of coherent light to produce chemical transformations in various chemical materials in their gaseous phases contained in an optical cavity. The intense monochromatic radiation from a high-power tuned carbon dioxide laser is playing an important role in the preparation of reactants that alter the course of chemical reactions. Such transformations find practical applications in the enrichment and separation of isotopes, such as hydrogen isotopes, uranium isotopes from the natural element for use as atomic fuel, and carbon-13 from carbon-12 for use as a marker in biochemical investigation; in the conversion of acetylene into butadiene in the presence of silicon tetrafluoride (SiF_4); and in other laser-induced chemical reactions such as alkali-metal-vapor interactions. In these reactions, the reactant molecules become excited and rapidly decompose and separate from the parent molecule. Laser-induced chemical transformation is a relatively new procedure for isotope enrichment and has a high potential to open new fields of investigation for commercial applications.

5-13 HYDROGRAPHIC LIDAR SYSTEM

An ocean depth–measuring system using laser beams has been developed by several technical firms. The system uses one of several green-colored lasers, such as frequency-doubled Nd-YAG, metallic vapor, metallic halide, or an excimer. A green-colored laser beam is used because of its efficient penetration of seawater containing the usual contaminants. The measurement involves projecting a laser beam at a slant to the surface of the water and measuring the time of return of the laser beam from the surface and from the bottom of the ocean. The optical transit time difference is used to compute the distance from the bottom of the ocean to the surface. There are many laser beam candidates for the hydrographic system, and the selection of the best laser beam has not been finalized at the time of this writing.

In operation, a narrow-bandwidth green laser beam from a frequency-doubled Nd-YAG laser emitting at 5320 Å is projected on the ocean surface in 2-mJ, 5-nsec pulses at a repetition rate of 400 Hz. Part of the beam is reflected from the surface and the other part from the bottom of the ocean. The difference between the aircraft-to-bottom distance and the aircraft-to-surface distance gives the depth of the water. A simple equation relating these distances to the numbers of pulses corresponding to the distances is

$$D = \frac{N_1 c}{2N_c} - \frac{N_2 c}{2N_c} \tag{5-1}$$

where D is the depth of the sea bottom from the surface of the water, N_1 is the number of clock pulses received by the laser system counter with respect to the sea floor, N_2 is the number of pulses counted during the time the beam impinges on the water surface and is reflected to the airplane, N_c is the clock frequency, c is the speed of light, and 2 is a factor accounting for the round-trip distance of the beam.

Joining the two terms in Eq. (5-1) and simplifying it, we have

$$D = \frac{c(N_1 - N_2)}{2N_c} \tag{5-2}$$

Example A laser depth-measuring system mounted on an airplane operates with a clock frequency of 100 MHz. If the beam round-trip distance between the airplane and the surface of the water, whose depth is to be measured, is 0.4×10^3 clock pulses, and 0.8×10^3 pulses have been received from the bottom of the water for a round trip, what is the depth of the water?

Solution

$$\begin{aligned} D &= \frac{3 \times 10^{10}(0.8 \times 10^3 - 0.4 \times 10^3)}{2 \times 10^8} \\ &= 0.6 \times 10^5 \text{ cm or 600 meters} \end{aligned}$$

5-14 LASER ROAD VISIBILITY SENSOR

A laser visibility sensor manufactured by International Laser Systems, Inc., designated as Model VS-1, is designed to be installed on highways and bridges for monitoring near-ground-level atmospheric visibility diminished by fog, snow, smoke, and dust that affect a car driver's view. The system consists of a transmitter and a receiver located 250 feet apart on one side of the road. The transmitter is provided with an amplitude-stabilized diode laser source for projecting a modulated 1000-Hz laser beam collimated by a collecting lens into a 1° beam for detection by the remotely located receiver. The operational scheme of the system is shown in the block diagram of Fig. 5-6.

In Fig. 5-6, the transmitter and the receiver are shown schematically stationed at their usual locations on the roadside where the atmospheric visibility is to be measured. The laser diode in the transmitter has a normal output of 20 to 30 W, but it uses only 5 watts stabilized by the power monitor, whereby additional power is fed as the diode ages; thus, compensation for a wide range of atmospheric temperature changes is

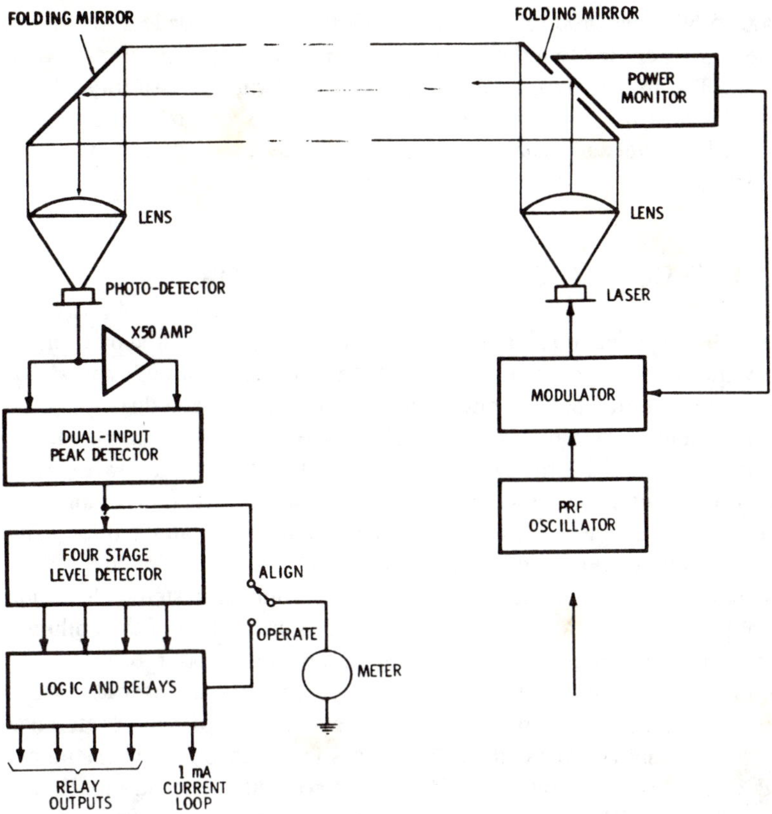

Figure 5-6 Block diagram of the laser visibility sensor Model VS-1 (ILS).

provided in the system. The receiver contains an optical receiving section, a peak signal detector, a level detector, a relay, a current-loop output circuit, and a power supply.

The two units, the transmitter and the receiver, are aligned in the azimuth plane by first sighting the receiver with a telescope mounted on the transmitter and then adjusting the direct field of view with jackscrews located on both of the units.

The modulated laser pulses in the transmitter are projected on a folding mirror, which bends the pulsed beam 90° and transmits it to the receiver 250 feet away. A field stop located at the focal point of the receiver lens establishes the 1.5° field of view of the receiver. An optical filter positioned behind the field stop blocks out the background radiation from the photodetector, which is a PIN (*p*-layer-intrinsic-*n*-layer) diode with a large optical receiving area.

The incident pulsed laser beam is routed in the receiver from the photodetector to two circuit channels; one channel receives part of the laser signal without amplification, and the other channel receives the other part of the signal with a gain of 50. The two signals drive a dual-input peak detector. The high-gain side of the peak detector responds only to weak signals corresponding to severe fog conditions. The output of the detector is parallel-fed to a four-stage visibility level detector adjustable to levels at 1000 feet, 1000 to 500 feet, 500 to 300 feet, and 300 to 100 feet or less. From the level detector, the signals are fed to a logic circuit that prevents operation of more than one relay at a time. The four relay outputs are provided for various visibility levels, each to display one level. The 1-milliampere current loop is used for remote monitoring of the discrete visibility levels and charting them on a single recorder. The system is energized by a 115/230-V, 50/60-Hz AC power source.

5-15 GENERAL ELECTRIC POLLUTION DETECTOR

An infrared laser pollution detection and measurement system manufactured by General Electric Company monitors a long path traversed by a laser beam through the atmosphere, integrating the pollution concentration along the entire path and displaying the average pollution content at ambient conditions. The measurement is based on the spectral absorption of the laser beam. The system uses relatively low and safe power levels over a range of several miles. Since long infrared wavelengths are employed, the radiation hazard to the eye is minimal, as claimed by the manufacturer. The transceiver optical system attenuates the laser power to less than 0.01 W/cm^2.

The system comprises a laser scanner, a transceiver optical system, electronic circuitry for signal processing, and a retroreflector to aid in integration of the pollutant particles along the laser beam path. The transmitter projects a laser beam to the retroreflector located at a remote point from a tunable carbon dioxide laser at various wavelengths. The returned beam energy is integrated and compared with a reference laser output. An error signal is produced by differences in the absorbed wavelengths, and this signal is converted into a direct readout of the pollutant concentration in the beam path. The carbon dioxide plasma is capable of giving off more than 50 wavelengths, and four wavelengths (95,050, 105,320, 106,750, and 107,190 Å) are selected for the pollution measurements to allow for a variety of weather conditions. Concentrations down to 14 ppb (parts per billion) of ethylene gas and 19 ppb of ammonia gas have been measured. As an extension of the principle used in this system, a circular scanning system with multiple retroreflectors located at various points on the circular area has been advanced as a laser monitoring unit for urban areas. One modification of this technique is shown in Fig. 5-4.

5-16 BELL LABORATORIES POLLUTION DETECTION SYSTEM

A team of scientists at Bell Laboratories have devised a pollution detection and measurement laser system that is capable of identifying concentrations of pollutant gases as small as 1 part in 10 billion. Similar in some respects to the General Electric system, the Bell laser-computer unit utilizes the affinity of different gases to absorb laser light at different wavelengths. Since different gases exhibit specific absorption characteristics at different absorbed laser wavelengths, the amount of absorption at each wavelength indicates the pollutant content in the atmosphere.

Air samples to be analyzed are held in an optoacoustic absorption cell to which the laser beam is directed. The laser beam energy absorbed by the air contained in the cell increases the temperature and pressure of the air sample in direct proportion to the amount of air in the cell. The pressure increase is sensed by a transducer, which converts the pressure change into an electrical signal. The computer receives the electrical signal and identifies the specific pollutant present by matching it with gas characteristic patterns stored in the memory. The computer can be programmed to process up to 20 different pollutant gases in this manner. The system is well suited to monitoring and identifying plant pollutants emanating from manufacturing processes, as illustrated in Fig. 5-7.

5-17 CONTROL LASER CORPORATION MARKER

Control Laser Corporation manufactures a CW Nd-YAG laser marker, designated as Instamark, which can engrave a wide variety of materials such as aluminum, titanium, steel, tungsten, ceramics, and polycarbonates. The surfaces of the parts can be marked before or after painting, anodizing, plating, or chemical etching. The marker can also scribe parts enclosed in glass containers capable of transmitting infrared radiation at 106,400-Å wavelength. Parts serializing can be accomplished by programming the microprocessor of the system. Even decorative designs can be produced on jewelry. The system is shown in Fig. 5-8.

The Instamark contains two galvanometer mirrors, which project an upcollimated laser beam on the workpiece to be marked. One mirror moves the laser beam along the *X* axis and the other moves it in the *Y* axis. A flat-field lens of 100-mm focal length focuses the laser beam on a plane surface 70 mm in diameter. As the mirrors deflect the focus of the beam in the *X* and *Y* axes, the desired character or symbol set from the microprocessor program is engraved on the workpiece. The memory capacity of the

Figure 5-7 Monitoring the effluent pollutants from a factory stack.

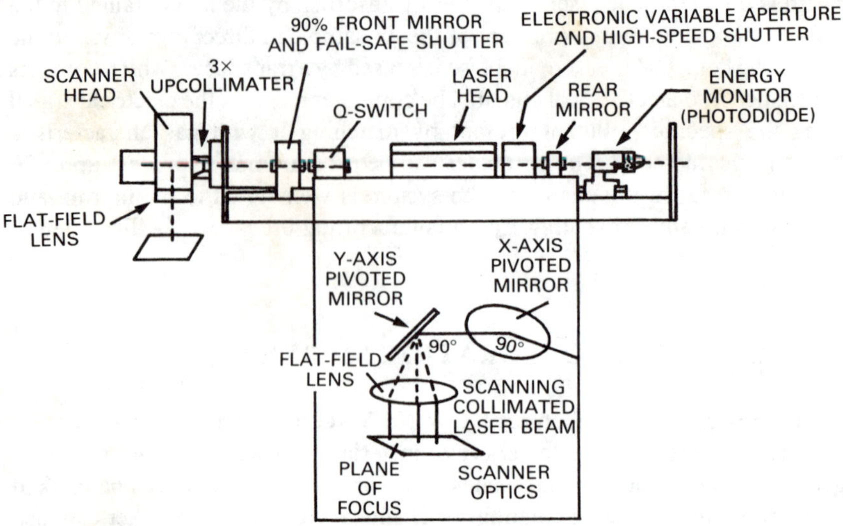

Figure 5-8 Schematic diagram of the Instamark optical train (Control Laser Corporation).

Instamark is 16,000 16-bit words, with provision for expansion to 32,000 16-bit words. For protection against malfunctioning hazards, the system is fully interlocked.

5-18 UNIVERSAL PRODUCT CODE CHECKOUT SYSTEM

A universal product code (UPC) checkout system is a supermarket cash register unit using a laser beam to scan the bar codes printed on merchandise to represent its name and price. The codes, known as universal product codes, appear on the container of the merchandise. They consist of black and white bars of different widths separated one from the other by specific distances that determine, when decoded, the name, price, and the inventory in general. As the merchandise moves on a conveyor belt toward the checker, the checker picks it up and passes the UPC symbol over a scanning window positioned horizontally in the plane of the conveyor belt. The cash register indicates the price of the merchandise while a printout tape indicates both the name and the price of the item. The machine also transmits the information to a central computer storage station that maintains records on the supermarket activity. When all the grocery items have been checked, the machine totals the price on the tape and the checker presents it to the customer for disbursement. The UPC symbols are shown in Fig. 5-9.

The supermarket checkout system consists of a cash register and a code scanning window positioned in line with the moving merchandise on the conveyor belt. A low-power laser source directs a pencil-thin beam from a He-Ne laser unit to a scanning mirror, which is located below the scanner window and projects the beam onto the bar codes at a given scanning rate. A photosensor receives the scanning laser beam reflected

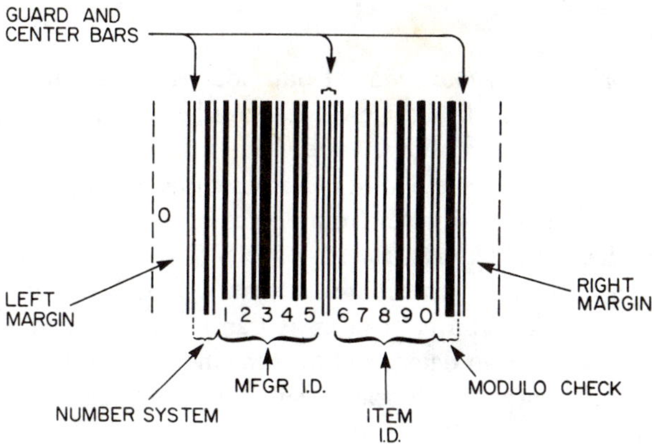

Figure 5-9 The categorized UPC symbols (Courtesy NCR Corporation).

from the merchandise code, develops a signal from the bars, and feeds the signal to an amplifier, as shown in Fig. 5-10. From the amplifier, the coded signal is transmitted to a coder/decoder section, which delivers the decoded information to the cash register to display it and simultaneously to print the name and price of the merchandise on the tape, which is ejected from one side of the cash register.

The principle of operation of the laser system consists of projecting a laser beam from a 1-mW He-Ne laser source onto the bar codes. The beam scans the wider bars for relatively longer periods than the narrower bars. The longer and shorter periods of scan time constitute various pulsed intensities, which develop specific signals in binary form that are amplified and then translated in the decoder into alphanumeric characters. These characters are then displayed on the cash register and simultaneously printed on the moving tape, which contains the name and price of the items and the tax on taxable items. This scheme is said to be three times faster than that employed by conventional checkers and is accurate.

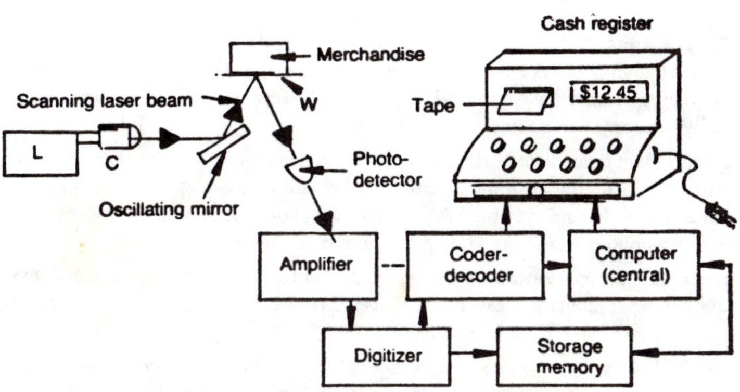

Figure 5-10 Schematic diagram of laser-scanned grocery checkout system.

5-19 FIBEROPTIC MONITOR FOR WATER CONTAMINANTS

The United States draws about one-half of its drinking water from underground aquifers, which are susceptible to contamination by industrial toxic wastes absorbed from chemical dumps. It is reported that about 50,000 hazardous waste sites in this country are contributing to this contamination. The current method of controlling this water pollution is by removing samples of the underground water from depths of up to 100 feet and testing the samples by spectrochemical methods. Because such efforts are tedious and time-consuming, researchers at Lawrence Livermore National Laboratory have developed a technique in which a 4880-Å argon laser beam is transmitted through a fiberoptic waveguide to the site of the water to detect and monitor the contaminants, which consist mainly of organic and inorganic chlorides, sulfates, aldehydes, and pesticides.

The technique consists of irradiating these contaminants with the argon ion laser beam to cause fluorescence of the pollutant particles. The fluorescent light from the particles is then transmitted to a spectrometer for identification and quantification of the particles in parts per million. To achieve this, a hole 1 to 2 inches in diameter is drilled into the ground and a glass fiberoptic cable with a 200-μm core and a numerical aperture of 0.4 is introduced through the hole into the water reservoir.

The 4880-Å beam from the argon ion laser causes fluorescence of the particles with emission of longer wavelengths. The fluorescent light is directed through the fiberoptic line into the spectrometer for analysis. To magnify the fluorescent effect, a sapphire ball is provided at the distal end of the optical fiber in contact with the water containing pollutants. The signals from the pollutants are passed through a multiplexer filter to isolate the different types of particles and determine their kind and amount. It is reported that the method can be used to analyze the contaminant contents of as many as 20 wells by means of a time-sharing arrangement.

QUESTIONS AND PROBLEMS

5-1 What is metrology? Why are lasers used in metrology? What are some of the applications of lasers in metrological work? What is a laser alignment system? What types of lasers are used in alignment work? Why is an expanded and collimated laser beam employed in alignment work? How is a laser alignment system constructed? How does the alignment system operate?

5-2 How is a laser gauging instrument constructed? How does the laser gauging instrument operate? What accuracy can be obtained from a laser gauging instrument? Draw a schematic diagram of a laser gauging and sorting system. How accurate is a laser sorting system? Why are the two photodiodes used in a laser sorting system matched? Explain the technique of using the laser gauging and sorting system.

5-3 What is a laser hardness-testing machine? How does it measure the hardness of the material being tested? How accurate is such a system? In what industry is this machine used? How does the instrument reject parts that do not meet the tolerance requirements? Explain.

5-4 What is a laser gyroscope? How is it constructed? How does it operate? On what systems is a laser gyroscope employed? How does the gyroscope develop an error signal? What is the significance of an error signal? What is the function of the error signal in controlling the attitude of a moving missile?

5-5 What is a laser velocimeter? Name a manufacturer of laser velocimeters. What does a laser velocimeter measure? How does it measure? What does a laser velocimeter comprise? How is an optical signal generated? What is the relation of the signal magnitude to the speed of the moving object? Can the velocimeter monitor any automated process? Explain how.

5-6 What is a pollution detection system? What does it detect? Can it measure the pollutants it detects? How accurate is the system operation? How is the Stanford University pollution detection system constructed? What type of laser is used in this system? Can the laser beam be tuned? What is the significance of having a tuned laser beam in this function? What gases can be detected by this system? What gas does the system sense remotely? Is remote sensing important physiologically to urban populations? Explain.

5-7 What is a laser anemometer? Why is it necessary to measure the optical frequency shifting of the laser beam in this system? What is the function of the radial diffraction grating in the device developed by the Atomic Energy Research Establishment of England? What is the range of velocity measurement in this device? How does Doppler frequency enter into the operation of this device? What is the importance of this device with respect to industrial gas burners?

5-8 What is a laser spectroanalysis system? What type of laser beam is used in this system? Describe the operation of this system. How does the system determine the character of the substance constituents? What are the advantages of laser-aided spectrochemical analysis?

5-9 What does a laser flaw finder detect in a semiconductor wafer? What type of laser is used in this device? What role does fiberoptics play in this flaw finder? What is the function of a photomultiplier in this device? What is the function of a microprocessor in this device? Discuss.

5-10 How does laser-induced chemical transformation occur in an optical cavity? What are the applications of such transformations? What is the mechanism involved in laser-induced chemical transformation? What are the potential applications of this technique?

5-11 What is a hydrographic lidar system? What types of lasers can this system employ? What is the principle of operation of the lidar system? How accurate are the results obtained with the system? Why should a green laser beam be preferred? What color is the laser beam from a frequency-doubled Nd-YAG laser? How is the ocean bottom measured from the aircraft? Explain.

5-12 A certain laser depth-measuring system operates with a clock frequency of 100 MHz. If the distance between the aircraft and the surface of the water whose depth is to be measured is 3×10^3 clock pulses and 4×10^4 pulses have been received from the bottom of the water for a round trip, what is the depth of the water? If the depth of the water is 800 m, how many clock pulses will be received one way from the aircraft?

5-13 What does the ILS Model VS-1 laser visibility sensor monitor? What is the significance of monitoring with this system? What does the system consist of, and how does each component enter into the operation of the system? Describe the manner by which the system operates and the results it produces. What are the ranges of visibility determination of the system? How are these ranges adjusted?

5-14 What is the constructional configuration of the General Electric pollution detection and measurement system? How does the system operate? On what principle is the measurement of pollution based? How is the integration of the pollutant particles accomplished? What role does an error signal play in the measurement of the pollutants? What is the relation of CO_2 wavelengths to the types of pollutants detected and measured?

5-15 How is the Bell Laboratories pollution detection and measurement system constructed? How does it operate? What accuracy does it possess? What is the principle of differentiating one gas from another in this system? What is the function of the optoacoustic cell in this system? How does the computer identify the different pollutants detected? Where can this system be used?

5-16 What does Control Laser Corporation's laser marker engrave? Can it scribe marks in glass containers? Describe the operation of the Instamarker. What is the memory capacity of the Instamarker? What lens focal length does the system employ? Why are two mirrors used in the system? Why is an infrared laser beam used in the system?

5-17 Draw a diagram of the UPC checkout system and describe its operation from the time the merchandise is pulled over the scanning window to the time the checkout slip is handed to the customer. Explain why this method is superior to conventional checkout systems.

SELECTED SOURCES OF RELATED DATA

Bonic, J. 1987. Airborne scanning lidar bathymeter measures water depth. *Laser Focus* February, pp. 48–52.

Camagni, P., and S. Sandrant, eds. 1984. *Optical Remote Sensing of Air Pollution*. New York: Elsevier.

Churnside, J. H., and H. T. Yura. 1982. Three-dimensional laser vector velocimetry. *Appl. Opt.* 21(5):845.

Cutler, P. H., and A. A. Lucas. 1983. *Quantum Metrology*. New York: Plenum.

Dimpfl, W. L., Chevron Research Corp. 1983. Removing Pollutants with Laser Radiation. U.S. Patent No. 4,397,823.

Divena, W. G., and S. M. Jarrett. 1982. Design and performance of a stabilized dye laser ring. *Rev. Sci. Instrum.* 53(9):1363.

Greenfield, E. 1986. Measuring laser power accurately. *Laser Focus* March, pp. 134–142.

Humphrey, L. M., et al. 1979. Doppler-Free Spectra from Dye Lasers. Bell Laboratories Technical Report. October.

Killinger, D. E., and A. Mouradian, eds. 1983. *Optical and Laser Remote Sensing*. New York: Springer-Verlag.

Krohn, D. A., and E. L. Vinarub. 1984. Fiberoptic sensors. *Photonic Spectra* February, pp. 51–57.

L&A Staff. 1983. Lumber measured with charge-coupled device using He-Ne laser. *Lasers and Applications* March, p. 33.

Laser-leveling used in produce fields. 1985. *Lasers and Applications* March, p. 42.

Lee, S. H. 1981. *Optical Information Processing Fundamentals*. New York: Springer-Verlag.

Levenson, M. D., and S. Kano. 1988. *Introduction to Nonlinear Laser Spectroscopy*. San Diego: Academic Press.

Pan, J. J. 1988. Integrated optics for microwave coherent detection systems. *Lasers and Optronics* March, pp. 87–93.

Parker, L. 1980. Measuring height with a position-sensing detector. *Laser Focus* July, pp. 40–46.

Smythe, R. A., et al. 1989. Interferometer Measures Linear Displacement. U.S. Patent No. 4,509,858.

Thermal remote anemometer device (laser). 1988. Langley Research Center, NASA Technol. Briefs, vol. 12, no. 6 (June).

CHAPTER

SIX

LASER BEAM COMMUNICATION

6-1 INTRODUCTION

The high coherence and directionality of laser beams have formed the basis for the development of serial laser communication systems capable of transmission of voice, data, and images, providing high fidelity and secure communication between several locations at the same time. Because of the extremely wide bandwidth of the laser radiation, up to 10^7 separate transmissions can be achieved without the risk of jamming by any known means. The transmission beam is 10^4 times narrower than radio frequencies. The beam can be modulated in amplitude, intensity, phase, frequency, and polarization formats, using analog, digital, or pulse transmission methods. At laser beam wavelengths, the frequencies are sufficiently high to obtain beam divergence in milliradians or microradians.

Laser communication is also possible through optical fibers. An optical fiber (fiberoptic) communication system is an acoustic or data transmitting and receiving apparatus in which the transmission line is a single optical fiber or a bundle of multiple optical fibers. The system comprises a voice or data transmitter, a fiberoptic line, and a terminal receiver. The transmitter is provided with an information-input link to convert the voice or data signals into electrical signals and with a modulator to modulate the electrical signals into a modulated current, which drives the laser beam. The modulated laser beam is interfaced with a fiberoptic cable through properly designed connectors to reduce the connector junction loss to a minimum. The fiberoptic cable terminates in a detector incorporated in the receiver in which the optical signal is converted back to the corresponding electrical signal current to drive a speaker or to be stored in the memory bank of a computer for future use.

6-2 AERIAL COMMUNICATION

In aerial communication, the transmitter and the receiver are generally in each other's field of view unless the beam is folded by reflectors. Several manufacturers are producing aerial laser communication systems for use in military field communication and ship-to-shore, ship-to-aircraft, and ship-to-ship intelligence transmission. The system, shown in Fig. 6-1, is also used by surveyors and at large construction sites to communicate with workers in the field of view. A well-collimated laser beam can be transmitted several miles on a clear day and to shorter distances when atmospheric conditions such as fog, mist, rain, or smog do not permit penetration of the laser beam to more than a few hundred feet. Also, the transmitted beam is attenuated by atmospheric pollutants, such as those mentioned in an earlier chapter, and by varying temperatures of the air. One other disadvantage is that the beam energy must be at a low hazard level for eye safety, which limits the distance to which the signal can be transmitted.

A laser beam traversing air suffers other losses, such as by refraction, scattering, absorption by water vapor, or other attenuation by aerial pollution. In spite of these limitations, a collimated laser beam can be transmitted to distances as far as the horizon on a clear day. Green laser beams can be made to penetrate to a distance of 150 to 200 feet in sleet, rain, or snow. Figure 6-2 illustrates the technique of communicating between several stations at the same time.

Other considerations of system design also enter into the application of laser beams for communication purposes. For instance, the detector at the receiving end must be compatible with the beam characteristics to eliminate or alleviate the degradation effects of the beam. To reduce costs, the beam must not require cryogenic cooling and constant operator attendance. The modulation speed of the beam determines the wavelength to be used. Since the best modulators at present operate in the visible spectral range, the

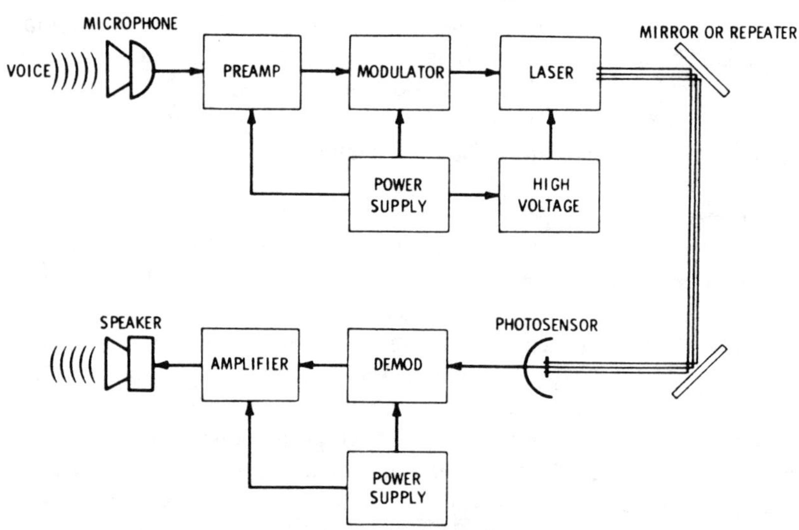

Figure 6-1 Block diagram of a laser aerial communication system.

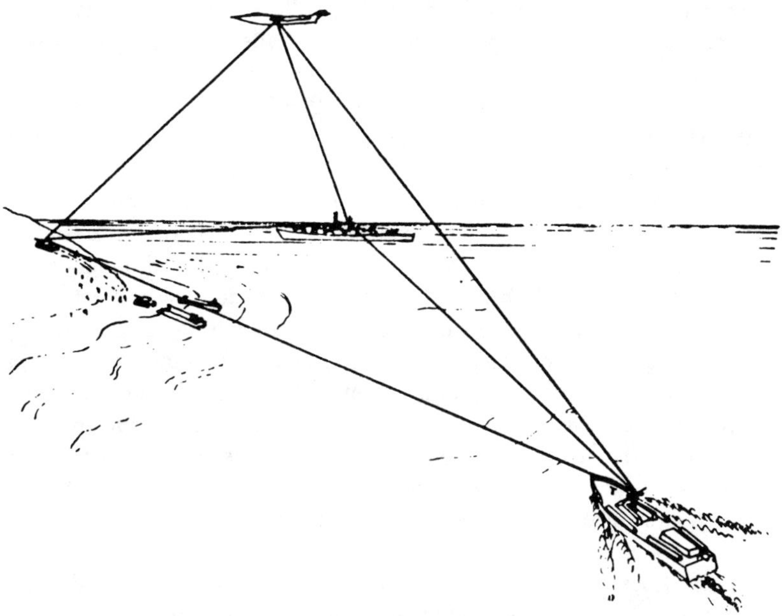

Figure 6-2 Laser aerial communication between stationary and moving stations.

power and efficiency of the laser are limited, which makes the transmission distances proportionately small. Among the laser candidates, carbon dioxide and Nd-YAG lasers furnish high power, but the carbon dioxide beam is difficult to modulate, which limits long-distance communication. Techniques have been developed for modulating He-Ne lasers and semiconductor lasers, but their applications have been limited to relatively short distances.

A few of the important advantages of laser communication over radio communication are as follows:

1. Laser communication permits all conversations to be private.
2. No jamming is possible, and the beam can be locked on the receiving system.
3. The radiation has extremely wide bandwidths, namely 10^{13} to 10^{15} Hz.
4. The beam can handle up to 10^{7} channels at the same time at different frequencies.
5. The laser method is economical to use.
6. Several conversations can be carried out at the same time among different stations.
7. Communication can be continued even when the parties are in movement, as the beam can be locked on the receiver.

6-3 FIBEROPTIC COMMUNICATION

For long-distance communication, or communication between two stations located at different parts of a ship, optical communication is carried out through light pipes, known as fiberoptic waveguides. For long-distance communications, repeaters (Fig. 7-

10) are inserted at various intervals in the fiberoptic line to amplify the transmission signals. Bell Laboratories has tried a single capillary conduit filled with tetrachloroethylene to transmit the laser radiation. The conduit has a minimum loss of 13.5 decibels (dB) per kilometer, using an infrared laser beam of 10,800 Å wavelength. Gallium arsenide, Nd-YAG, He-Ne, and other lasers are also employed for transmission of light beams. The outside diameter of the quartz capillary is 95 μm and its inside diameter is 65 μm. The index of refraction of the quartz capillary is 1.457 and that of the liquid tetrachloroethylene is 1.5, which satisfies the conditions for total internal reflection for a laser beam. This scheme has been superseded by other more efficient techniques.

The present fiberoptics are solid fibers and are invariably used in laser communication systems. Such a system comprises principally an information transmitter, a fiberoptic line, and a terminal receiver. The transmitter is provided with an information input link containing a modulator, which converts the input signals into an electrical signal output. The modulated electrical signals are fed into the laser generator to modulate the laser beam. The modulated beam is then projected into the proximal end of the cable, which conducts the beam by total reflection to the distal end of the cable, at which a receiver-detector receives the modulated beam and demodulates it to the original signals.

The length of the fiberoptic cable may vary from a few meters to hundreds of kilometers. As in electrical conductors, there is a transmission loss in the optical fiber material as well as in the interfacing optical connectors. The loss depends on the distance through which the signal is transmitted; the larger the distance, the greater the optical loss in the fiberoptic material. The loss in optical intensity also varies with different fiber materials and is in the range of several hundred to less than 1 dB/km. Furthermore, if the transmission consists of voice data, the system is capable of transmitting from 1 to 100 Gb/sec (gigabits per second).

The optical signal travels through the fiberoptic cable by total reflection, as stated earlier. Accordingly, a portion of the beam reflects out of the fiber walls by dispersion, and this loss partially constitutes the fiber transmission loss. Another cause of attenuation is impurities such as transition metal ions (Fe, Cu, Mn, Co, and Cr) in the fiber material. The traveling beam suffers an additional loss at each fiberoptic-link interface, and this loss is further augmented at the fiberoptic-detector linkage. The total transmission loss is measured in decibels per kilometer. Figure 6-3 illustrates the optical intensity loss through the fiberoptic cable, at the link interfaces, and at the fiberoptic-detector linkage in the receiver.

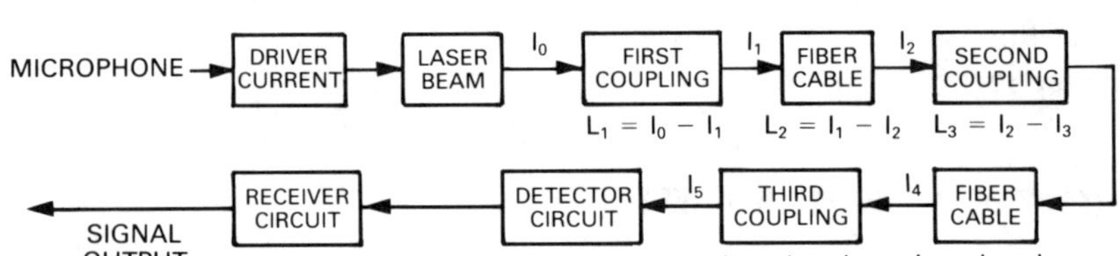

Figure 6-3 Functional block diagram of a fiberoptic transmission system.

The optical transfer-function loss in decibels between the transmitter and the receiver may be expressed as

$$L_t = L_1 + L_2 + L_3 + L_4 + L_5 + \cdots + L_n \tag{6-1}$$

where L_t is the total linkage and transmission loss, and L_1, L_2, L_3, etc. are the individual transfer-function losses at the increments shown in Fig. 6-3.

The optical intensity loss I_t in decibels may be given as

$$I_t = I_0 - (L_1 + L_2 + L_3 + L_4 + L_5) \tag{6-2}$$

and

$$I_d = 10 \log\left(\frac{I_0}{I_t}\right) \tag{6-3}$$

where I_0 is the initial laser intensity (in dB) at the exit port of the laser system and I_d is the intensity (in dB) at the detector.

The principal advantages of using fiberoptic cable rather than metal cable for data transmission are that the fiberoptic cable is lighter in weight than copper wire of equal data transmission capability; is immune to interference by radio waves or electromagnetic waves in general; offers information transmission security; has a broad bandwidth, offering greater signal transmission capacity than metal cables of equivalent size; eliminates cross-talk problems; and has surprisingly lower signal attenuation than metallic conductors transmitting over the same distance. A large variety of lasers can be used for data transmission, including the less costly light-emitting diodes (LEDs). Also, for long-distance communication the fiberoptic cables can be linked together by optical connectors, or repeaters, which are optoelectronic devices that receive, amplify, and regenerate data signals in transmission from one cable link to another, as shown in Fig. 7-10. The fiber transmission line can also transfer forensic and tactical information from a main data storage computer to other terminals located a few feet or many miles away. A block diagram of the fiberoptic transmission system is shown in Fig. 6-4.

Transmitter Laser Sources

Both visible and invisible laser light can be transmitted through fiberoptic cables. For invisible light, such as infrared or ultraviolet laser light, special light-transmitting cables, such as those made from quartz, sapphire, or Pyrex, are used. This radiation can be modulated because of its monochromaticity, and the cable can be impressed with any optical data. Radiation from a gas or ion laser, solid-state laser, liquid dye laser, or semiconductor laser can be employed as the optical data carrier. Dye lasers, however, are not adapted to long-distance data transmission because of their low-level optical intensities. Accordingly, for remote communication purposes, a He-Ne laser, argon laser, Nd-YAG laser, GaAs laser, InAlAsP laser, or CO_2 laser would be more suitable, although the last laser is somewhat difficult to modulate. For medium-distance

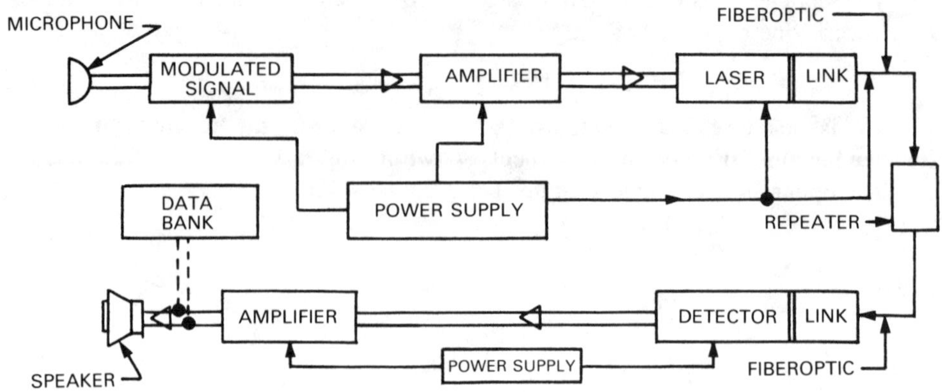

Figure 6-4 Simplified block diagram of a typical fiberoptic transmission system.

transmission and low operational cost, AlGaAs LEDs emitting at 8200 and 8400 Å are specified in practice.

LEDs are favored for optical data transmission because of their small size, high optical intensity with radiance up to 100 W/cm^2-steradian, wavelength range, and operation at low voltage and because their output radiation can easily be controlled by varying their bias current. They also have a half-life in excess of 1 million hours at 20°C. The LED lasers used for fiberoptic data transmission can produce more power output than LEDs. The LED lasers must operate slightly above their threshold current level in order to retain their lasing efficiency and coupling with the fiber; since they are time- and temperature-sensitive, they lose their output efficiency at temperatures above the specified threshold level and degrade with time. Here the LEDs offer a broader temperature range, longer operating life, and easy modulation; these characteristics make them attractive candidates for optical fiber communication.

An ideal semiconductor laser for use in fiberoptic communication systems has been developed by the Institute of Applied Physics, Hamburg University, Germany. The laser is a neodymium–aluminum borate crystal of high purity and between 10 and 100 μm in length. It emits a single spectral line of 10,600 Å because of a proprietary process of doping the aluminum borate. The laser crystal can be tuned to the minimum absorption characteristic of the transmission fiberoptic by varying the degree and nature of the proprietary dopant. Thus, it yields an attenuation of about 1 dB/km. The new laser is excited to radiation by optically pumping it with a krypton ion laser emitting at 8000 Å.

Recent semiconductor diode laser developments include gallium-aluminum-phosphide (GaAlP), gallium-indium-arsenic-antimonide (GaInAsSb), indium-gallium-arsenic-phosphide (InAsGaP-GaP), and many other gallium arsenide derivatives. Diode lasers usually have greater radiant emission efficiencies than any other solid-state or gas ion lasers. This efficiency represents the number of photons produced at the *p-n* junction with a given number of electrons (current) that have crossed the junction. A simple equation defines this relationship as

$$Q_{\text{eff}} = \frac{P_t e \lambda}{hcI} \tag{6-4}$$

where Q_{eff} is the quantum emission efficiency, P_t is the total photonic power, e is the charge on the electron, λ is the wavelength of the emitted photons, h is Planck's constant, c is the speed of light, and I is the current across the *p-n* junction.

In modulation of the diode, the optical response to the signal may be expressed approximately as

$$R_o = \frac{I}{[1 - (w^2/w_0^2)^2 + w^2 j^2]^{1/2}} \Delta I \qquad \text{(6-5)*}$$

where R_o is the optical response to the signal, w is the frequency of modulation current ΔI, w_0 is the resonance frequency at the junction, and j is the damping constant, which is equal to the carrier lifetime t_1 plus the photonic lifetime t_2, respectively 10^{-9} and 10^{-12} second.

Other laser sources that are being investigated are metal vapors such as copper vapor and He-Cd vapor. These vapor lasers are competing with inert-gas or ion lasers, and the manufacturer of one type claims to possess a better laser source than another manufacturer; this remains to be seen when their applications become practical.

Fiberoptic and Connector Characteristics

There are three general types of fiberoptic materials. One type consists of a silica core with plastic cladding of a smaller refractive index than the core material and is used in multimode optical data transmission. A second type, known as a graded-index fiber, has a doped silica core fiber whose refractive index varies from the center of the core to the outer surface in a nonlinear gradient. The third type consists of a doped silica core and doped silica cladding of lower index of refraction than the core.

Of the three types of fibers, the graded-index fibers exhibit the lowest radiation dispersion, although the other two types are also low-loss fibers. The graded-index fiber operates in multimode with an LED input light, but when a Nd-YAG laser is used the fiber can operate in single mode at up to 100 GHz bandwidth. Typical fiber diameters range from 0.05 to 0.125 mm, and the fiber transmission loss varies from typically 100 dB/km down to 0.5 dB/km, and even less with certain recent developments. The numerical aperture varies from 0.24 to 0.67 typically. (The numerical aperture of a fiber is its ability to capture the maximum amount of light incident conically on its cross-sectional surface, with the apex of the cone incident on the fiber surface.) Table 6-1 illustrates some of the characteristics of optical fibers and cables.

Silica or quartz fibers are used singly or in a bundle. In either case, coupling the radiation from an LED or a laser diode to the fiber poses some problems, which are

*H. Kressel, *Topics in Applied Physics,* Vol. 39, Springer-Verlag, Berlin, 1980; H. Kressel and J. K. Butler, *Semiconductor Lasers and Heterojunction LEDs,* Academic Press, New York, 1977.

Table 6-1 Typical characteristics of optical fibers*

Fiber type	Core and cladding	Attenuation, dB/km	Fiber core diameter, μm	Numerical aperture	Bending radius, cm
PS-05-10	Silica, plastic	10	125	0.3	0.5
GG-02-05	Doped silica, borosilicate	3	50	0.25	0.5
GS-02-05, multimode	Doped silica, borosilicate	3	50	0.25	0.5
ISM-6-GG-(6), 6 fibers	GG-02-05	6	Cable diam., 6 mm	0.25	5.0
S-6-GG-(1)	GG-02-05	6	Cable diam., 2.5 mm	0.25	2.5
ESM-6-GG-(7)	GG-02-05	6	Cable diam., 6.0 mm	0.25	5.0

* Manufactured by Electro-Optical Division of ITT, Roanoke, Virginia.

not unsurmountable. Optical connectors are used to interface the light source to the end surfaces of the linking fibers. Also, in coupling the radiation to the fiber or the fiber to the terminal detector, an optical connector or coupling is used. The LED is inserted into one end of the coupling to link its light to the fiber attached to the other end of the companion coupling, as illustrated in Fig. 6-5. For fiber-to-fiber coupling, both ends of the coupling are precision made so that the surfaces of the fibers are aligned accurately. Another type of fiber-to-fiber linking is the splice-coupling method, in which two ends of the fibers are permanently joined with an optically pure bond interface; this type will exhibit a minimum loss of radiation by dispersion.

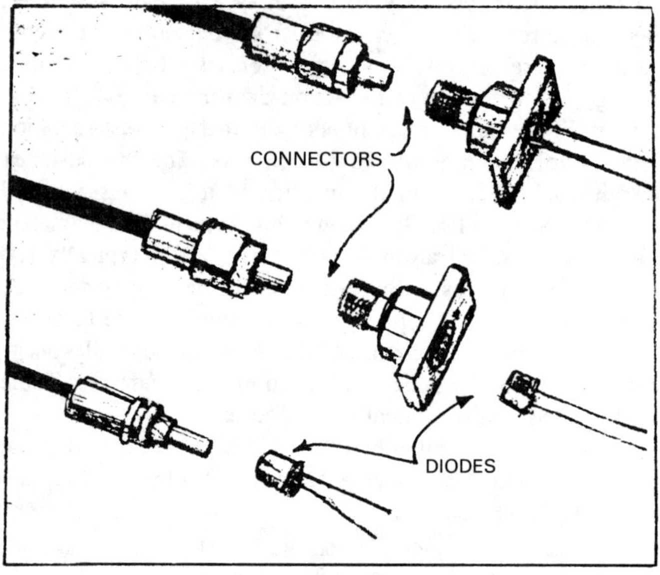

Figure 6-5 Examples of optical couplers.

Dispersion Loss in Fibers

As the data bits propagate through an optical fiber, they become blurred before they reach the receiving end, and the information may even be lost if the blurred data pulses become incoherently superimposed on each other. One of the main reasons for this is the fact that the traveling bit pulses make zigzag patterns along the interior of the fiber core. The rays that have become incident on the fiber aperture at the transmitting end at a larger incidence angle make more zigzag paths and thus take longer to reach the receiving end of the cable. The difference between the rays that take a shorter time and those that take a longer time to travel the length of the cable accounts for the spreading or divergence of the laser beam pulse. This dispersion is a function of the difference in index of refraction between the core material and the cladding of the core and is augmented by the distance the pulse information travels in the fiber.

To overcome the dispersion problem, several optical fiber manufacturers produce fibers whose index of refraction changes with radial distance from the fiber optical axis, as described in a previous paragraph. Furthermore, the pulse dispersion varies with the wavelength of the light transmitted, which changes the index of refraction of the fiber. With GaAs LEDs, which emit in the visible spectrum and contain incoherent wavelengths, the dispersion effect is prominent, whereas with GaAs lasers, which have a hundred times smaller linewidths, the dispersion effect due to wavelength variation is practically nil.

6-4 FIBEROPTIC WAVEGUIDE PRINCIPLES

Laser waveguides are light pipes that transmit laser radiation from a laser source to a destination. Laser waveguides are ordinarily made of fibers of glass, plastic, or quartz, depending on the type of light wavelength they will conduct. The fiber diameter varies typically from 2 to 250 μm, depending on the type of material and service for which they are intended. Usually, the fibers are gathered into a bundle and enclosed in a polyvinyl chloride sheath; a combination of kevlar and polyurethane is also used for the sheath. The ends of the fibers are polished and bonded, for long-distance use, with an epoxy material. The terms waveguide and fiberoptic conduit are practically interchangeable, but the word waveguide ordinarily refers to short fiberoptic pieces, such as those used in medical work, temperature sensing, welding, and the like.

It has long been known that as light travels through any transparent medium, there is a loss of brightness due to several factors. These factors are the angle at which the light enters the transparent medium, the length of the fiber, the absorption characteristics of the fiber, the reflection loss at the proximal end face, the index of refraction of the fiber core, and the wavelength of the radiation transmitted. Like an optical lens, the fiber has a numerical aperture that determines the light-gathering property of the fiber surface. Also, the numerical aperture for a straight fiber is slightly different from that for a curved fiber. However, because of the very small diameter of the fiber used for laser transmission, this difference is not significant.

There are two general types of fiberoptic conduits—coherent and noncoherent. In the coherent type, the fibers along the entire length of the bundle are arranged so that their relative positions at both ends are the same. This type of fiberoptic bundle is employed as an image conduit because the image incident on the face of the proximal end is transmitted unchanged to the other face of the fiber at its distal end. This type is used extensively in medical diagnosis applications. In the noncoherent type, the fibers are randomly grouped together but have relatively the same light transmission efficiencies as the coherent type. Fiberoptic laser conduits have numerous uses in industrial, medical, dental, and military applications, in addition to metrological instrumentation.

Since the discovery of fiberoptic material as a light conductor, many improvements have been made to increase the efficiency of transmission by refining the purity, density, and transparency of the material and improving the index of refraction of the fiber. One of these improvements involves coating or cladding the fiber strands with a glass or plastic of lower refractive index than the core in order to increase the transmission by total reflection of the laser beam from one surface of the fiber wall to another during its travel through the fiber. Any impurity or contaminant within the core or the surface of the fiber affects its transmission efficiency. Even fingerprints on the fiber surface affect laser transmission. If the surface is painted with some material having a relatively high refractive index or absorption properties, such as paint, total extinction of the light through the fiber occurs; this effect is known as light frustration.

The spectral transmission of the fiberoptic includes ultraviolet, visible, and infrared radiation, and there are specific fibers for transmitting radiation in each of these categories. For instance, quartz fiber may be used to transmit ultraviolet as well as visible wavelengths and to some extent infrared radiation; germanium and lanthanate glass fibers can transmit infrared radiation wavelengths. Many other materials are transparent to these types of radiation, and they will be discussed under pertinent headings.

Numerical Aperture

The amount of light accepted by the fiberoptic cross-sectional surface depends on the numerical aperture (NA) of the fiber. The numerical aperture is the maximum light-admitting measure of the fiber and is numerically equal to the product of the refractive index N_a of the medium and the sine of the angle ϕ between the normal to the fiber face and the incident light beam. This relation may be expressed as

$$\mathrm{NA} = N_a \sin \phi = (N_1^2 - N_2^2)^{1/2} \tag{6-6}$$

where N_1 is the refractive index of the fiber core and N_2 the refractive index of the fiber cladding. For the greatest efficiency of laser transmission through the cladded fiber, the refractive index N_1 of the core must be greater than the refractive index N_2 of the cladding, C, shown in Fig. 6-6.

As stated earlier, a ray of laser light travels through a fiberoptic strand by total reflection. This occurs when the laser ray, for instance R in Fig. 6-6, becomes incident on the end surface of a fiber core having a refractive index (N_1) higher than that of the

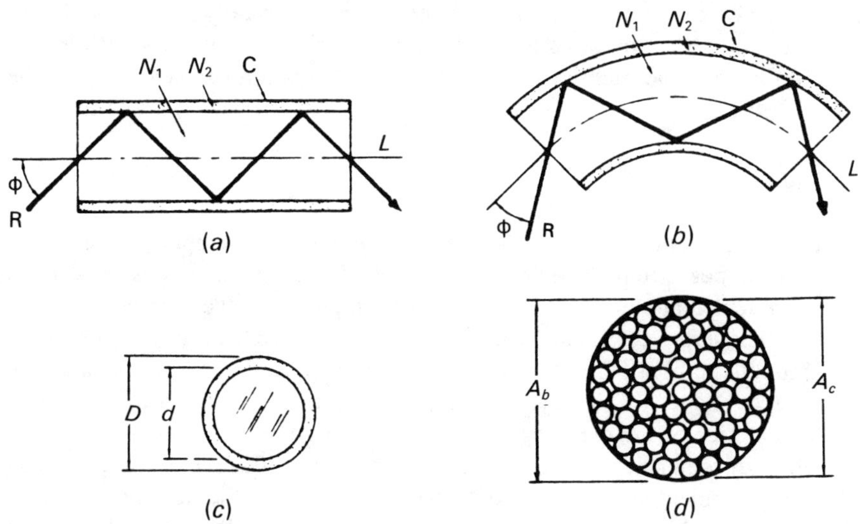

Figure 6-6 Laser ray transmission through an optical fiber.

cladding (N_2) and when the laser ray is within the critical angle, shown in Fig. 6-7. The ray then undergoes multiple reflections from the interface between the cladding and the core until it reaches the opposite end of the fiber, where it projects out at the same angle of incidence ϕ. If the ray is incident on the fiber end surface at an angle greater than ϕ, it will reflect out of the cladding. Equation (6-6) ordinarily is applied to a straight fiber and will change for a curved fiber of relatively large diameter. However, for very fine-diameter fibers, within the range of about 40 μm, the deviation will not significantly affect the relation in Eq. (6-6). Most fibers used in practice will be confined to small diameters.

The refractive index of a fiber core made of glass is typically 1.65, and the cladding index of refraction is 1.53. This proportion changes with different types of fiber and cladding materials, and this change in index ratio may cause the numerical aperture to vary. For plastics, the indices of refraction of the core and the cladding are lower than that of glass. The fiberoptic bundle diameter for applications discussed in this book

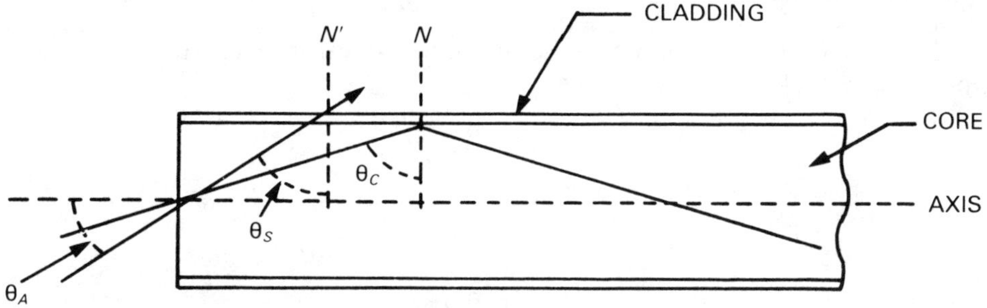

Figure 6-7 Diagrammatic illustration of the critical angle.

varies from 1 mm up to 1 cm. The numerical aperture may typically range from 0.2 to 0.8. By judicious combination of core and cladding materials, it is possible to vary the numerical aperture, the thermal properties, and the transmission characteristics of the fiber.

Critical Angle

One of the most important factors in radiation transmission through optical fibers is the angle at which the beam reflects within the core. This angle is known as the critical angle; a beam reflected at any angle within the core smaller than the critical angle will leave the fiber. The larger the critical angle, the greater will be the transmission of the laser beam through the fiber. Conversely, the smaller the critical angle, the smaller the transmission because some of the rays from the radiation at a smaller critical angle will emerge to the exterior of the cladding and be lost.

The relation of the critical angle to fiber transmission is illustrated in Fig. 6-7, in which a light ray incident at angle θ_i with respect to the axis of the fiber reflects from the interface at the cladding to the opposite side, making an angle θ_c with the normal N. This angle is the critical angle, and when a light ray impinges on the core-cladding interface at an angle equal to or greater than θ_c, total reflection in the fiber occurs. The light ray then continues to be reflected at the same angle throughout the entire length of the fiber until it emerges from the opposite end. However, any light rays that enter the core of the fiber at an angle θ_s that is smaller than the critical angle θ_c pass out of the cladding, as illustrated in the figure.

The critical angle θ_c may be expressed by

$$\sin\theta_c = \frac{n_2}{n_1} \tag{6-7}$$

where n_2 is the index of refraction of the cladding and n_1 the index of refraction of the fiber core.

Example A certain optical fiber has core index of refraction 1.546 and cladding index of refraction 1.450. A ray of laser radiation becomes incident on the cross-sectional surface of the fiber at an angle of 50°. (a) Determine the critical angle in the core material. (b) Calculate the numerical aperture of the fiber. (c) Calculate the angle of refraction of the core.

Solution (a) In accordance with Eq. (6-7), the critical angle of the ray in the core will be

$$\begin{aligned}\sin\theta_c &= \frac{n_2}{n_1}\\ &= \frac{1.450}{1.546}\\ &= 0.9380\end{aligned}$$

and the angle θ_c will be 69.7°.

(b) The numerical aperture of the fiber will be

$$\begin{aligned} \text{NA} &= (n_1^2 - n_2^2)^{1/2} \\ &= (2.39 - 2.10)^{1/2} \\ &= (0.29)^{1/2} = 0.54 \end{aligned}$$

(c) The angle of refraction of the ray in the core will be obtained from

$$n_1 = \frac{\sin i}{\sin r}$$

$$\begin{aligned} \sin r &= \frac{\sin 50°}{1.5460} \\ &= \frac{0.7660}{1.5460} \cong 0.5 \\ &= 30° \end{aligned}$$

Thus, it is seen that whether a ray of a laser beam will be conducted through the fiber or will project out of the fiber depends not only on the critical angle but also on the index of refraction of the core material as well as that of the cladding. Accordingly, these factors are taken into consideration in designing a fiber transmission system.

Packing Density of Fiber Bundle

The packing density of a fiberoptic bundle is the ratio of the area of the total number of fibers in the bundle to the area of the overall bundle. This relation can be explained by using the illustration in Fig. 6-6*a*, in which d is the diameter of the core and D is the diameter of the fiber with its cladding. For instance, if there are N fibers in the bundle shown in Fig. 6-6*b*, the total area of the fiber core, A_c, will be given by

$$A_c = \frac{N\pi(d)^2}{4} \tag{6-8}$$

and the area of the bundle, A_b, will be given by

$$A_b = \frac{N\pi(D)^2}{4} \tag{6-9}$$

The packing density P_d of the fiberoptic bundle will be

$$P_d = \frac{A_c}{A_b} \tag{6-10}$$

or

$$P_d = \frac{d^2}{D^2} \tag{6-11}$$

The percent utilization P_u of the total fiberoptic bundle will be the total core area divided by the total bundle area and multiplied by 100, as shown in Eq. (6-12).

$$P_u = \frac{d^2}{D^2} \times 100 \tag{6-12}$$

Equation (6-11) is simplified by canceling the π and 4 from each term that constitutes the area.

Example Let us assume that a certain fiberoptic bundle has 400 fibers and that each fiber has a diameter of 60 μm and a cladding thickness of 5 μm. What will be the packing density of the bundle? What will be the percent utilization of the fiberoptic bundle?

Solution The packing density will be

$$F_d = \frac{d^2}{D^2}$$

$$= \frac{60^2}{70^2}$$

$$= 0.73$$

and the percent utilization will be

$$P_u = 0.73 \times 100$$

$$= 73\%$$

From these relations it is readily apparent that the larger the diameter of the fiber, the greater is its utilization percentage. However, other optical factors enter into the use of these equations in laser waveguides. For instance, the numerical aperture will change and the resolution will suffer when coherence is the prime requirement, as in medical diagnostic work. Also, to achieve flexibility, the fiber diameter must be reduced and thus the number of fibers increased in the bundle.

Valtec Fiberoptic Waveguide

The Valtec Corporation manufactures fiberoptic instruments for various applications. One of these instruments is a product code scanner consisting of a tubular stylus at the terminal portion of the waveguide that transmits the scanning light to and from the product code, as shown in Fig. 6-8. In operation, the stylus is passed over the product code bars to transfer the product information from the bars to a computerized checkout and inventory system for retail merchandising. The point of the stylus has a spherical lens that rotates over the bars as the stylus is passed over them. The device has a resolution of 0.006 inch at a scan rate of 25 inches per second, and its lifetime is claimed to be 10,000 hours. It operates from a 15-V DC current using 0.625 W.

One other versatile instrument manufactured by Valtec Corporation is a medical implementation device that supplies a bright and pure light at ambient temperature. The instrument is used principally in delicate surgical work where intense light with

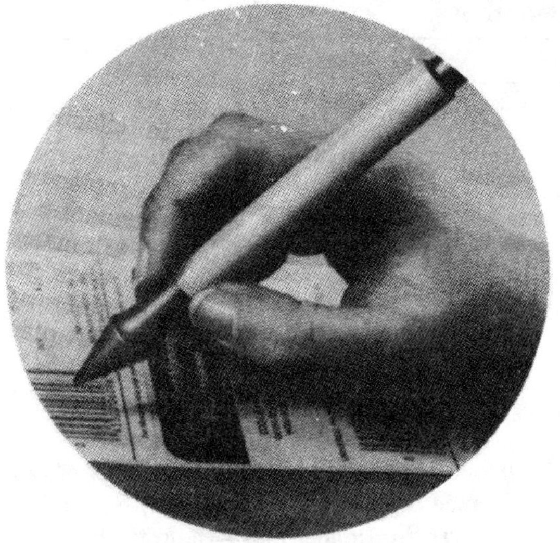

Figure 6-8 Fiberoptic product code scanner (Valtec).

practically no heat is essential. The device is claimed to be fully autoclavable and offer complete electrical isolation.

6-5 RECEIVER CHARACTERISTICS

The receiver of the fiberoptic transmission line is provided with a photosensitive detector at its front end to which the distal end of the transmission fiber is linked. The detectors most suitable for the fiber-detector link are PIN diodes (*p*-layer-intrinsic-*n*-layer diodes); avalanche photodiodes with high numerical apertures to accept light from different angles; and phototransistors, which also amplify the transmission signal. The avalanche photodiode exhibits high speed and high receiver sensitivity because of its gain due to increased thickness of its avalanche site. Table 6-2 lists various photosensitive detectors.

The amplifier for the front end of the receiver with any of these detectors has a low signal-to-noise ratio because the amplifier is designed for a smaller bandwidth than that of the optical signal. There are other types of amplifiers: (1) When the transimpedance type is used in the front end, the detector load resistance functions as an effective feedback resistor. (2) In the receiver end design, the capacitive effect of the detector, its temperature stability, and its low-frequency signal cutoff must be considered. Thus, the photodetector must match the optical design in the application to minimize the background noise and quantum noise due to photons (phonons) and to compensate for amplifier noise. The designer of an optical transmission system must start with the manufacturer's specifications for the materials and electrical parameters. A well-designed receiver must have a front end designed for the specific bandwidth at which the transmission system is to be used.

Sufficiently high voltage gain should be incorporated in the receiver to raise the

Table 6-2 Various types of photodetectors

Silicon	Germanium	Others*
PIN diodes	Photovoltaic diodes	Lead sulfide diodes
Avalanche photodiodes	Photoconductive diodes	Lead selenide diodes
Doped diodes	Doped diodes	Mercury-cadium telluride (HgCdTe)
Photovoltaic diodes	Avalanche diodes	Indium antimonide (InSb) Indium arsenide (InAs)
Photoconductive diodes		Lead-tin-telluride (PbSnTe)
Hybrids and FETs		Gallium-arsenide-phosphide (GaAsP) Indium-gallium-arsenide (InGaAs and many others)

* The diodes in this column are also made as photovoltaic and photoconductive types.

signal level to a usable output voltage, which can be determined by the receiver power, diode responsivity, and effective load resistance. Further voltage gain may be obtained by cascading additional amplifier stages. The transimpedance may be adjusted by means of an automatic gain control to a level sufficient to convert the anticipated input signal to the desired output voltage. When cascaded amplifiers are inserted into the circuit, the overall rise time will be approximately 1.1 times the square root of the sum of the squares of the individual amplifier rise times. The system rise time, however, can be reduced by proper design of the receiver bandpass to equalize dispersion due to multi-mode operation and optical material.

Laser Photodiode Technology

The latest development report from Bell Laboratories on photodiodes for long-wave-length communication systems with low-dispersion silica fiberoptic lines operating in the wavelength range 1.3 to 1.55 μm presents indium-gallium-arsenic-phosphide (InGaAsP) alloy material epitaxially grown on InP substrate as the most promising lattice-matched photodiode constituent. This material offers broad spectral sensitivity, and the InP substrate can be employed as the back window material for admitting light into the *p-n* junction, as shown in Fig. 6-9. The back window illumination of the diode minimizes losses due to surface recombination of the carriers generated by the incident light. A 2000-Å-thick layer of silicon nitride (SiN_x) is applied as diffusion mask for Zn or Cd formed adjacent to the *p-n* junction. A centrally located top metallization and an annular metallization formed peripherally to the back window provide contacts for external leads. This combination of diffusion materials offers high transparency to long-wavelength incident illumination.

PIN photodiodes have the property of high detection sensitivity, operation at low voltages such as 5 V, and high quantum efficiency (see Eq. 6-4) in the neighborhood of 100%. Photons are absorbed in the depletion region of the InGaAs/InP interface adjacent to the n^- InP buffer substrate. The percentage of total quanta absorbed is a function of the internal quantum efficiency. Also, the sensitivity of the optical receiver is influenced by the dark current and the detector capacitance—the smaller these quantities, the lower the noise. The dark current increases exponentially in this type of detector; therefore, the device is made to operate below its breakdown voltage. Because of the high-doping character of the InP layer, the diode must not be operated at higher voltages than specified. Since these devices have high-speed response and high reliability, they are capable of functioning in the gigabits per second range.

Avalanche photodiodes have the same internal quantum efficiencies as PIN diodes but they also have sufficient gain to increase the device sensitivity considerably, and their tolerable dark current is much lower than that of PIN photodiodes; this is due to excessive avalanche noise arising from the ionization process dependent on the gain. This limits the transmission of a 1.3-μm-wavelength beam to about 15 km between any two repeater stations. To increase the speed of the photoresponse, the depletion region of the avalanche diode must penetrate about 2 μm into the InGaAs layer, and the light-absorbing layer must be *n*-type for low-noise operation. A barrier to the application of the avalanche photodiode for long-wavelength communication is its rapid avalanche-forming condition at excess noise with increased voltage and resultant carrier multiplication and initiation of ionization.

Added to these problems is lack of knowledge about the reliability of operation of avalanche photodiodes. The passivating layer is contiguous with the semiconductor gain region held at a high electric field level, which may result in large dark currents or in degradation of the diode quality. These problems, however, do not seem insurmountable, and when they are overcome the high sensitivity offered by the device will make it potentially attractive for long-wavelength communication service. While there are other devices that also amplify the signal, such as modulated barrier photodetectors, photoconductors, and phototransistors, they all draw high current with resultant high noise that has a tendency to degrade the performance of the device. It is possible that further study of the diode composition GaAlAsSb will uncover an ideal material of high sensitivity and reduced dark-current properties for transmission of long-wavelength radiation in laser communication.

Recent developments of GaAs PIN photodiodes or germanium avalanche photodiodes for use with single-mode fiber systems have found some success. Other detectors such as GaAsIn/InGaAsP/InP avalanche photodiodes have also been tried and have superseded the former diodes; these diodes have low dark current with high gain in

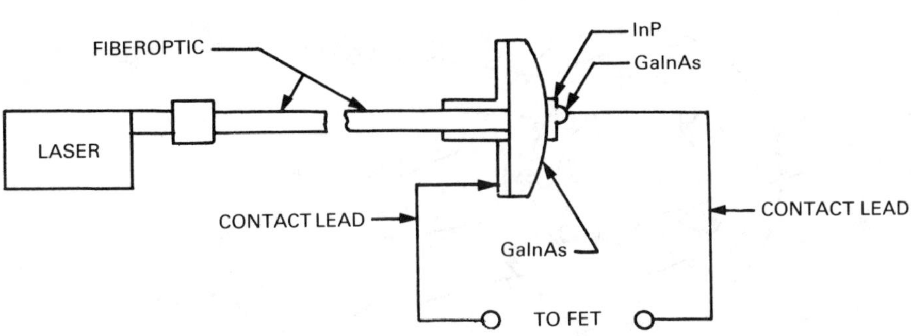

Figure 6-9 Schematic diagram of a planar InGaAs PIN photodiode device.

excess of 10 GHz, but as yet they have failed to meet the ideal photodetector requirements. It appears that the potential improvements will come from monolithic integration of laser sources, detectors, and associated circuitry through reduced parasitic capacitances and reduced device dimensions with high reliability. Some of the projected applications involve the use of cleaved-coupled cavities (see Section 9-4), frequency-selective feedback schemes, or injection locking to force the laser system to operate at a single frequency and obviate the use of repeaters for long-distance transmission of data.

General Design for Communication Lines

In a commercial communication line in which repeaters are used and the cables are stressed during installation, the cables are usually provided with steel wire reinforcement and electrical conductors to transmit boosting power to the line repeaters. Such a cable, shown in Fig. 6-10, consists of a steel wire embedded in a silestic material. Peripheral to it are two copper wires to conduct current to the repeaters. Around the silestic cable is tubular hytrel, which contains a number of fiber strands for optical transmission. Surrounding the hytrel are at least two other plastic sheathings, kevlar and polyurethane. These sheathings prevent the cable from being chewed by rodents in the ground and also offer tough resistance to abrasion, oxidation, and other environmental attack from the ambient surroundings. At various distance intervals, the cable is linked with electronic components to give a boost to the transmission of the data or voice signals. This is only a general description of the construction of a typical cable; different manufacturers have different proprietary designs and construction. However, the design shown in Fig. 6-10 may be considered as basic format.

Glass Fiber Drawing Technique

Although there are several methods for manufacturing single waveguide fibers, one method developed by Galileo Electro-Optics Corporation deserves discussion here. The process consists of melting a batch of core material (lead silicate) in a platinum crucible and floating the cladding material of low index of refraction over the molten lead

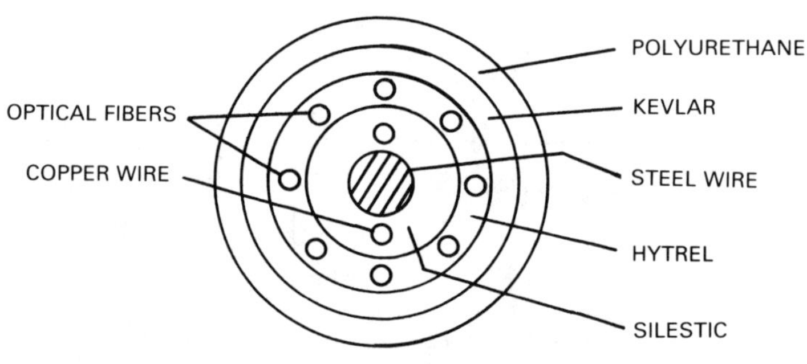

Figure 6-10 A typical design for a communication cable.

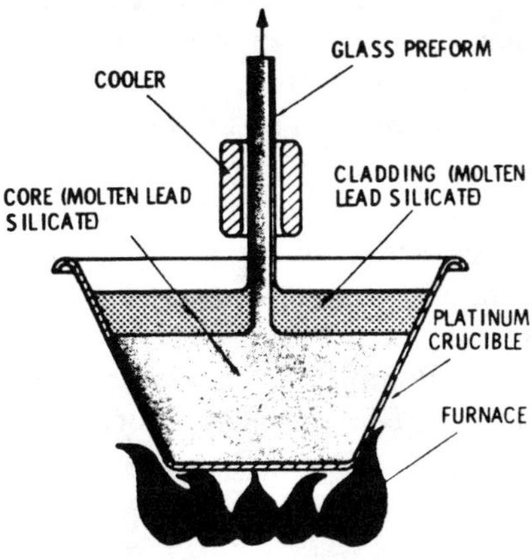

Figure 6-11 Galileo stratified melt process for producing fiberoptics. (Courtesy of Galileo Electro-Optics Corp.)

silicate. The two materials are heated to their fusion point, and a glass preform is inserted through the two layers and then slowly withdrawn through a cooler, forming a fiber with lead silicate cladding on its periphery. It is claimed by the maker that this process produces significantly lower radiation loss than other methods using plastic-clad fused silica. The finished product has a transmission attenuation of 30 dB/km. However, because of its relatively high numerical aperture of 0.48 and high radiation transfer index, the material is preferred for applications in which distances do not exceed 200 m, and for medical work the loss rate is insignificant because of the short length of the fiber cable. The process is shown in Fig. 6-11, with various essential components labeled.

6-6 FIBEROPTIC APPLICATIONS

Fiberoptic cables are employed in numerous applications, and new uses are being developed for them as the technology advances. Some of the prominent applications of fiberoptic transmission systems are in private broadband networks, facsimile transmission, intercomputer links, process control systems, automation of industrial systems, patient-monitoring devices, military ships and aircraft, microwave-to-optical intercommunication, memory delay lines, inspection of hazardous situations, security surveillance systems, and many others.

Among the many systems presently in use, a laser system for transmitting Wirephotos has been developed by the MIT Lincoln Laboratory. It uses a He-Ne laser beam of 1-mW power, which is intensity-modulated by a distant transmitter signal scan across a silver-emulsion photographic paper developed by 3M Corporation. The paper records the varying signal intensities in various shades of gray. The paper is then subjected to a thermal environment to develop the image, and a high-gloss photographic-quality image is obtained. The prints are said to be suitable for storage for long periods of time without deterioration.

For short-range line-of-sight applications, semiconductor laser communication systems are becoming increasingly practical, especially for field use and military intelligence service. Likewise, fiberoptic transmission systems for shipboard use, from one cabin to another, are becoming very common on new ships as well as on some of the old ships, which are being reequipped with fiber transmission systems because of the immunity of their signals to extraneous electromagnetic interference.

Fiberoptic information transmission technology has already penetrated local communication networks using data rates up to 100 megabits for distances from several miles up to 10 miles. Because of the high bandwidth of present optical fibers and the potential application of wavelength-division multiplexing (see Chapter 9), bidirectional transmission formats have become possible. In these efforts, LEDs and laser diodes will play important roles in both single-mode and multimode transmission lines.

Laser Systems Communicator

A laser communicator manufactured by Laser Systems, Inc., generates 10-kHz pulses at 10 W to transmit information at 10-kilobits per second to a distance of 25 miles in clear weather. It operates from 12- to 24-V DC battery power or a 115-V AC power supply to drive a gallium arsenide diode. The pulse rise time is 100 nsec at a 9040-Å wavelength. The pulses are transmitted and received through dual telescope channels using 10-power lenses. The telescopes are built side by side in the unit to provide a compact and portable laser communicator. Available beam divergence ranges from 1.3 to 10 mrad with interchangeable lenses provided with the unit. The developer claims that the system can detect signals as small as 0.58 nanowatt through a 50-mm optical aperture up to a 10-mrad field of view.

A fiberoptic communicator of almost the same characteristics has also been developed to transmit both digital and analog optical data to distances up to 111 km without use of a repeater. Multiplexing the information to several stations has already been demonstrated.

Central Telephone Communication System

A 4.2 km-long fiberoptic communication system was installed between MGM Grand Hotel and Central Telephone Company in Las Vegas, Nevada. The system, which was said to be the fourth of its kind in the world, uses laser diodes in its transmitter. For its receiver, the system is provided with photodetectors manufactured by Valtec Corporation. The existing telephone communication ducts have been used for the fiberoptic cables, which have been identified by red polyethylene sheathing to differentiate them from other copper wire cables in the duct. To give strength to the fiberoptic cable, a 3/32-inch-diameter insulated steel wire has been disposed centrally to the cable for pulling and handling during installation. The cable consists of six graded-index fibers of low decibel loss. Each fiber strand is covered with polyethylene sheathing and with kevlar, a flame-resistant plastic material.

The channel capacity of a fiberoptic line is directly proportional to the bandwidth, which can range to more than 100 MHz. An expression for the information capacity of the line is

$$C = B \log\left(1 + \frac{P_a}{N_w}\right) \tag{6-13}$$

where C is the information capacity of the signal, B is the bandwidth of the channel, P_a is the average power, and N_w is the white-noise power of the given channel.

Other Fiberoptic Applications

In addition to the various applications of fiberoptic materials already discussed, the following application areas are important:

Intra- and intercomputer links
Telephone loops and trunks
Private broadband networks
Facsimile transmission systems
Memory delay lines
Ground station links for satellites
Submarine cables
Sonobuoy systems
Patient monitoring
Electronic instruments
Microwave-to-optical interconnections
Fusion and nuclear operations
Telemetering
Process control systems
Automation of industrial inspection
Lightning arresting systems
Hazardous situation inspections
Civilian ships and aircraft

QUESTIONS AND PROBLEMS

6-1 What is meant by laser beam communication? How many separate conversations can be carried out with a laser beam at the same time? In how many formats can the beam be modulated for communication? What is optical fiber communication? What does such a system consist of? What is the function of the voice transmitter? What is the function of the detector in the receiver?

6-2 Can several aerial communications be made at the same time even when an aircraft is flying? Explain how this can be accomplished. At what power level should an aerial communication beam be retained? Why? How far can aerial communication be carried out? What wavelength of the laser beam should be used for optimum modulation speed? What type of laser beams can be modulated for communication? Can frequency-doubled Nd-YAG be employed for speech modulation?

6-3 State several important advantages of laser communication. Give the component parts of a fiberoptic communication system. Is there any limit on the distance to which optical communication can be made? What types of lasers can be used for optical communication? What is the minimum signal intensity loss that can be achieved? Up to how many bits can be transmitted through optical fibers? Up to how many megahertz can be transmitted through optical fibers?

6-4 Which transition metal impurities cause loss of signal intensity in the transmitting fiber? State several factors that reduce in signal intensity in the fiber. Explain what is meant by optical transfer-function loss between the transmitter and the receiver. What is the advantage of fiberoptic transmission versus metal-conductor transmission? Can light-emitting diodes (LEDs) be used in fiberoptic transmission of data? Can invisible laser light be transmitted through the fiberoptic? Explain how it can be.

6-5 In transmitting a data signal on an infrared laser beam, what types of fibers can be used? What types of laser beams can be employed for remote transmission of data and speech? Why are LEDs favored for data transmission? At what level with respect to threshold current do LEDs and LED lasers operate? Explain why a neodymium-aluminum borate crystal is used in optical communication. What optical pump is used to excite this crystal? Why do diode lasers have greater emission efficiencies than any other type of solid-state lasers? What is meant by emission efficiency of the diode?

6-6 Calculate the quantum emission efficiency of a diode with a wavelength of 9050 Å and a photonic power of 50 W per square inch, if the diode is forward biased with a current of 10 A.

6-7 What are metal vapor lasers? Why should they be used in fiberoptic data transmission? What general types of fiberoptic materials are used? Which of these has a doped silica core and a doped silica cladding of lower index of refraction? Which fibers exhibit the lowest radiation dispersion? How can graded-index fiber be made to operate in a single mode? How can it be made to operate in a multimode format? What are some typical fiber diameters?

6-8 In the case of light source-to-fiber coupling, how are LEDs incorporated in the coupling interface? How is fiber-to-fiber coupling accomplished? Explain how dispersion loss is incurred in a fiberoptic line. How can dispersion in the fiberoptic be reduced? Why is the dispersion effect due to wavelength practically nil in GaAs lasers? Why is the dispersion phenomenon in LEDs prominent?

6-9 Why does a fiberoptic have a core of greater index of refraction than the cladding? State four factors that are responsible for the loss of brightness of the beam as it travels through the fiberoptic cable. What does the numerical aperture of a fiberoptic determine? What is the difference between a coherent and a noncoherent fiberoptic? When is a coherent fiberoptic used? Should a coherent fiberoptic be employed for data transmission? Explain why.

6-10 Explain what type of fiberoptic is used for transmitting infrared, visible, or ultraviolet radiation. What is numerical aperture as applied to an optical fiber? In what configuration is a laser ray transmitted through an optical fiber? Calculate the numerical aperture of an optical fiber whose core index of refraction is 1.657 and cladding index of refraction is 1.584.

6-11 What is the packing density of a fiberoptic cable? What is meant by percent utilization of a fiberoptic bundle? Calculate the packing density of a fiberoptic cable if the diameter of the fiber core is 40 μm and the diameter of the fiber is 50 μm, assuming that there are 600 fibers in the cable.

6-12 How does a Valtec waveguide operate? What does it accomplish in its operation? Describe how a typical fiberoptic strand is fabricated. What is the critical angle as related to a ray passing through an optical fiber? If the critical angle is large, what happens to a ray that enters an optical fiber? If the critical angle is small, what happens to a ray incident on the end of the optical fiber?

6-13 A ray of laser radiation impinges on the cross-sectional surface of a fiber whose core index of refraction is 1.582 and cladding index of refraction is 1.460. If the angle of incidence of the ray is 45°, calculate (a) the critical angle in the core material, (b) the angle of refraction at the core, and (c) the numerical aperture of the fiber.

6-14 Describe the structure of a fiberoptic transmission line receiver. What types of detectors are used at the receiver end of a fiberoptic cable? What is a transimpedance-type photoamplifier? What electrical and physical parameters should be considered in the design of the receiver end? Why should the photodetector characteristics match the optical design of the receiver? To raise the signal level to a usable output voltage, what parameter should be increased in the receiver? What advantage can be gained by cascading amplifiers? What is the function of automatic gain control in the receiver?

6-15 What is the importance of the use of an InGaAsP photodiode developed by Bell Laboratories? Why should InP be used as the rear window for this photodetector? What is the purpose of back window illumination of the diode? What is the purpose of a diffusion mask for the diode? State some of the important characteristics of PIN photodiodes used in the receiver end of a fiberoptic line.

6-16 How is the sensitivity of the optical receiver influenced and how does it affect the noise? What is the effect of the high-doping character of the InP laser in the diode? What are avalanche photodiodes? How does the device sensitivity of an avalanche photodiode compare with that of a PIN photodiode? How is the photoresponse speed of an avalanche photodiode increased? Explain. What barrier is imposed on the application of an avalanche photodiode for long-wavelength signal communication?

6-17 Describe a typical design for a communication cable. What material in the cable prevents attack by environmental conditions? What is kevlar's importance in the cable construction? How do repeaters receive their energizing current in the fiberoptic line?

6-18 State some of the important applications of fiberoptic cables. Describe the MIT Wirephoto transmitter and its unique features as differentiated from conventional phototransmitters. For short-range line-of-sight application, which types of laser systems are more practical? Can fiberoptic transmission systems transmit data or voice to short distances as effectively as they do to long distances? Are there any barriers to optical data transmission at short distances? If so, what are they?

6-19 Describe the Laser Systems laser communicator. Is a pulse rise time of 100 nsec faster than the conventional diode rise times? What significance has rise time for the efficiency of transmission of optical data? In fiberoptic cable transmission, what important role does beam divergence play? Discuss.

SELECTED SOURCES OF RELATED DATA

Bigelow, H. 1985. Fluoride glass fibers in telecommunications. *Lasers and Applications* September, pp. 52–56.

Bigelow, H. 1986. Tunable WDM from GTE doubles fiber capacity. *Lasers and Applications* November, pp. 36–40.

Bigelow, H. 1987. High-speed laser brings fiber and satellites together. *Lasers and Applications* January, p. 28.

Camagni, P., and S. Sandroni, eds. 1984. *Optical Waves in Crystals*. New York: Elsevier.

Carts, Y. A. 1988. Long-wavelength diode lasers provide fast transmission. *Laser Focus/Electro-Optics* June, pp.136–139.

Chaffee, C. D. 1988. *The Rewiring of America: The Fiberoptic Revolution*. San Diego: Academic Press.

Channin, D. J. 1982. Emitters for fiberoptic communications. *Laser Focus* November, pp. 52–56.

Dawson, L. R., et al. 1980. LEDs for 32-megabit communication. *Bell Syst. Tech. J.* 59(2):161.

Fridman, J. D., and J. Svacek. 1983. Atmospheric optical communication. *Laser Focus* March, pp. 96–97.

Gallawa, R. L. 1979. *Optical Waveguide Communication*. Boulder, CO: U.S. Department of Commerce.

Hodara, H., ed. 1988. *Fiber and Integrated Optics,* quarterly journal. Washington, D.C.: Taylor & Francis.

Hunsperger, H. G. 1984. *Integrated Optics: Theory and Technology*. New York: Springer-Verlag.

International Laser Systems, Inc. 1983. GaAs Aerial Communication, technical bulletin. Orlando, Fla.

Izawa, T., and S. Sudo. 1987. *Optical Fibers: Materials and Fabrication*. Boston: Reidel.

Kressel, H., ed. 1982. *Semiconductor Devices for Optical Communication*. New York: Springer-Verlag.

L&A Staff. 1985. Fiberoptic connector specification table. *Lasers and Applications* September, pp. 124–127.

Lamrow, C. M., and V. A. Bhagavatula. 1985. Advanced fiber designs. *Laser Focus* March, pp. 82–92.

Lee, S. H., ed. 1981. *Optical Information Processing Fundamentals*. New York: Springer-Verlag.

Lovely, P. S. 1984. Measuring fiber performance parameters. *Laser Focus/Electro-Optics* May, pp. 98–110.

Mack, R., sr. ed. 1987. InGaAs photodiodes for long-wavelength fiberoptic communication. *Laser Focus/Electro-Optics* December, pp. 136–141.

Midwinter, J. E. 1979. *Optical Fiber for Transmission*. New York: Wiley.

Nagel, S. R., ed. 1987. Optical Fiber Materials and Properties, Materials Research Society Symposium Proceedings. Pittsburgh.

Pan, J. J. 1988. Integrated optics for microwave coherent detection systems (communication). *Lasers and Optronics* March, pp. 87–93.

Suematsu, Y., et al. 1982. *Introduction to Optical Fiber Communications*. New York: Wiley-Interscience.

Tunable Lasers. 1987. Appl. Phys. Ser., Vol. 59. New York: Springer-Verlag.

Willardson, R. K., et al. 1985. *Lightwave Communications Technology*. New York: Academic Press.

Young, M. 1984. *Optics and Lasers,* 2d ed., Vol. 5. New York: Springer-Verlag.

CHAPTER

SEVEN

MILITARY LASER SYSTEMS

7-1 INTRODUCTION

The technological potential of laser beams in military tactical systems is varied and numerous. Only a few representative systems are taken up in this chapter. They are confined to both the nondestructive and destructive aspects of the system technology. In the nondestructive category, laser beams are being employed to guide missiles, aircraft, drones, and other moving projectiles to their targets with high precision. The beam is also employed in battleground training of soldiers, who, instead of using blank ammunition, project laser beams in shooting maneuvers from rifle-configured devices. The rifle that houses the laser generator fires a pencil-narrow laser beam up to range of 1000 feet.

Each soldier is equipped with a uniform containing radiation sensors fixed at various points. When a sensor receives a laser beam, the soldier equipped with that sensor is considered to be hit at the body portion containing the sensor, which produces an electronic signal. An antenna erected in the soldier's helmet transmits the signal to a computer in the central station at the field. The signal informs the computer memory center whether the soldier was hit in the head, body, or extremities so that he can be eliminated from the field; the soldier is also made aware of the hit. The laser generator is generally a GaAs diode operating at a wavelength of 9040 Å and is energized by an electric battery source.

In other nondestructive applications, ruby laser rangefinders are being used on M-60 forward air controllers for artillery rangefinding and on AH56A Cheyenne helicopters as part of the navigation system. Battery-powered laser scanning devices are also being employed by the military and more recently by the police in many states of the country.

These devices can detect light invisible to the naked eye and amplify it 45,000 times or more so that objects and the enemy can be seen in the dark.

Although various methods have been devised for laser rifles to simulate the firing of the explosive weapons, one unique method is the use of a time-code modulation technique based on time synchronization, pulse position, and coding time slots. The laser employed in this technique is a gallium arsenide diode, which is temperature-sensitive during operation. The diode threshold current decreases at low temperatures and increases at elevated temperatures. When excessive lasing occurs below room temperature, failure results from current limitation. Above room temperature, excessive current density of the diode causes the failure. Accordingly, close control over the temperature is achieved by provision of a thermal control loop comprising a modulator, a thermistor, and a regulator. The regulator controls the DC current to the modulator in order to generate a supply voltage proportional to the current for lasing. The laser beam is collimated in the transmitter and has a beam divergence of 1 mrad; a simple converging lens of 3-inch focal length is used.

In the destruction application of laser beams, energy from a high-power laser is directed in a pencil beam at targets. The laser beam then destroys strategic areas on the target or drills holes in fuel tanks to explode the targets. On the battlefield, with a forward air controller or a ground observer maintaining a laser beam on a coordinate, a missile from a single fighter bomber can home in on a target with no trouble. Laser-guided bombs have already proved to be most effective counterweapons against 130-mm guns. It has been stated in a number of publications that a devastating laser beam weapon that will make nuclear weapons obsolete is in the development stage. Although such reports require serious scrutiny, it is well known to the scientific world that sufficiently high-powered laser beams have been developed to destroy nuclear warheads in moving missiles. Materials for generating high-energy laser radiation have been developed to the point where they may soon be used for triggering the explosion of nuclear warheads or for initiating controlled nuclear fusion for peacetime atomic energy.

Lasers generally employed in military applications include the ion lasers, molecular lasers, semiconductor lasers, liquid dye lasers, Nd-YAG or Nd-glass lasers, and for some purposes excimer lasers. The most powerful laser among these is the molecular or carbon dioxide laser. The Nd-YAG or Nd-glass laser is also used where destructive laser energy is specified. Ion lasers and semiconductor diode lasers are generally used in rangefinders, target designators, missile guidance systems, and battlefield simulation systems.

7-2 NONDESTRUCTIVE LASER SYSTEMS

Rangefinders

Laser rangefinders are standard equipment in many types of military artillery. A rangefinder is a device that uses a laser beam to measure distance. This is accomplished by projecting a pencil beam to a target whose distance from the laser source is to be measured. The timing system in the rangefinder determines the travel time of the beam to and from the target and electronically calculates the distance traveled one way to the target. For timing purposes, the projected laser beam is pulsed, since it is easier to count the pulses per second by a counter in the rangefinder than to use a continuous-wave laser beam. The counter counts from the first pulse leaving the instrument to the first pulse returning to it from the target. Since the total number of pulses is equal to twice the distance to the target, the total pulses are divided by 2 in the instrument computer, and the result is converted by the computer into kilometers, meters, and centimeters, or miles, feet, and inches, for whichever unit the range instrument is calibrated.

Example Suppose a rangefinder operates with a clock frequency of 100 MHz and the counter has counted 4×10^3 pulses. If the rangefinder is calibrated in centimeters, what distance will it measure? A simplified relation may be given as

$$D = \frac{N_1 C}{2N_2} \tag{7-1}$$

where D is the distance from the laser source to the target, C the speed of light, N_1 the number of clock pulses counted by the counter, and N_2 the clock frequency or pulse repetition rate per second. Substituting the numerical values of the quantities in Eq. (7-1), we obtain

$$\begin{aligned} D &= \frac{(4 \times 10^3)(3 \times 10^{10})}{2 \times 10^8} \\ &= 6 \times 10^5 \text{ cm or } 6 \text{ km} \end{aligned}$$

In the above example, the speed of light is given in centimeters; if the distance is to be measured in miles, the speed of light should be given in miles.

The signal from the target is inversely proportional to the fourth power of the target-to-transmitter range. The retroreflected energy from the target is uniformly distributed over the cross-sectional dimension of the signal bundle. This energy is confined to the solid angle limited to the receiver optical aperture and may be expressed as

$$N = \frac{E}{hf} \frac{1}{S^4} \frac{A_r A}{\phi_r \phi_e} J^2 \tag{7-2}$$

where N is the number of photons entering the receiver telescope, E is the transmitted energy in joules, S is the range from the laser system to the tracked object (target) in meters, A_r is the effective reflecting area of the object in square meters, A is the effective cross-sectional area of the entrance pupil of the telescope in square meters, J is the atmospheric attenuation of the laser beam (in joules), ϕ_r is the solid angle through which the energy is reflected and is equal to A_r/S^2, and ϕ_e is the solid angle of the beam reflected from the object. The number of photons entering the telescope and hence being detected by the photodetector may be converted into watts per square centimeter.

The laser beam employed in rangefinders may be from one of the following sources: diode laser, He-Ne laser, Nd-YAG laser, ruby laser, or dye laser. Recently, carbon dioxide laser beams have also been tried, and some problems due to their excessively high power levels are now being resolved by several manufacturers of carbon dioxide lasers. Furthermore, since the carbon dioxide laser operates in the infrared spectrum, a suitable photosensor sensitive to CO_2 wavelengths is now being developed for use on a commercial scale. Although the Nd-YAG laser also operates in the infrared, it can be frequency-doubled to a visible 5320-Å wavelength and is in extensive use for rangefinders.

A rangefinder, as shown in Fig. 7-1, consists of three principal components: the transmitter, the receiver, and the electronic signal-processing sections. The latter sections display the target distance in digital format, in either metric units or British units, depending on the type of military service. Because each manufacturer has a proprietary method of rangefinder construction, we shall discuss several major commercially manufactured rangefinders.

A typical rangefinder, for example, operates from a He-Ne laser emitting at a wavelength of 6328 Å. The unit is mounted on a tripod and has a telescope for sighting the target. Such a system will preferably be driven by a clock frequency of 98 to 100 MHz and will contain a frequency counter to count the number of pulses of the beam from the time it leaves the rangefinder to the first pulse from the target that reaches the counter and stops it. The counter processor then integrates all the pulses during a round trip, divides the result by 2, and multiplies it by the speed of light. The count translated into a digital format is displayed on the data readout instrument panel.

Subsequent to the discovery of the ruby laser by Hughes Research Laboratories, one of the first applications of the ruby laser beam was the development, by that company, of a laser ranging system designated as COLIDAR, an acronym for coherent light detection and ranging. This system operated Q-switched and provided pulses of several megawatts peak power with pulse duration in the nanosecond range. In operation, the Q-switched laser beam was projected on a retroreflector positioned at the proximal point of the distance to be measured. Counting of the pulses started simultaneously with the triggering of the laser and continued until the beam returned to the receiver from the target. The counted pulses were translated to a one-way distance measurement in digital form.

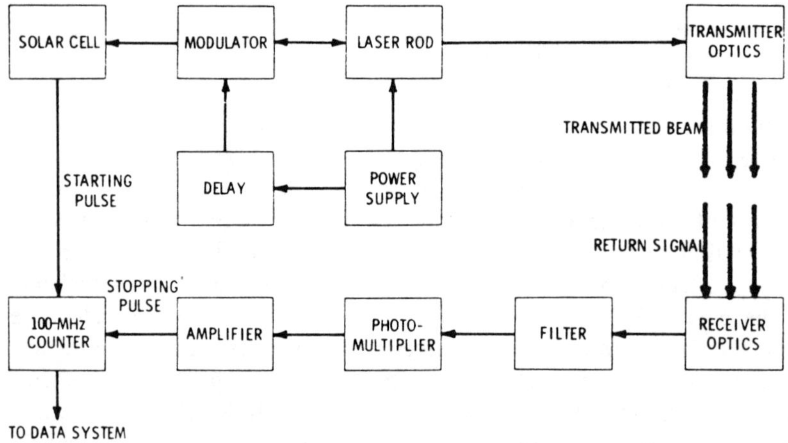

Figure 7-1 Block diagram of a typical rangefinder circuit.

Figure 7-2 A laser rangefinder in operation. (Courtesy International Laser Systems, Inc.)

The principle of operation of such a system may be described by means of the block diagram shown in Fig. 7-1. As the laser beam leaves the optical aperture in the transmitter, a small portion of the beam is sampled by a silicon photodetector (solar cell), which starts a 100-MHz pulse counter. The laser signal returning from the target is detected by a photomultiplier, amplified by it, and then fed to a counter, which then stops counting. The distance is determined by the one-way pulse durations, each pulse duration being equal to 10^{-8} second, as given by Eq. (7-1). A photograph of a typical rangefinder is shown in Fig. 7-2.

The output laser beam is processed through a collimating telescope to form a parallel pencil beam. The beam signal returned from the target is gathered by a second telescope at the front end of the receiver, the noise is filtered from the signal, and the filtered signal is impressed on the photomultiplier, amplified, and then fed to the counter.

Many improvements have been made in rangefinders, so that at present a laser rangefinder is a reliable and accurate ranging tool. The equipment is compact and easily portable. Certain rangefinders are installed on many military weapons and artillery. Some systems are made to track rockets, aircraft, and missiles. Other systems can lock onto a moving target so that continuous intelligence is available about the position and distance of the target. The laser beam is amplitude-modulated to achieve better resolution than that of a Q-switched optical laser system. Also, the use of a telescope with a relatively large optical aperture makes the fine laser-processed signal resolution superior to that of microwave radar.

A laser rangefinder developed by International Laser Systems (ILS) is a lightweight, portable, and easily maintainable military system. The rangefinder is mounted on a tripod and has a goniometer (angle-measuring meter) and a transit case. In addition to measuring range, the unit can measure the azimuth and elevation angles respecting target attitudes. The rangefinder contains a transmitter, control and ranging circuits, a sighting telescope, and a receiver.

The transmitter operates with a Q-switched Nd-YAG laser that generates 2-MW peak power at a pulse repetition rate of 30 pulses per second and a pulsewidth of 12 nsec. The emergent laser beam has a diameter of 2.8 cm and beam divergence of 0.5 mrad. Thus, the intensity of the retroreflected laser beam from a fast-moving target at a maximum range of 9995 m is sufficiently high for acquisition by the photodetector and the sighting telescope. The goniometer measures the azimuth angle on a full-scale 6400-mil vernier and elevation angle on a ±50-mil angular scale.

The receiver is provided with a silicon avalanche photodiode detector connected to a video amplifier having a variable sensitivity to minimize backscatter effects during short-range measurements. The range is indicated on an illuminated numeric display positioned adjacent to the sighting eyepiece. All targets at a distance shorter than 250 m are discriminated by the provision of a minimum range gate; other targets in the field of view are indicated on the range receiver. The receiver optical aperture is 50 mm and it has a field of view of 2 mrad.

Because of its easy portability and lightweight structure (about 8 kg), the ILS Model LRR-101 rangefinder can be used with small field artillery and infantry units, for target observation and localization. The data received by the rangefinder can also be utilized in computerized fire-control systems. The ranging unit operates from a 24-V rechargeable nickel-cadmium battery.

Another manufacturer's rangefinder employs a GaAs laser and measures ranges varying from 50 to 10,000 feet against a cooperative target. A cooperative target is provided with a reflector to retroreflect the laser beam to the receiver of the rangefinder. Accordingly, a retroreflector can project the greater part of the beam incident on it to the receiver; this property of the system enables the unit to measure longer distances using a small-power laser diode. The GaAs diode operates at 9040 Å in the infrared spectrum. The beam divergence at the exit aperture of the system is 5 mrad and varies up to 50 mrad within the range of the rangefinder. The diode emits 5 W and is operated at 2000 pulses per second with a pulsewidth of 40 nsec. Its rangefinding and processing logic is the first pulse. At the receiving end, an avalanche detector with a sensitivity of 10^{-9} W/cm^2 detects with a fast response. The optical aperture of the rangefinder is 5 inches and has a field of view of 5 mrad. The unit operates from a 24-V current source.

Laser rangefinders are standard equipment on many types of military artillery and on laser-guided air-to-ground missiles, which have already demonstrated the effectiveness of the laser systems. The information received is relayed to centrally located station computers. Rangefinders of this type require acquisition of the information within tens of centimeters at distances from 5 to 8 km. The trend is toward using space as an operational environment with satellites as headquarters for information transceivers. For this type of application, a chemical laser emitting at 27,000 Å in the infrared has been chosen because of its high system weight-to-energy ratio. Other high transmission laser candidates for this purpose are CO_2, CO, and deuterium fluoride lasers. Ultraviolet excimer lasers have also been tried with relative success. Along with these, krypton and xenon lasers emitting in the green wavelengths are being used for communication from a satellite to submarines, since green light can penetrate water with higher transmissivity than light of other colors.

Based on the rangefinder principle, laser radars have been constructed and used successfully for detection and measurement of shorter distances at lower altitudes than can be achieved with microwave radar. Both the transmitter and the receiver are built into a single unit.

MIT's Laser Radar

The MIT Lincoln Laboratory has developed a laser radar with an accuracy of 14 inches at a range of 50 miles. It is well known that the information processing capability of microwave radar is dependent on the frequency of the broadcast beam. Since laser radar has a frequency approximately 10^4 times that of microwave radar, laser radar is 10^4 times more accurate in both angular and Doppler resolutions. Thus, the distance of an object 3½ miles away from the laser system can be determined within 1 inch. The system can also measure the velocity of a moving target to within ¼ m/sec, as reported.

This radar uses a carbon dioxide laser generator operating with a 10-kHz frequency at a 5-μsec pulsewidth. The laser beam returning from the target is detected by means of a mercury-cadmium telluride photodetector. The aerial attenuation of the system beam ranges from ¼ to 2 dB/km due to the carbon dioxide and water content of the atmosphere. In a routine application and in a cloudy atmosphere, the signal will be further attenuated; however, in combination with a microwave radar, the unit can provide target acquisition as well as detailed data. Because the backscatter signal characteristic from the target depends on properties different from those of microwave radar, an object visible by laser radar is not necessarily visible by microwave radar.

Target Designator

One manufacturer produces a rangefinder that can also operate as a target designator. A target designator is a military implement that pinpoints targets for laser-guided weapons. This system is equipped with a Q-switched Nd-YAG laser, which dumps its energy in short, intense pulses at the target. The beam reflected from the target impinges on a silicon PIN or avalanche diode in the receiver section counter. The signal is transformed in the counter computer into digital data and displayed on the display panel of the system. The divergence of the beam is 3 mrad, and the beam diameter may be increased, when necessary, by means of a beam expander and collimated.

The avalanche photodiode is preferable for this system because of its noise-free operation and its sensitivity 100 times greater than that of a PIN photodiode. To reduce spherical aberration, the optical system of the rangefinder/designator contains a three-element lens array provided with an optical filter to minimize the background radiation. The Nd-YAG laser may also be frequency-doubled to 5320 Å to make the beam visible

Figure 7-3 A hand-held target designator in operation. (Courtesy of Hughes Aircraft Co.)

to the operator. A target designator, when built without a rangefinder, can be small and compact for hand-held operations. A typical target designator is shown in Fig. 7-3.

Military Television Transmitter

In addition to rangefinders, the military agencies frequently use television transmission systems for strategic communication. Such a system is manufactured by American Laser Systems, Inc., Santa Barbara, California, for transmitting television images to line-of-sight distances. The system consists of a semiconductor injection laser operating in the infrared and transmitting television signals through wideband optical links to an avalanche photodetector located in the front end of a receiver. The transmitter is aligned by means of an internal telescope with the receiver, which may be located as far away as ½ mile in daylight conditions and nearly 1 mile during night transmission. Since the optical bandwidth in the electromagnetic spectrum is not restricted by Federal Communications Commission (FCC) regulations, no license is required for the image transmission. The audio subcarrier operates at 10.7 MHz, which has the inherent advantage of wide frequency separation between the video and sound subcarriers. For faithful reproduction, a nonlinear compensation network is used to eliminate nonlinearity due to the feedback network operating in conformance with diode characteristics.

By use of the recently developed silicon avalanche photodiode for direct modulation, the signal-to-noise ratio of the system is greatly improved over that of other brands; at a distance of 1500 feet in sunlit daylight this ratio is 48 dB, increasing to 52 dB on an overcast day. To overcome the atmospheric scintillations due to hot air rising from the surface of the earth, the system has a reference signal in the video section and a demodulator to automatically eliminate the atmospheric disturbance from the video signal. A most useful application of this system is in tracking military test projectiles; in this application it is impossible to perform the transmission with a microwave system due to electromagnetic interference, especially at low atmospheric levels.

Eye-Safe Detectors for Rangefinder/Designators

Conventional rangefinder/designators use a He-Ne laser or a frequency-doubled Nd-YAG laser. Although these laser beams are effective to long distances, they are hazardous to the operating personnel. The most compatible photodetectors for these laser wavelengths, 6328 and 5320 Å, respectively, are silicon avalanche photodiodes. Since the eye can focus wavelengths up to 10,600 Å, an operating wavelength that has been standardized by the American National Standards Institute (ANSI) is 14,000 Å or longer. The selection of erbium-doped or holmium-doped laser elements with emission wavelengths between 14,300 and 21,000 Å appears to be preferable. An excellent photodetector for these wavelengths would be HgCdTe, but it is difficult to fabricate. The usual detectors chosen are GaInAs/InP and GaInAs/GaAs, which have quantum efficiencies ranging from 40 to 90%. The latter detectors are epitaxially grown on GaAs substrates and have a high spectral response that increases with increasing indium content, whereby the detector can be made to respond up to 17,300-Å wavelengths. Other detectors can be used for longer-wavelength reception, but their excessive signal-to-noise ratio precludes their use commercially.

The signal-to-noise effect is reduced with an increase in gain by use of a PIN GaAs-FET preamplifier at the front end of the receiver; this approach has been found to be superior to the use of a germanium avalanche detector, whose spectral response ranges up to 17,000 Å with high gain characteristics. This technique was developed after this arrangement was used in fiberoptic transmission lines and found to perform successfully. Thus, the combination of a GaAs-FET preamplifier with a GaAsIn/GaAs detector is the preferred system arrangement for rangefinders/designators at present.

War-Game Simulation with Lasers

For battlefield maneuvers in war-game simulation with laser beams, the government has assigned various military experimental ground facilities supported by electronic instruments and computers. One such laser simulation device is a rifle-like laser target illuminator manufactured by International Laser Systems, Inc., and shown in Fig 7-4. The device contains a harmless laser beam furnished by a gallium arsenide diode and collimated into a pencil-thin radiation beam. The laser beam is projected at soldiers equipped with military outfits containing silicon sensors, or PIN photodetectors,

Figure 7-4 Soldier with complete battleground outfit. (Courtesy International Laser Systems, Inc.)

mounted on their helmets, arms, bodies, and legs. When a laser beam impinges on a sensor, an audio alarm located within the soldier's helmet sounds, and simultaneously the soldier's weapon (laser rifle) is disabled. The electronic signal from the "hit" is transmitted by means of a microwave antenna in the helmet to a remotely located central intelligence station, where computers process the signal to cancel out the soldier from the field as a casualty. A schematic illustration of the battlefield simulation conditions is shown in Fig. 7-5.

The gallium arsenide laser diode operates in the pulsed mode, flashing one optical pulse per trigger pull, and is effective to a distance of 1000 m. The energy of the laser

Figure 7-5 A battleground scene, with a soldier using a GaAs weapon simulator. (Courtesy International Laser Systems, Inc.)

beam is less than 1 mW, so any accidental exposure of the eye to the beam is not generally hazardous. He-Ne lasers have also been used in place of GaAs lasers. The detector in the helmet produces an output when illuminated by an optical pulse. This output is amplified to operate the audio alarm in the helmet as well as an optionally available flashing light.

Laser beam-equipped weapons are rifles, machine guns, tanks, and other movable artillery. The lasers are pulse coded and operate at 3000 Hz; thus, the weapon used in a casualty can be identified by the pulse rate. It must be emphasized here that the training of the soldiers is intended solely for realistic battlefield simulation and not for marksmanship, as illustrated in Fig. 7-5.

ILS Marksmanship Trainer

An International Laser Systems (ILS) marksmanship trainer, the Model LT-100 Lasertrain System, shown in Fig. 7-6, consists of a rifle, a target assembly, and a scoring console and is used in a target training range inside a building. The system includes a replica of an M-16 rifle with a laser transmitter mounted on the barrel. The rifle "shoots" laser beam pulses to simulate bullets, using the light from a gallium arsenide laser diode, which operates in the infrared spectrum at 9040 Å. The laser unit is boresighted to the rifle, and when the rifle trigger is squeezed a laser pulse is emitted and projected

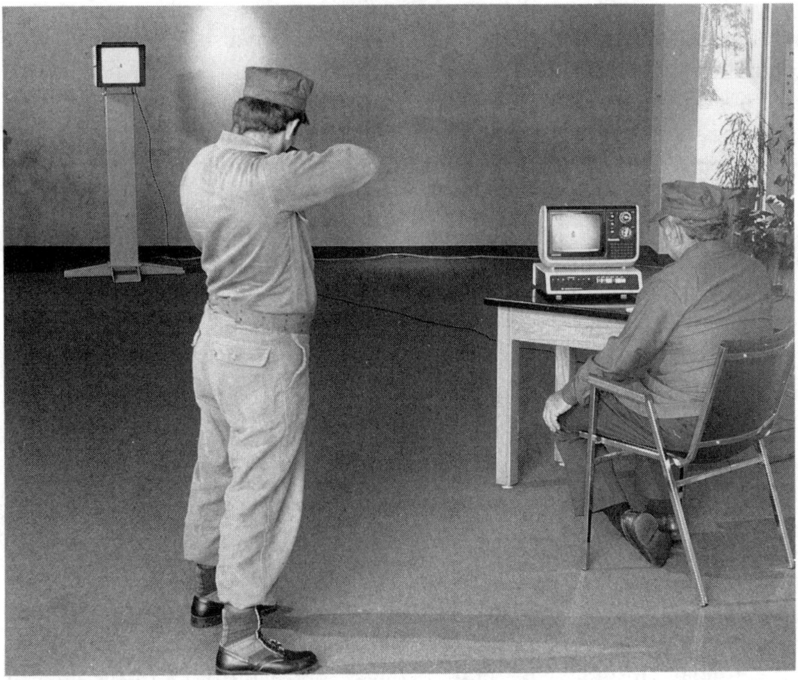

Figure 7-6 Indoor marksmanship training with Lasertrain Model LT-100 (ILS).

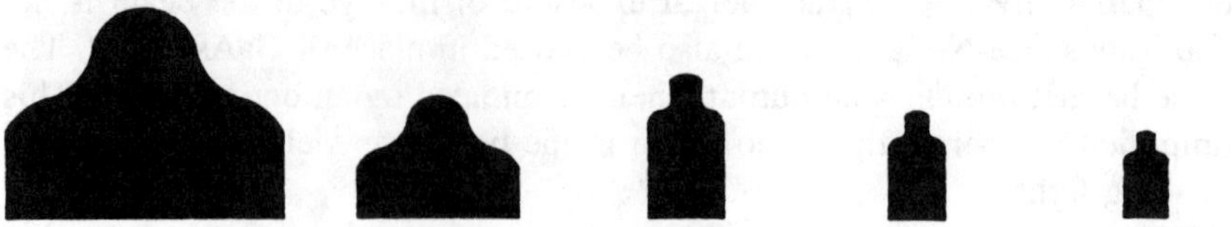

Figure 7-7 Simulated ranges for target training.

to the target. If the target is hit, a sensor in the target assembly displays the location of the hit while a sound simulating a rifle shot is simultaneously produced.

For different target ranges, from 3 to 300 m, different target silhouette sizes simulating increased distances are used, as shown in Fig. 7-7. A switch on the replica rifle is used to change the system operation from the marksmanship to a diagnostic mode. The audible rifle sound can be varied with a volume control provided on the control console. The system operates from a rechargeable battery providing approximately 2000 rounds of ammunition flashes per full charge.

Simulaser Marksmanship Trainer

Recently, some improvements have been made in the marksmanship laser trainer by Simulaser Corporation of Los Angeles. These improvements consist of increasing the number of detectors on the soldier's military outfit to 34, making the probability of a hit approximately the same from all angles. Additional protection from electromagnetic interference from nearby power plants is also provided. Threshold module data are decoded and processed by a microcomputer in the soldier's vest pocket; this device indicates whether the hit is a near miss or a kill and records the time. As many as eight events can be stored in the computer's memory section. A near miss causes the soldier's harness horn to sound intermittently, and a kill produces a continuous sound. The device automatically disables the "killed" soldier's equipment from further operation. This technology is said to have been applied to a hand-grenade code generator without serious injury to combatants. The Simulaser marksmanship trainer is shown in Fig. 7-8.

Automated Aircraft Tracking Laser System

In a Precision Automated Tracking System (PATS) developed by GTE Sylvania Corporation and accepted for ground use by the Federal Aviation Administration, a laser-mounted television camera system initially acquires the aircraft and the laser beam locks on it and automatically tracks the craft's course. The acquisition of the aircraft (target) is performed by an operator, who locates the target on the television camera

Figure 7-8 Simulaser marksmanship laser-aided trainer.

screen by means of a joystick or by use of the autotrack mode of the computer-controlled camera, which is mounted, for versatility, on a cinetheodolite.

With manual acquisition, the angular rate is 100 mrad/sec for both azimuth and elevation; in automatic tracking, the angular rate increases to 500 mrad/sec. The rangefinder field of view is 3 mrad. Once the target is acquired, the laser comes into play by projecting a pulsed beam from a Nd-YAG laser to the target; the beam reflected from the target returns to the locked-on subsystem, which collects the strategic information. A photograph of PATS is shown in Fig. 7-9*a*.

The aircraft or target is tracked by means of a servo system and the reflected laser beam is received by the data processing unit, which forms a closed loop with the servo system. The elevation and azimuth signals are created by the returned laser signals, and the resultant error signal is amplified and fed to the servo mechanism to sustain the beam on the target. The tracking accuracy of the system is said to be ± 1 foot at a distance of 20 nautical miles. At this range, the accuracies of the elevation and azimuth measurements are 0.005°. The Nd-YAG laser beam comprises 15-nsec bursts. The target range is measured by one-half of the transmission and reception time of the laser pulses. The entire system is mounted on a mobile van so that target acquisition can be made at any desired location.

Sandia's Laser Tracking System

An advanced laser tracking system developed by Sandia National Laboratories is designed to monitor flying vehicles at speeds of 13,600 miles per hour. The system is mounted on a 40-foot tractor trailer and uses a 5-W argon laser for the tracking mechanism. The beam is directed to the flying test target by means of an 18-inch-diameter mirror gimballed to rotate and direct the beam to a wide angle of view. The beam can be made to lock on the moving target to acquire data accurately and with fine temporal resolution and to analyze a wide range of missile data.

The retroreflected beam from the target is acquired by the same mirror (mentioned above) and split by it into two beams. One of the split beams is transmitted to the recording camera and the other is fed to an electro-optic tracking system. The received signal from the target is processed in the tracking subsystem, which contains a synchro mechanism to motivate the azimuth and elevation motors to lock the beam on the target at all times. Missile flight data are analyzed by the combined information on distance and tracked data to form three-dimensional coordinates giving the pitch, yaw, and roll. The tracking system is shown in Fig. 7-9*b*.

This system is akin to the automatic aircraft acquisition system suggested in Chapter 14. However, there is a distinct difference between the two systems. The suggested system would be provided with a divergent laser beam projector; upon automatic acquisition of the target, the beam would lock on the target in a collimated mode and the servo-computer system would automatically take over to track the target and record its progress. The entire system would have a small volume so that it could be mounted on a tripod as well as on a missile base.

(a)

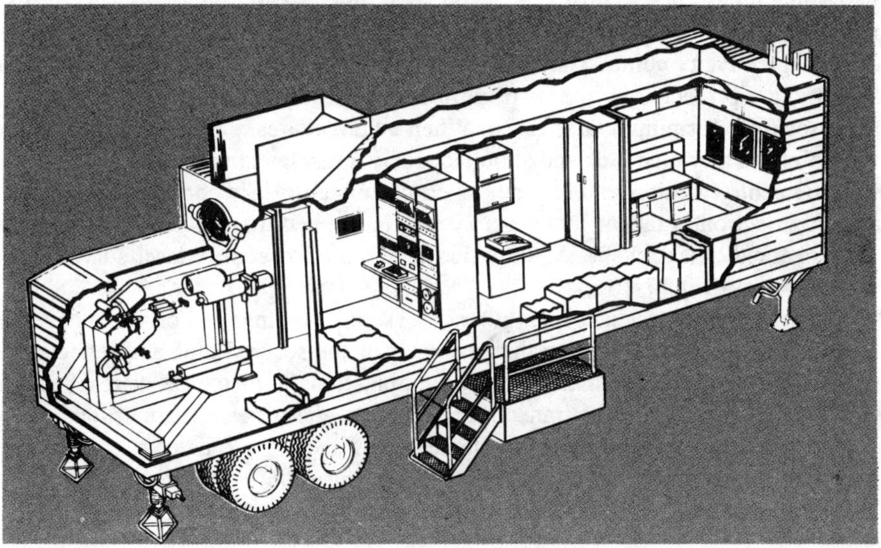

(b)

Figure 7-9 (*a*) The GTE Precision Automated Tracking System (PATS). (*b*) Sandia's computer-controlled trailer-mounted target-tracking system.

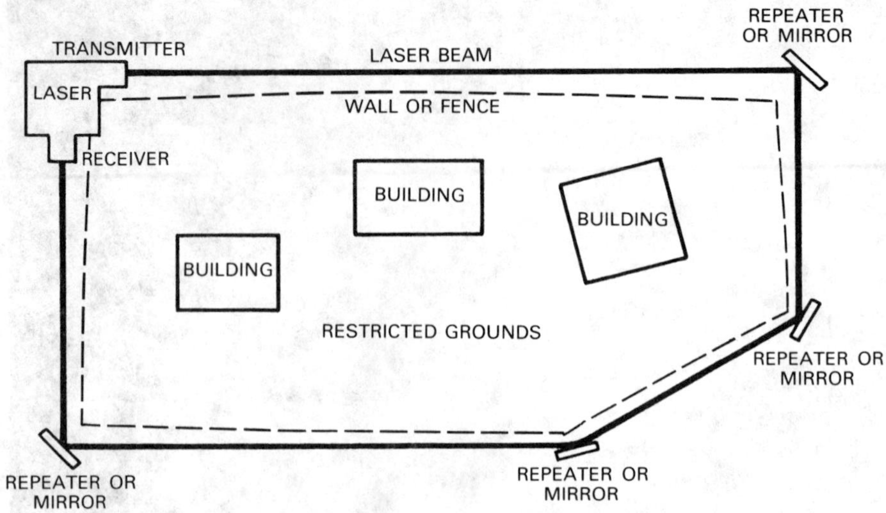

Figure 7-10 An installation with security surveillance by laser beam.

7-3 MILITARY SECURITY SURVEILLANCE SYSTEMS

A security surveillance laser system is a combination of detection, acquisition, and alarm equipment designed to guard restricted areas and buildings against intruders and escapees. The laser system comprises a transmitter and a receiver, both of which are located strategically at a central station. One or more laser beams, preferably of infrared wavelengths, are projected by means of mirrors or repeaters so that they surround the area to be secured, forming a laser fence. When a human breaks the beam by crossing it at any point, an alarm sounds and simultaneously a display is projected on a wall or screen at the central station to alert authorized personnel to take corrective measures. When such an action is taken, the alarm system is stopped manually by resetting the laser system for the next event. A simplified diagram of a security laser surveillance system is shown in Fig. 7-10.

Variations of laser security surveillance systems are many. For instance, one system uses a single laser beam on the perimeter of the secured area and reflects the beam from one mirror to another at the corner junctions until the beam reaches the receiver, located adjacent to the transmitter. When the laser beam is interrupted by a person crossing it, an alarm system is actuated. However, this type of system places everyone in confusion because the location of the intrusion is unknown. Thus, this system is not any better than the old burglar alarm that uses a visible light beam.

Segment-Locating Laser System

Another type of laser system projects a laser beam through several isolated detectors from which the beam returns to the control station, where the location of the intrusion is determined from the frequency of the projected beam. Still another system is provided with a number of sensors located in the path of the laser beam, and the location of the intrusion is determined by the particular sensor that has been actuated. One other laser system, which possibly transcends all others, projects split laser beams at various sections of a wall or fence. This system can determine whether a person is entering or escaping from the grounds and can indicate the exact location of the interrupted beam. Also, this system actuates a "talking alarm" that audibly informs the authorized personnel that an intrusion or a getaway from the grounds is in progress.

A segment-locating intrusion alarm system is shown diagrammatically in Fig. 7-11. In the figure, a laser beam from a master transmitter located at a central station is directed around the perimeter of the protected area by means of repeaters located at various points in the path of the laser beam. When a laser beam between the repeaters is interrupted by a person crossing it, the repeater closest to the central station produces a logic signal characteristic of that repeater. The signal passes through the other repeaters without transformation and is received at the central station by a silicon photodetector. The beam signal is decoded by the receiver at the station and the segment interrupted is identified. This action actuates an alarm to alert the authorized personnel. When the event is over, the system is reset to a quiescent operational state.

The repeater consists of a receiver transducer that receives a laser signal from the main transmitter at the central station or from the preceding repeater during the quiescent operational state of the system—that is, when there is no beam interruption. The transducer output is partially coupled to an OR gate logic circuit and partially coupled to a ramp generator. The OR gate feeds its output to the transmitter transducer containing a gallium arsenide laser diode, which emits pulsed radiation. The ramp generator output is coupled to a signal-level detection circuit in which the signal from the ramp generator is compared to a preset reference DC voltage. When the ramp (reset) signal is equal to the reference voltage, an output pulse is produced and coupled to a pulse shaper. The output from the pulse shaper is connected to the second input side of the OR gate. The resultant signal output of the OR gate is coupled to the transmitter transducer, as illustrated in Fig. 7-12.

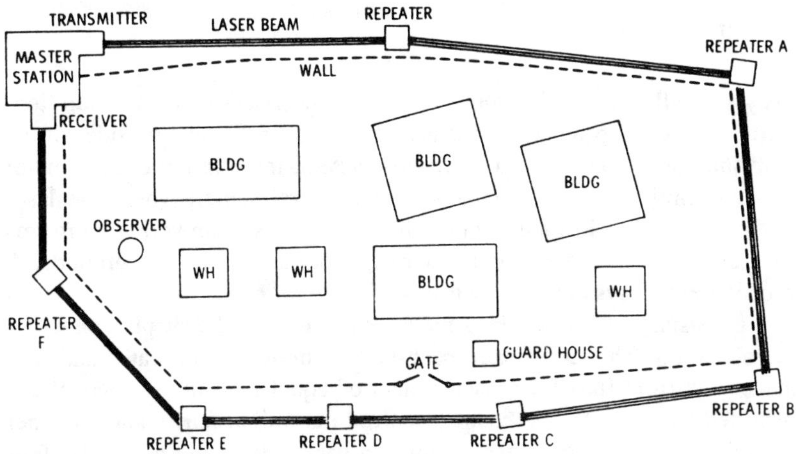

Figure 7-11 Segment-locating intrusion alarm system.

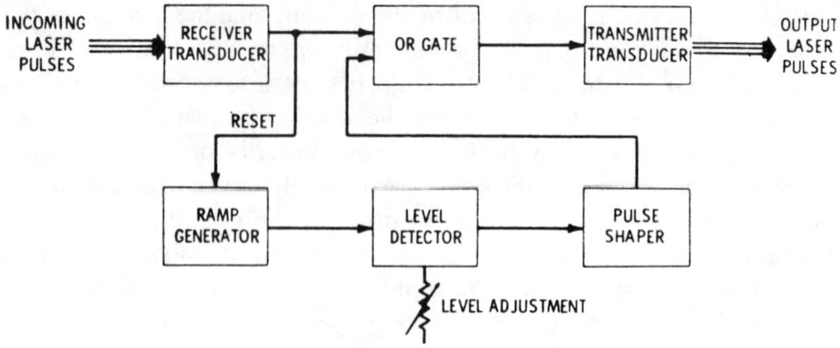

Figure 7-12 Block diagram of a repeater system.

When the receiver transducer at A in Fig. 7-11 receives an input laser signal, it feeds a pulse through the OR gate to the transmitter transducer, which generates a laser pulse from the gallium arsenide diode with minimum delay in relation to the OR gate signal received by it. This condition sustains the repeater at a normal operational state when there is no interruption in the laser beam. The level-detection voltages that are applied to the level-detection circuits of the other repeaters from B to E are successively higher, so that each repeater station has its own "signature" pulse.

As the laser beam is interrupted at a segment (e.g., between A and B in Fig. 7-11) by a person crossing the beam, the segment signature pulse is produced. This pulse is increasingly displaced in time from the missing transmitted pulse. For the next repeater, the pulse time is further increased, and so on for the other repeaters. Thus, each repeater produces characteristically a longer pulse than the preceding one. In this way, the pulse time in the segment at which the intrusion has occurred can be decoded by the master receiver, which then indicates the precise segment area where the intrusion has occurred. One principal disadvantage of the system is that it cannot indicate whether the person is escaping from the facility or entering it.

Four-Beam Security System

A unique security surveillance laser system, developed by Laserkinetics Corporation, secures a restricted facility by projecting four laser beams over the top and sides of the wall or fence surrounding the guarded area. In this system, interruption of any one of the four laser beams initiates visual displays in the central station where the laser transmitter and the receiver are located. At the same time, the system verbally informs the security personnel whether the person is escaping from or entering the grounds. A block diagram of the entire system is shown in Fig. 7-13.

The system, consisting of a laser transmitter, a receiver, and a display panel, is located at a central station. The laser beam from the transmitter is collimated and split by half-transmissive mirrors into four laser beams of equal intensities. Two of the beams are transmitted on each side on the upper part of the wall or fence, and the other two are transmitted a foot or so from the ground on each side of the wall. The four laser beams are projected along the entire perimeter of the secured area by using either mirrors or repeaters at the points where it is necessary to bend the beams. The beams reach the master receiver at the central station and are sensed by four photodetectors, which feed the laser beams to their respective amplifiers. The output of each amplifier is coupled to a spring-biased latching relay, which is retained at an open-circuit state during the continuous photodetection of the beams by their respective photodetectors.

When one of the laser beams, for instance, the top outer beam A, is interrupted by a person crossing the laser beam (entering), photodetector 1 fails to receive any photosignal and no photocurrent is produced in amplifier 5. As a result, the spring-biased latching relay in channel 9 closes a circuit that feeds current to indicator 13 and to alarm-signal amplifier 17, which then energizes the acoustic system to broadcast the event. Acoustic system 21 is provided with four sound tracks, one for each channel. Since the laser signal was received, in this case, from the laser beam covering the outer top section of the wall, the signal indication on the wall panel will be ENTER and

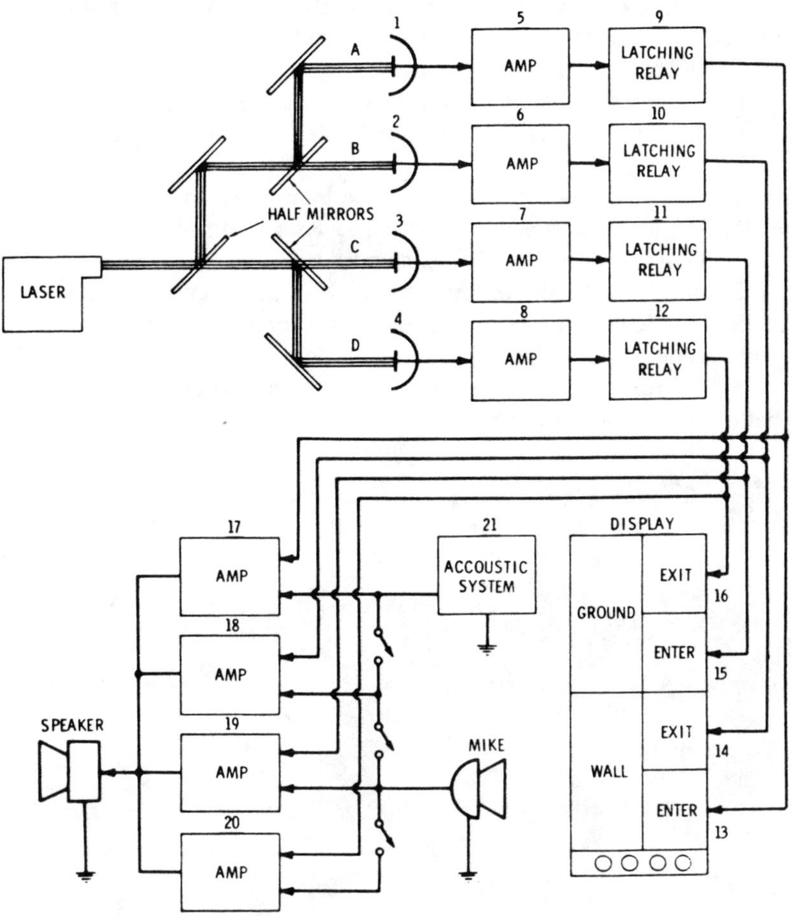

Figure 7-13 Block diagram of a facility-monitoring laser system.

every 2 seconds the acoustic channel will play the word "entering." Had the laser beam signal come from the inside section of the wall, the acoustic system would have played the word "escaping." In the latter case, the signal would have been processed through photodetector 2, amplifier 6, relay channel 10, and amplifier 18 before it entered the second track of the system. The display panel would indicate EXIT at point 14.

The Air Force Laser Fence

A portable security surveillance laser system has been developed by the Mitre Corporation and the Air Force Systems Command, Hanscom Field, Maine. The system, designated as the Intrusion Detection and Identification System (IDIS), detects and identifies the location of an intruder in the restricted area where aircraft and military weapons are kept. The equipment, shown in Fig. 7-14, resembles six vertically mounted traffic-light tubes attached to a 6-foot post. Each tube is provided with a horizontally mounted laser generator, and the tubes are positioned 1 foot apart, starting 6 inches from the ground. Each laser generator is a gallium arsenide diode, operating at 9050 Å, and is encased in a transmitter with a focusing output lens. A companion post, provided with matching tubes, contains the receiver, its electronic subsystem, and a lens at the output end of each tube. This post can be located at distances up to 500 feet from the transmitter.

In operation, the six pulse-coded laser beams are transmitted to their respective receivers at the 500-foot distance. Each pulse of ½-W laser energy is projected at a repetition rate of 1000 pulses per second. When an intruder crosses one or more of the beams at the same time, some of the pulses from each of the laser transmitters are interrupted. The detector located in each receiver senses the missing counts. These missing counts are decoded to the receiver, and the location of the intruder is immediately determined by the reading of the receiver counter. The system operates from either a set of batteries or a local AC power source.

Figure 7-14 The Air Force portable laser fence (IDIS).

Intercontinental Laser Security System

An extension of the Laserkinetics security surveillance laser system is a facility that provides continuous monitoring and detection along the border between two countries, such as Mexico and the United States. The projected development program consists of laser detectors placed a specific distance apart along the entire border separating the two countries. The detectors monitor the activity of the traffic day and night. The intelligence signals transmitted from the various segments of the laser fence are decoded and displayed on a display panel. Times and locations of intrusions are constantly recorded on a chart, which displays the results of the surveillance for each event separately.

The system includes laser transmitters, receivers, pulse counters, decoders, and computerized information-charting equipment coupled together to form a central intelligence system that can receive a maze of signals from the approximately 1500 miles of border. The circuitry rejects precarious signals, such as those created by moving animals, birds, windblown objects, and the like. The intrusion signal is picked up by a processor, which discriminates between human and nonhuman bodies and triggers a sensor transmitter, which transmits a coded infrared signal to the central station either directly or through a repeater. The signal at the receiver is either processed immediately or stored for complete-event occurrence.

The decoded and processed signal is printed out, showing a transmitter identification code, the segment code, the time of occurrence, and the character of the event. A panel chart indicates by a lighted map display the area and the point of intrusion. If this action is progressive, the map indicates the progress by means of the extended light path over the intruded area. The system is complemented by an alarm system that is activated in emergency cases to alert the patrol stations along the border.

7-4 DIRECTED-BEAM LASER RIFLES

A number of laser weapon systems that have been developed or are under development cannot be discussed here because of their classified nature. These systems are projected for destructive application and their power and use vary according to their applications. A few of the systems include directed-beam devices, target penetration beams, free-electron energy beams, and other radiant beams, which will be described in this section.

Directed-Beam Laser

A directed-beam laser rifle developed by Arthur M. Muncheryan of Laserkinetics Corporation, shown in Fig. 7-15, consists of a portable laser system built into a rifle configuration with a self-charging power supply. The device consists of a six solid-state laser rods interfaced with coniform fiberoptic amplifiers. The laser rods are simultaneously activated by flashlamps, which receive high-voltage pulsed current from a capacitive network contained in the stock of the rifle-configured system. The capaci-

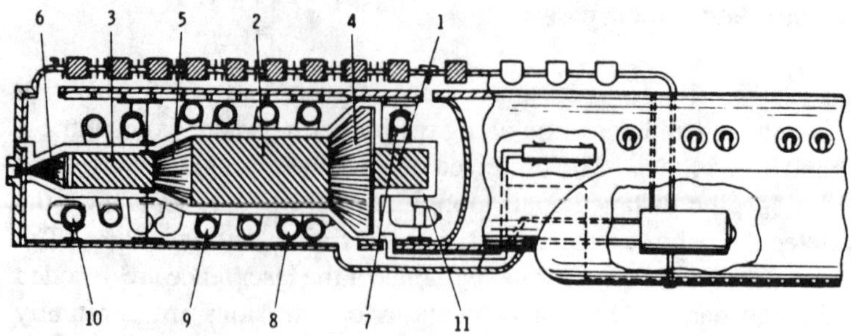

Figure 7-15 Directed-beam laser rifle.

tors are energized by a self-charging battery network, which obtains its energy from a radioactively excited battery of cells and from silicon solar cells illuminated by the flashlamps during operation of the laser system. To reduce the size, weight, and cost of the system, a vibrator mechanism is used to convert the low-voltage direct current to a pulsing high-voltage current for energizing the flashlamps. The lasing elements are either ruby or Nd-YAG rods operating in the pulsing mode.

In Fig. 7-15, the laser generator consists of three Nd-YAG rods, designated by 1, 2, and 3, positioned alternately between the fiberoptic cones 4, 5, and 6. The cones amplify the laser beam by condensing it before transmitting it to the next rod sequentially. The rods are excited to emission by flashlamps 7, 8, and 10. Between the flashlamps and the laser rods separated by fiberoptic cones is a glass envelope, 11, through which a cooling fluid circulates to cool the lasing elements. When the laser triggering mechanism is actuated, the fluid is simultaneously supplied from a compressed-fluid reservoir in the system. For emergency action, the batteries may be charged from an external 115-V AC source.

The system can provide a low-power laser beam for battlefield maneuvers, as discussed earlier, or it can provide a high-power laser beam for drilling holes in objects or for destroying the delicate working mechanism of a target system located at distances up to 100 to 200 m, depending on the power supply provided within the laser system or isolated from it. Because of its high-power capability, the device may be used to dig holes in concrete, metal-ore mines, and granite. Also, the device may be employed as a portable welding system that can weld aircraft-fuselage skin and structural parts in the field without disassembling them from the craft.

CW/Pulsed Directed-Beam Lasers

For increased penetration of metal targets, the United States Army, Fort Belvoir, Virginia, has combined a continuous-wave carbon dioxide laser beam with a pulsed Nd-YAG laser beam. It is reported that the time of penetration has been reduced 90% from that of the continuous-wave carbon dioxide laser beam alone. In tests in which this technique was used on unpainted steel, the improvement in penetration doubled.

When the technique was used on painted steel, the penetration time was reduced five times compared to that for unpainted steel.

The combination of continuous and pulsed laser beams serves two purposes related to coupling efficiencies with the target: (1) it overcomes the high reflectivity of the metal surface, and (2) it disposes of the molten metal accumulation at the focus of the beam and thereby facilitates beam penetration at the focal area. This is explained by the fact that when the metal surface is irradiated with a continuous-wave carbon dioxide laser beam, the focal area melts, and the simultaneous incidence of the pulsed Nd-YAG laser beam removes the molten metal from the focus cavity. No significant difference has been found between the effect of a 1.3×10^8 W/cm^2 Q-switched laser beam and that of a 3.7×10^4 W/cm^2 conventionally pulsed laser beam used in the tests.

The cavitation of the focal area is due to the impulse (force $\times$ time) of the ejected molten metal. The penetration increases with increases in impressed laser power and coupling efficiency of the radiation with the metal surface—the rougher the surface, due to painting or irradiation of the surface with the continuous-wave laser, the higher the coupling efficiency. Irradiation of an aluminum sheet surface with a pulse power of 2.5×10^3 W/cm^2 permits the continuous-wave carbon dioxide laser beam to produce a hole that cannot be made with a continuous-wave or pulsed laser beam alone. This investigation and the conclusions reached indicate that it is highly possible to penetrate flying targets encased in aluminum or steel materials by use of the substantially increased laser powers available with present techniques.

Free-Electron Excited Beams

A high-energy laser weapon system using free electrons to excite laser radiation is reported to be under development by the Department of Defense. The new laser is thought to have three times more emission efficiency than a conventional laser, and with a more lethal striking power than ever before. The radiation is produced in association with a spiraling beam of free electrons at relativistic undulations. The system is continuously tunable from ultraviolet to infrared. The potential service of the system will be on aircraft, satellites, and naval ships, as defense against enemy attack.

Other high-power laser equipment for military use could be developed by increasing the size of the lasing element, which is optically pumped by continuous flash light from a mercury arc or carbon arc system similar to those used in aerial advertising and in military searchlights. Such a system would be capable of generating megajoules of laser energy pulses. When the beam is collimated to a 1-cm diameter and directed at hostile targets, it can instantly cut large areas of metal or drill holes through strategic mechanisms to deactivate or destroy the mechanisms. The laser beam could first be produced at diameters of several feet, collimated, and then concentrated to a suitable diameter and directed to a point of interest. The resulting concentrated laser energy could be made equivalent, at the point of contact, to atomic fission energy. The system could also be made to trigger a hydrogen bomb without the use of an atomic bomb as detonator.

7-5 TACTICAL CO_2 LASER SYSTEMS

Carbon dioxide laser systems have found distinct applications in military tactical environments. These systems are capable of sensing targets, identifying them, and then acquiring them. They also have precision guidance capabilities to home on targets such as munitions and bombs. Such systems form video images of target thermal emissions on screens, so that even in complete darkness the acquisition of targets becomes possible. Use of a long-wavelength CO_2 laser at 90,000 to 110,000 Å may supplant the use of the shorter-wavelength Nd-YAG laser. For large-field-of-view laser systems, the signal is focused on the photodetector by means of a receiving convex lens whose image collection area may be given as

$$A = \frac{\pi d^2}{4} \tag{7-3}$$

and

$$\phi_d A = \frac{\pi D^2}{4F} \tag{7-4}$$

where ϕ_d is the range angle, A is the lens area, d is the effective diameter of the lens, F is the f-number of the lens, and D is the diameter of the detector. The focal length of such a lens is equal to its effective diameter times its f-number.

With this system, the lowest signal that can be detected is in the neighborhood of 10^{-8} W/cm^2, using HgCdTe or PbSnTe photodetectors. The signal-to-noise ratio of the system may be given as

$$R^* = \frac{\mu P_r}{\sqrt{A_c \beta}} \tag{7-5}$$

where R^* is the signal-to-noise ratio of the voltage, μ is the detector sensitivity, P_r is the signal power from the target surface, A_c is the circular area of the detector, and β is the receiver bandwidth for a silicon photodetector.

In Doppler velocimetry measurements, the signal-to-noise value may be given as

$$R_d = \frac{\eta P_0}{\Delta f}\left(\frac{D_d D_e}{R_a f}\right)^2 \times 4 \times 10^{11} \times a^2 \overline{G}\, \overline{V^2} \tag{7-6}$$

where R_d is the signal-to-noise ratio for the moving target, η is the quantum efficiency of the photodetector, P_0 is the power in the multiple beam, Δf is the bandwidth of the beam (in MHz), $\overline{G}$ is the scattering function of the radiation, and $\overline{V}$ is the visibility factor of the projected beam.

In these types of systems, large fields of view would be required to search, acquire, and lock on the target. One efficient way to achieve this is to project the laser beam divergently and, subsequent to detection and acquisition of the target, to narrow down the beam to a pencil-thin ray and lock on the moving target. When using this technique, the retroreflected signal from the target must be received by a large-diameter signal-collecting (convex) lens, or preferably by a reflective-type optical element. Since the received energy will be in the infrared range, scattering due to atmospheric contaminants or countermeasures will be smaller than when shorter-wavelength laser beams are used. Both pulsed and continuous-wave CO_2 laser beams can be utilized in this technology and have high performance unsurpassed by equivalent electro-optical systems.

7-6 ANTIBALLISTIC LASER DEFENSE

Antiballistic laser defense systems have a great advantage over conventional antiballistic missile systems because of the enormous speed with which the laser defense systems can operate to long-distance lethal ranges. In order to be effective against enemy intercontinental ballistic missiles, the laser beam must hit the target when it is above the atmosphere, since the exhaust of the target could be tracked readily and the booster lifting the target into space would be easily destroyed by the laser beam. The energy required to perform this function will involve megawatts of laser power. A candidate for such an amount of power is the chemical laser consisting of hydrogen and fluorine, whose reaction gives off enormous amounts of laser radiation, without the hazards of such reactions if the combination occurred on earth. Other advantages of a chemical laser are its operation at the low temperatures prevalent in space and the requirement of a vacuum for its performance.

Although chemical lasers possess considerable advantages, they also have certain inexpediences that prevent their use in space-based ballistic missile defense, at least for the present. For instance, the enemy could cover the missile body with thermal insulation or a high reflective coating to reflect the infrared wavelengths from the beam. Furthermore, since chemical lasers operate at long infrared wavelengths, a mirror more than 10 feet in diameter would be required to collect and precisely deliver a large amount of focused beam to the target sufficient at least to drill a hole at a strategic target site. Even a minute displacement of the aiming mechanism at a range of 500 km or more would miss the target. Thus, a highly reliable laser source stabilizer would be required, although a lock-on mechanism would partially overcome this disadvantage.

With the preceding drawbacks in mind, the development of short-wavelength lasers with higher emission energy would seem to be preferable to the use of long-wavelength infrared beams. The short-wavelength laser beams would operate in the pulsing mode, with each pulse delivering a powerful impulse to cavitate the surface of a strategic missile area. In addition, the shorter wavelengths would be more readily absorbed by the missile metal to vaporize it, and thermal shielding would be ineffective against short-wavelength pulses. The focusing mirror size could be made smaller for rays such as those in the ultraviolet range. However, short-wavelength laser beams also have their drawbacks. The power levels required to produce the ultraviolet rays will militate against the use of short-wavelength lasers in space defense effort.

Another technique under consideration for space defense by means of laser beams involves the use of nuclear-powered sources to produce x-rays by the explosion of low-yield hydrogen bombs. The x-rays thus produced can be amplified by laser beams to enormously high power for projection to the target. A laser station of this type might consist of dozens of Nd-YAG rods, each generating its own laser beam and directed to and locked on a given target up to 4000 km from the laser source. The nuclear power would optically pump the laser rods simultaneously to produce the radiation beams directed to different targets preliminary to the detonation of the bomb material. However, the laser rod is also vulnerable to vaporization by the detonation of the bomb. Furthermore, if the oncoming missiles are in multiples of dozens or more, the laser station with its nuclear power must be destroyed prior to the disposal of the lasing elements in the manner described. Apparently, research in the United States is in progress along these lines regardless of the various drawbacks of the individual techniques described above.

One other method has been advanced in which ground-based laser systems project powerful laser beams to space-based mirrors, which focus the laser beam on the oncoming target to destroy it. In this type of facility, infrared and visible laser beams would be employed, since they can penetrate the space medium more readily than ultraviolet or x-ray beams; they also can be located at levels above the clouds to obviate their absorption. The laser sources for these beams may be located at ground level and the energy from them transmitted through large-diameter fiberoptics to the top of a high mountain, from which the beams may be projected by means of a local mirror to the relay mirror in space. Both mirrors may be made to lock on each other for at least the time necessary to deliver the laser beam to the flying target.

Other methods, including launching the laser system by means of a missile from a submarine or earth-based station, have been advanced but have not received support from the scientific world. Considering all these various methods of utilizing laser beams against ballistic missiles, the most tenable and possibly practical method would be the use of space-based laser sources with adaptive optics to maneuver automatically the trajectories of the laser beams to their target. What if the enemy deploys the same types of systems in space to destroy the laser sources? It is very possible, and in this event the idea of destroying ballistic missiles by means of laser beams at any convenient location would be discarded. It further appears that under these conditions the most logical expedient for defense is strategic stability of diplomacy achieved by the opposing parties.

QUESTIONS AND PROBLEMS

7-1 What are nondestructive and destructive military laser systems? Give examples of each. What types of lasers are utilized in each of the nondestructive and destructive systems? Which is the most powerful laser used in the military field? Can a Nd-YAG laser be used in both destructive and nondestructive applications? Explain how.

7-2 What is a rangefinder? How is it constructed? How does it operate? What types of lasers can be used for a rangefinder? On what one principal factor (parameter) does a rangefinder operate? Is there any other method by which this factor can be measured? Name it.

7-3 A certain rangefinder operates at 98 MHz, and its counter has counted 6×10^2 pulses. What is the distance measured by the rangefinder? If the distance measured by the rangefinder was 2.4 km, how many pulses would the counter count one way?

7-4 Can a diode laser be used in a rangefinder? Can a CO_2 laser be used in a rangefinder? Can a CO_2 laser beam be modulated? Can Nd-YAG be used frequency-doubled in a rangefinder? In what spectral range would Nd-YAG be when frequency-doubled? Which of the two lasers, a He-Ne laser or a Nd-YAG frequency-doubled laser, is preferable for long-distance measurement? Explain why.

7-5 What is a COLIDAR? What type of laser does it use? In what operational mode does it operate? What is the principle of operation of a COLIDAR? At what megahertz does this system operate? How is the noise filtered from the signal? On what type of military artillery is a COLIDAR employed? What type of beam modulation does this system employ in its distance measurement?

7-6 How is the ILS rangefinder constructed? At what peak power does it operate? What type of lasing element is used in the ILS rangefinder? In what mode does it operate? What is its maximum range? What type of photodiode is used in the ILS rangefinder? Why are distances shorter than 250 m discriminated by the rangefinder?

7-7 How is MIT's laser radar constructed? What is its accuracy? How accurate is it in both angular and Doppler resolution? What type of laser system does this radar employ? At what frequency is this radar modulated? Do atmospheric conditions affect this radar?

7-8 What ranges can a GaAs diode laser measure against cooperative targets? Why is a retroreflector used in measuring distances against a cooperative target? What is the beam divergence of this laser system? How many watts of laser power can this laser generate? What type of photosensor does it use? What is the sensitivity of the photosensor? What size optical aperture does it have?

7-9 Why is space projected to be used as an operational environment for rangefinders? Why is a chemical laser emitting at 27,000 Å chosen for this type of application? What types of lasers can be employed for such a system? What types of lasers are now being used for communication between a satellite and a submarine?

7-10 What is a target designator? How does it operate? How is the reflected signal from the target transformed into digital data? Can this system be used as a rangefinder? Why is an avalanche photodiode used in this system? How can the spherical aberration in this system be reduced? How can the system operate in the visible spectrum? Discuss.

7-11 Describe the operation of the American Laser Systems television image transmission system. How far can it transmit the image? At what frequency does the audio subcarrier operate? For what purpose is a nonlinear compensation network used in this system? How is its signal-to-noise ratio improved? How is its atmospheric scintillation reduced? What is its most useful application?

7-12 Describe the ILS battlefield simulation rifle. Why are silicon sensors used for the battlefield simulation? Can PIN diodes be used instead of silicon sensors? What is the advantage of such a laser system over actual, conventional battlefield operations? On what military equipment can laser beams be used for battlefield simulations? Discuss. Give the application of ILS's Lasertrain.

7-13 What improvements have been made by Simulaser Corporation over the ILS Lasertrain? How does the simulator system avoid electromagnetic interference? How are hits identified? How are threshold data decoded? How many events can be stored in the computer's memory section? When does a horn sound in the soldier's harness?

7-14 State some of the laser weapon systems. Describe the directed-beam laser rifle developed by Laserkinetics Corporation. How is the laser beam amplification accomplished? Are the system lasing elements pumped electrically or by flashlamps? If the system is battery-operated, as claimed, how does it generate a high voltage for its operation? How does the self-restoring power supply function? Can this laser rifle be used in battlefield simulations? How?

7-15 How does the United States Army achieve increased penetration of metal targets with its destructive laser systems? How are the metal coupling efficiencies obtained from the laser systems used by the Army? How is the high reflectivity of the target-metal surface overcome? How is target penetration or drilling a hole in the target accomplished? What type of laser is used in the system? How does the cavitation in the focal area occur?

7-16 What is the high-energy laser weapon system being developed by the U.S. Department of Defense? How powerful is that system in comparison with the presently developed laser systems? Through how wide a spectral range is the system tunable? What is its potential application? Describe a possible method of producing enormously high laser powers. Could such high powers detonate a nuclear reaction?

7-17 What is GTE Sylvania's PATS laser equipment? How is the acquisition of a target accomplished? How does the system lock on a moving target? In automatic tracking, what is the angular rate of acquisition? What type of laser does it use? How are the elevation and azimuth signals created by this system? What is the tracking accuracy of the system? How does this system compare with Laserkinetics' suggested system?

7-18 What is a security surveillance laser system? What subsystems does it have? How many methods are there for determining the location of an intrusion into the restricted area? Is there a system that can determine whether a person is entering or escaping from a restricted area? Whose system is it?

7-19 What is a repeater? How is it used in a security surveillance system? Describe the operation of a repeater. When the ramp signal is equal to the reference voltage, what happens to the resultant output pulse? In a surveillance laser beam line, what happens in the repeater when a laser beam is interrupted? What is the function of the master receiver in this surveillance system?

7-20 Describe the construction and operation of the Laserkinetics security surveillance laser system. Can this system identify the location of the intrusion? Can it determine whether the interruption of the laser beam is due to intrusion or escaping from the restricted area? Describe how it can accomplish this latter.

7-21 How does the Air Force's portable laser fence operate? What type of laser generator does it have? How far can the pulse-coded laser beams from the laser fence be transmitted? At what repetition rate is the energy projected from the laser system? What does the detector in each receiver sense? How are the decoded counts in the receiver translated into location information?

7-22 State the purpose of the improved and expanded Laserkinetics security surveillance system. What does the projected development consist of in the entire system? How extensive a border can this system protect from intruders? How can the system discriminate between human beings and other animals? In your response, make an attempt to utilize your knowledge as gathered from this book. Can this system indicate the progress of the intrusion at any given point? How does it display this information?

SELECTED SOURCES OF RELATED DATA

Button, K., et al. 1984. *Optically Pumped Far Infrared Lasers*. New York: Plenum.

Carter, A. B., and D. N. Schwartz, eds. 1984. *Ballistic Missile Defense*. Washington, D.C.: Brookings Institution.

Divens, W. G., and S. M. Jarret, 1982. Design and performance of a stabilized dye laser ring. *Rev. Sci. Instrum*. 53(9):1363.

Dubinsky, R. N. 1988. Lidar moves toward the 21st century. *Laser Focus/Electro-Optics* April, pp. 93–106.

Frame, W. F. 1984. Boresight stabilization in military laser systems. *Lasers and Applications* March, pp. 87–93.

Fraser-Volpe Corp. 1979. Military Laser Systems, technical bulletin. Warrington, Pa.

Gialloranzi, T. G., and J. M. McMahan. 1984. Naval Research Laboratory laser sensors. *Laser Focus/Electro-Optics* May, pp. 75–84.

Hecht, J. 1984. *Beam Weapons*. New York: Plenum.

Hill, R. A., et al. 1982. *Development of Flyable CO_2 Laser Beacon*. El Segundo, Calif.: Hughes Aircraft Company.

Hinman, W., and W. Cannizo. 1983. Military ground-based laser designators. *Lasers and Applications* February, pp. 38–40.

Hora, H. 1981. *Physics of Laser-Driven Plasmas*. New York: Wiley.

Johnson, A. M. 1988. Eyesafe lasers for military use. *Lasers and Optronics* June, pp. 59–63.

L&A Staff. 1982. Battlefield simulation system. *Lasers and Applications* November, pp. 18–22.

L&A Staff. 1983. Tunable filters for detection of lasers on battlefield. *Lasers and Applications* January, pp. 87–93.

L&A Staff. 1983. Laser weapons strategy. *Lasers and Applications* December, pp. 24–30.

L&A Staff. 1984. Medium power lasers penetrate the military. *Lasers and Applications* January, pp. 28–30.

L&A Staff. 1984. Pentagon shrouds plans for blinding lasers. *Lasers and Applications* February, pp. 38–40.
Laser Focus Staff. 1984. Laser Weapons. *Laser Focus/Electro-Optics* February, pp. 18–26.
Laser Focus Staff. 1987. Laser instrumentation: High altitude monitor, and ultralow loss measurement. *Laser Focus/Electro-Optics* February, pp. 22–26.
Laser guides electron beam in particle-beam weaponry. 1985. *Lasers and Applications* May, pp. 54–56.
Lasers in strategic defense. Distinguished panel discussion. 1986. *Laser Focus* February, pp. 124–130.
Muncheryan, A. M. 1968. Directed-Beam Portable Laser System (Handgun). U.S. Patent No. 3,404,350.
Muncheryan, A. M. 1969. Directed-Beam Portable Laser System for Military. U.S. Patent No. 3,478,278.
Papayoanou, A. 1982. CO_2 lasers for tactical military systems. *Lasers and Applications* December, pp. 49–55.
Payne, K. B. ed. 1983. *Laser Weapons in Space*. Boulder, Colo.: Westview Press.
Schomisch, J. 1987. *Guide to Strategic Defense Initiative (SDI)*. Arlington, Va.: Pasha Publications.
Simons, B. S. 1982. Laser-Light Detection System. U.S. Patent No. 4,366,378.
Tebo, R. 1985. Military electro-optics for flat panel electroluminescence. *Laser Focus* March, pp. 45–52.
Tomasi, G. P. 1986. Fiberoptics for military service. *Photonics Spectra* August, p. 73.
Tucker, J. B. 1984. The fallacy of laser defense. *Technol. Rev.* April, pp. 30–52.
Yevick, G. J. 1982. High-Power Window Mirror for Laser. U.S. Patent No. 4,357,077.

CHAPTER
EIGHT

MATERIALS PROCESSING WITH LASER BEAMS

8-1 INTRODUCTION

Laser processing of materials involves laser systems and procedures for welding, drilling, cutting, scribing, thermal treatment, and annealing. To accomplish these processes successfully, acquaintance must be made with certain essential requirements related to the laser beam and the equipment utilizing the beam. Any significant deviation from these requirements may lead to total failure in achieving the purpose and the many advantages of laser processing over other types of analogous functions. The following are some of the requirements:

1. The laser beam must be of sufficient energy to transmit heat to the material to be processed.
2. The time of processing (pulsewidth) of the laser beam must be relatively long (several milliseconds) to conduct the heat into the material.
3. The laser beam wavelength must be compatible for optimum absorption by the particular material.
4. Very close contact must be maintained between the contiguous surfaces if welding is to be performed.
5. The laser beam pulse shape must be controlled precisely for the particular job.
6. The beam focus must be adjusted for the radiant intensity to permit proper penetration into the material.
7. Human error in focusing the beam must be eliminated if possible.
8. Ideally, the process must be conducted in an enclosure so that no restriction is imposed on the operator for safety purposes.

9. The equipment must be provided with an automatic line-power regulator to eliminate capricious variation of laser power during operation.

Numerous types of laser systems can be employed in materials processing of metals, ceramics, glass, plastics, textiles, and the like. They are Nd-YAG lasers, carbon dioxide lasers, alexandrite lasers, and sometimes argon and excimer lasers. Certain of the laser beams used in industry are visible and others are invisible. If visibility of the beam is a requirement, the processing should be accomplished by use of a visible laser beam or the beam should be made visible by mixing it with a visible low-power laser beam, such as that from a He-Ne laser, as illustrated in Fig. 8-18. For deep penetration into a metal, the beam may be employed in the pulsing mode, and for continuous cutting a continuous-wave laser beam may be used. For cutting thin sheets of ceramics, pulsed laser beams are preferable. For cutting and drilling thick sheets of metals, a carbon dioxide laser may be used in the continuous mode. For processing bright and shiny surfaces, a coarse black coating should be used for effective absorption of the laser beam heat. Other methods suitable for the workpiece in hand may be devised by the operator after he has gained some experience with metals processing.

8-2 LASER BEAM WELDING OF MATERIALS

A typical laser beam welding machine consists of a laser generator, a power supply to energize the generator, an optical system to shape and focus the beam on the workpiece, an electrical control panel with control instruments, a workbench or console on which to perform the laser welding, and frequently a cooling system to keep the laser generator temperature at a level favorable for efficient laser emission. Other auxiliary equipment may also be added as the need arises.

Laser Welding Systems

The diagram of a typical laser welding system shown in Fig. 8-1 illustrates a work console, a cabinet containing a power supply section, and a water-cooling system. The water-cooling system is connected through water pipes to the laser head, which is constructed integrally with an electric control panel section supported on the console. The laser head contains a laser rod that is optically pumped (illuminated) by means of a flashlamp, which receives the energizing power through a cable from the power supply section. The power supply, in this case, consists of a capacitive bank whose output is adjusted by means of a capacitor control and a voltage control.

To operate the equipment, the main switch is turned on, which energizes the power supply section and the cooling system pump. The workpiece is positioned on the *X-Y* translator. A foot switch triggers the flashlamp to illuminate the laser rod and to excite it to emission of laser radiation. The radiation is projected as a pencil beam from the laser rod into a chamber containing a mirror positioned at 45° with respect to the beam

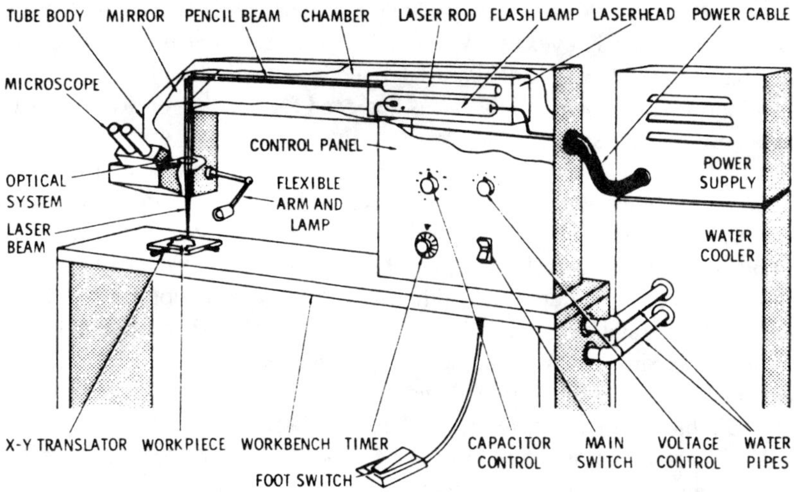

Figure 8-1 A typical simple laser welding system.

optical axis to bend the beam 90° and to project it on a focusing (convex) lens. A binocular microscope, attached to the laser head, is used for observing the position of the workpiece with respect to the converging laser beam. The workpiece is illuminated by a lamp with a flexible arm for adjusting the light. The time of exposure of the workpiece to the laser beam is controlled by means of a timer located on the control panel, the time of exposure (milliseconds) depending on the type of material to be welded.

A schematic circuit diagram of a simple, pulsing laser generator is shown in Fig. 1-20, which illustrates generally the arrangement of the different component sections corresponding to those shown in Fig. 8-1. In the circuit diagram, a current rectifier receives a high-voltage current from an external 115-V alternating current and converts it into a direct current to charge the pulse-forming capacitor network. The water pump circulates deionized distilled water through the laser head prior to the triggering of the laser generator and sustains its operation to normalize the temperature in the laser head.

It will be noted from Figs. 8-1 and 1-20 that there is a considerable distance between the lens and the workpiece, and the welding is accomplished without contact between the laser head and the workpiece. The separation from the workpiece is due to the particular focal length of the lens; in this case a long-focal-length lens has been chosen for illustration. The workpiece must also be stabilized by means of jigs or a pair of tweezers manipulated by the operator. Since both the primary (direct) laser beam and secondary (reflected) laser beams are visible to the eye in such equipment, laser safety goggles must be worn by the operator during the operation of the laser system. In the old designs of the laser welder, the laser beam traversed the microscope-tube body before it was incident on the workpiece. A shutter was provided in the microscope tube to shut off the view of the operator from the laser focus area just before triggering the laser generator. Recent designs have eliminated this feature by the use of a television screen, which is eyesafe as well as more convenient for the operator; this subject will be discussed more fully in a later section of this chapter.

The laser welder/driller shown in Fig. 8-2 is designed for welding or drilling small precision parts at production speeds of up to 50 welds per minute, each pulse time being from 1 to 6 msec. It has a collinear, binocular viewing system with a fail-safe shutter. Sheet stocks as thick as 0.05 inch and wires with diameters up to 0.07 inch can be welded with a focus spot size varying from 0.001 to 0.06 inch. The system can drill holes in materials up to 0.3 inch thick, and the hole size can range from 0.0005 to 0.06 inch in diameter; a hole depth-to-width ratio of 26:1 can be obtained with the equipment. The unit operates from a 220-V, 60-A, 50/60-Hz, single-phase power source. It is cooled by means of a water-to-tap-water heat exchanger requiring 1 gallon per minute (gpm) of tap water at 20 pounds per square inch (psi). Refrigerated-type cooling is also available when requested from the manufacturer.

A laser welding system manufactured by GTE Sylvania, Inc., Model 1610, is rated to operate at an average power of 150 W, using a Nd-YAG laser with pulse energies of 3 to 20 J at pulsewidths of 0.1 to 8 msec. It has a pulse repetition rate of 100 Hz for the lowest energy level and 15 Hz for the highest energy level. The system can be operated either in single mode or repetition-pulse mode, with manual or automatic remote control for on-line production. Coaxial alignment is achieved by means of a binocular microscope having alignment cross hairs for viewing and focusing on the workpiece with white light. Its integrated power meter gives a continuous power reading from 0 to 200 W. The system is capable of welding microcircuit leads, with a spot diameter varying from 0.01 to 0.04 inch. The equipment operates from a 220-V, 50-A, three-phase, 60-Hz power source and has a cooling system using 2 A from the same

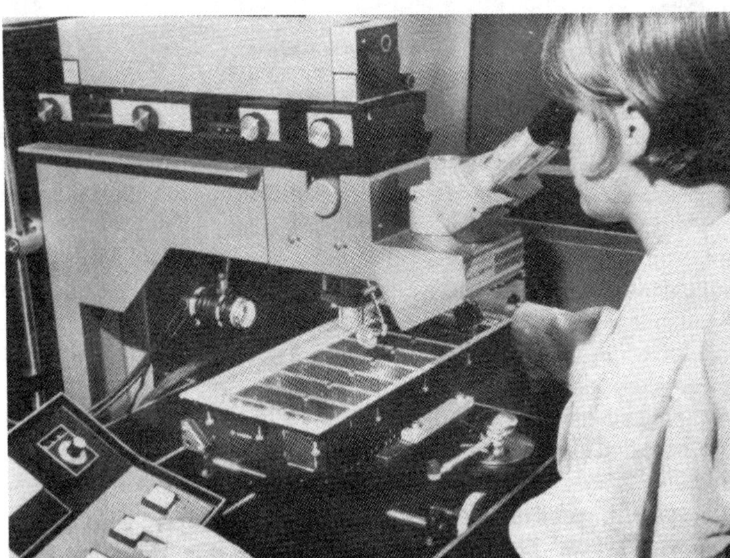

Figure 8-2 A commercial laser welder/driller system. (Courtesy of Holobeam, Inc.)

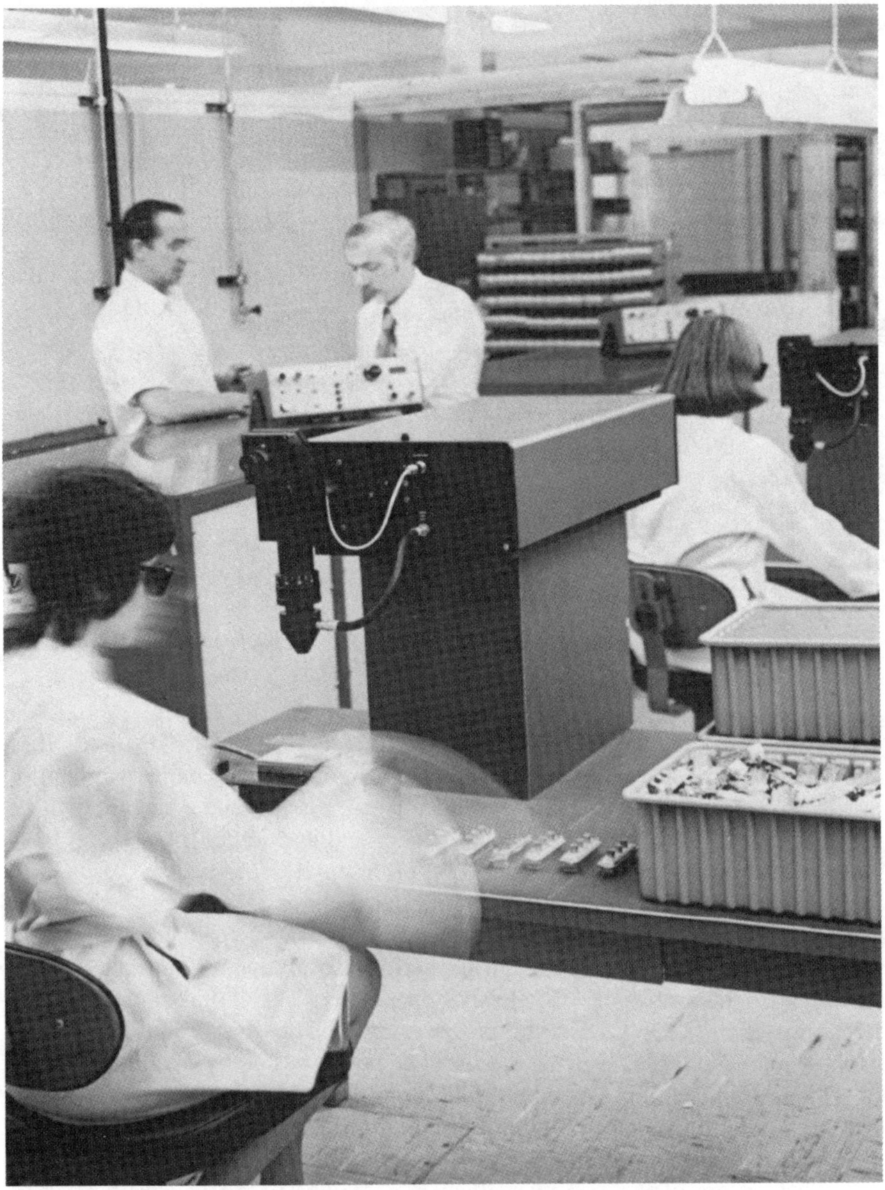

Figure 8-3 View of laser-welding operations (GTE).

power source. The cooling system is a closed-loop type using 4.5 gpm of clear filtered water at 20 psi. Figure 8-3 is a general view of the laser welding operation showing one of the welders in the foreground.

A high-power laser welder manufactured by Control Laser Corporation is shown in Fig. 8-4. The system, Model 2660-6R, is a continuous-wave Nd-YAG laser with a

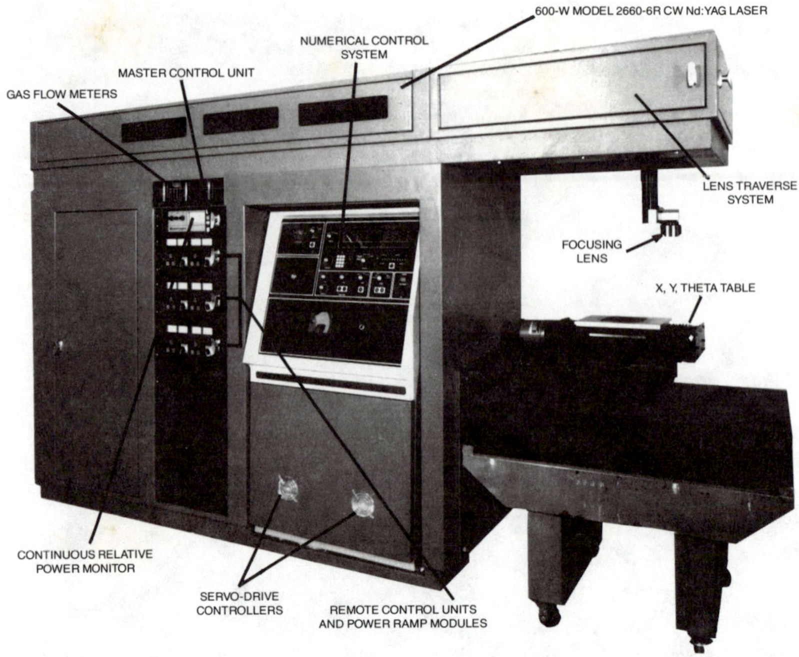

Figure 8-4 High-power laser welder, Model 2660-6R, by Control Laser Corporation.

maximum power output of 600 W and a beam divergence of 14 to 18 mrad. It is provided with a relative power monitor, a power ramp module, a numerical control system with an *X-Y* and theta translator, a lens traverse system, and a cabinet with an integral frame assembly to support the laser system and its components.

Unlike conventional lasers, the Model 2660-6R laser welder is made up of six laser heads, each of which generates 100 W of laser output in a continuous-wave mode. Each head has its own power module, water-to-water heat exchanger, and remote control unit. The remote control units are slaved to a master control unit that permits separate operation of each unit or collective operation of all the units together. To provide control of the laser power, the power ramp module operates in conjunction with the numerical controller, which can control all operations in the welding sequence except for loading and handling of parts. The beam is upcollimated and is focused with a 50-inch focal length lens, which is motivated by an *X-Y* traversing system. In the cascaded module assembly, the system can generate up to 800 W of laser power at 10,600 Å. The system is powered by a 40-kVA, three-phase, 60-Hz, 460-V source.

Formation of Weld

With a given thickness of material to be welded, several factors enter into the procedure. The specific parameters of the laser emission energy are set by means of panel controls, shown in Fig. 8-1. The lens focus is adjusted with the optical subsystem, and the laser generator is triggered with the foot switch to project a laser pulse on the workpiece. The focusing of the laser beam on the workpiece is performed indirectly in some (earlier) machines; that is, a light source, not shown in Fig. 8-1, is directed through the laser optical subsystem, before triggering the laser beam, and is focused on the workpiece by use of the microscope. A chart, such as that shown in Fig. 8-5, can be prepared for routine work, and the voltage is set in accordance with the thickness of the material to be welded.

During laser triggering, the light beam is cut off from view and simultaneously an optical path for laser transmission is opened so that when the laser beam is triggered it is exactly in focus on the workpiece. Although this operation may become cumbersome at times, it is obviously necessary to protect the operator's eyes from exposure to the laser beam. In a procedure of this type, it is improbable that the incident laser beam will have a precise and identical diameter each time an operator sets the focus with a visible light. Furthermore, it is doubtful that different operators will obtain the same size focus on the workpiece. In other words, precise and identical focus size is imperative for obtaining uniform weld quality.

With a given lens focal length at a constant laser emission rate, experience has shown that the distance between the lens and the workpiece directly influences the size of the weld, the depth of the weld nugget, and the metallurgical quality of the weld structure, as illustrated in Fig. 8-6. This implies that an improperly distanced weld zone may result in the formation of a hole (A), a concave weld with a puncture in its center (B), or a convex weld (D). Part (C) of Fig. 8-6 illustrates material that is properly welded because of the correct position of the beam focus. In part (E), only the top layer of the material has fused, without affecting the bottom layer. These conditions point to the critical nature of the relationship between the optical positioning and the resultant thermal transition of the heat-affected area. The need for a self-focusing, self-distancing, and self-power-setting means of laser beam adjustment on the workpiece is obvious.

Since a system such as that described above is not available in current laser equipment, the operator must still rely on the charts and graphs that he has prepared

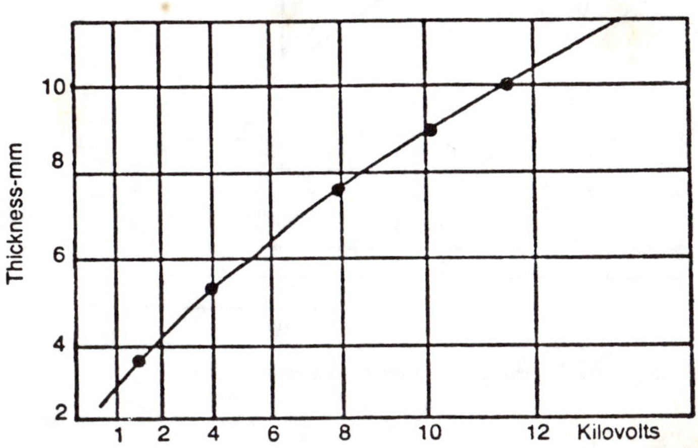

Figure 8-5 Chart showing the thickness-kilovoltage parameters for welding nickel sheets.

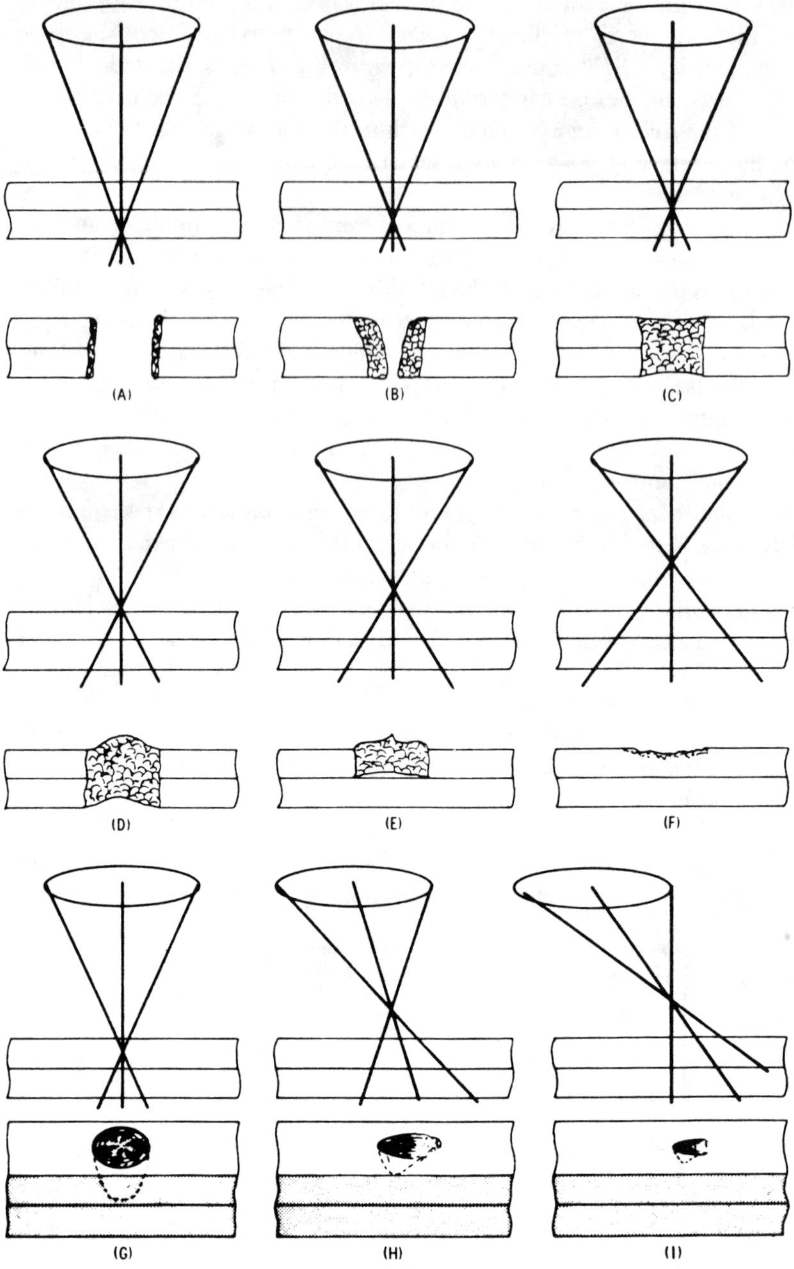

Figure 8-6 Various focus conditions due to beam focus locations in a workpiece.

for each type and thickness of material routinely processed in the plant. A few of such graphs are shown in Fig. 8-7, which gives examples of different welding techniques employing several variations of pulse repetition rates for a carbon dioxide laser beam used at various speeds. This type of parameter determination is current with most conventional laser welding equipment.

Electrical parameters must also be chosen so that the thermal energy at the focal area is sufficient to melt the two parts being welded together and the weld area has the same diameter as the beam focus. The two faying surfaces of the parts being welded must be in good thermal contact with each other, with minimum or no air space between them, so that the radiant heat will be completely absorbed by the weld area. To establish intimate contact, pressure may also be exerted on the parts being joined. Furthermore, the operator may occasionally prefer a weld size larger than the focus waist to obtain a stronger weld joint. This may be achieved by moving the lens system slightly toward the workpiece and increasing the pulsewidth.

Self-Focusing Laser Welder

A laser welding system has been developed by Laserkinetics Corporation to eliminate human errors and to incorporate as much flexibility of use as possible in the laser welder. The laser system, whose circuit diagram is illustrated in Fig. 8-8, comprises

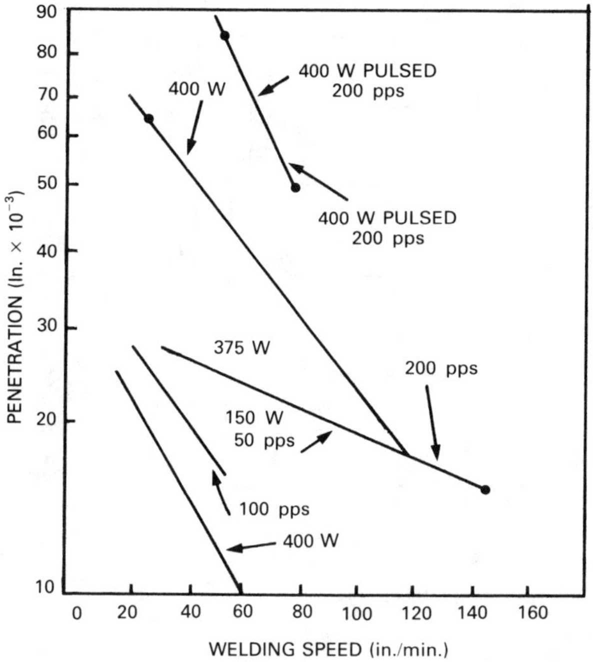

Figure 8-7 Graphs showing weld penetration versus welding speed for pulsed laser beams.

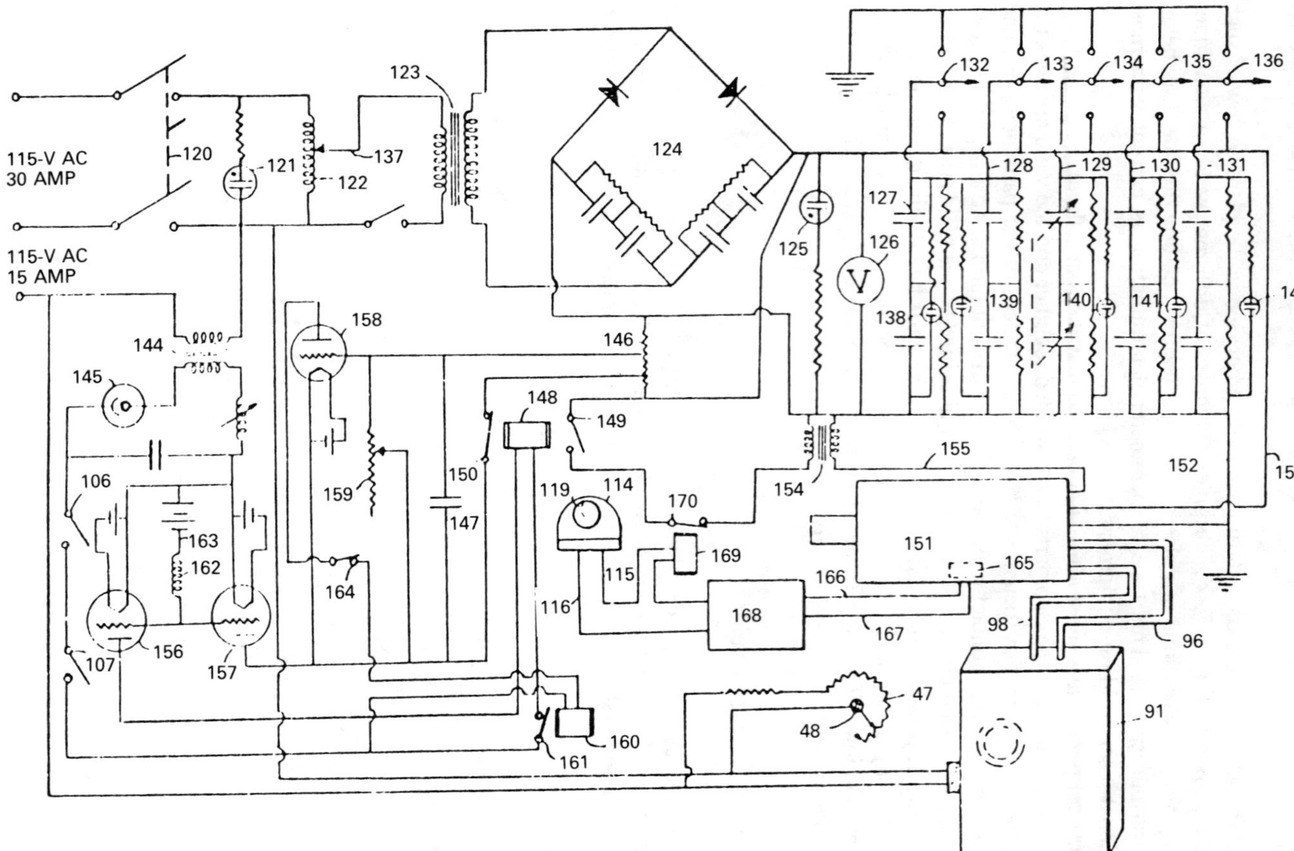

Figure 8-8 Schematic circuit diagram of the self-focusing Laserkinetics welder/driller.

a single-unit console containing a laser generator, a power supply unit with a pulse-forming network, a closed-circuit water-cooling subsystem, a laser head embodying the optical mechanism, controls for laser intensity and pulsewidth adjustment, and electrical meters to indicate the various parametric settings and the temperature condition of the laser head. A foot treadle is used to move the laser head to and from the workpiece and simultaneously trigger the laser radiation when the head contacts the workpiece. The latter feature protects the operator from laser reflections and eliminates the shutter mechanism provided on conventional, manually operated laser systems. This scheme also eliminates pressurization of the faying surfaces. An auxiliary waveguide of special construction is provided for welding, cutting, and drilling to be performed remote from the console.

The laser system is capable of welding, drilling, cutting, scribing ceramic, and trimming resistors to tolerances. It can also be used for welding sophisticated circuit parts, bonding and cutting plastics, and cutting other synthetic materials, such as garment materials, with slight component modification. Some of the metals for which this system can be employed are gold, silver, stainless steel, nickel, titanium, aluminum, copper, tungsten, zinc, tantalum, beryllium, and alloys of these metals. The unit is especially adapted to weld semiconductor devices. The thickness of the metals that the system can handle is dependent on the density, thermal conductivity, reflectance, and energy diffusivity of the metal in consideration. Generally, thicknesses between 0.001 and 0.4 inch can be processed with the system, using 1 to 40 J of laser energy. Higher-energy systems are also available, with circuit variations.

Some of the principal advantages of the system are as follows:

1. The laser focus is automatically adjusted on the workpiece.
2. Because the laser head contacts the workpiece, no stray radiation is visible to the operator.
3. No microscope is necessary for focusing the beam, so human error is eliminated.
4. When necessary, the laser beam can be directed by means of a proprietary waveguide to a workpiece up to 8 feet away from the console.
5. The laser beam can be transmitted to areas inaccessible by conventional methods.
6. Operator training is reduced to a minimum, eliminating further errors.
7. The laser beam can be focused above the workpiece, within any desired depth of the material, or on the surface of the workpiece.
8. The laser head can rotate 360° in the vertical plane.
9. A binocular microscope is used only to locate and magnify the work area.

Semiconductor devices and integrated circuits can be welded with this system just as by a resistance-welder method. Only one or two welds per lead are sufficient to attach the leads of the integrated circuits; switch terminals, capacitors, and other microcircuit components can be welded to the printed circuit board. Experience with printed circuit component attachment by resistance welding has shown that more than 50% of the component leads develop microcracks. Some of these microcracks are harmless to the device, but others, such as those installed on equipment on shipboard, gradually grow because of the interaction of the metal (kovar) with the moist sea air.

Eventually, the lead material develops corrosion, as shown in Fig. 8-9, which results in opening of the circuit and consequent malfunctioning of the circuit board assembly. Such a condition rarely occurs in laser-welded device components.

In Fig. 8-9, the gold-plated kovar leads are lifted and pushed aside to reveal the badly degraded and corroded weld interfaces. This is one of many examples of similar failures that occur in computer equipment installed aboard ship. For reasons of this type, all device components are currently solder-joined to the printed circuit boards, with a solder material having a melting point of 450°F or over. Solder-joined component leads are not favored because when the equipment operates air-cooled and at relatively elevated temperatures the solder melts and opens the microcircuits, resulting in failure of the system.

In contrast to resistance welding, a specimen of laser-welded kovar is shown in Fig. 8-10, which indicates the uniformity of weld size that can be achieved with properly chosen laser parametric settings.

Figure 8-11 shows a top view of a kovar weld, and Fig. 8-12 shows a cross-sectional view of the same weld after etching with nital etchant. The diameter of each weld is about 15 mils. By pulsing at least two welds on each microcircuit lead, the weld interface area increases with an increase in holding strength. The welding schedule for each weld shown is 1.4 J per pulse of laser energy, at a pulse duration of 3 msec, using a planoconvex lens of 25-mm focal length. A tensile or pull test on a single weld

Figure 8-9 View of corroded integrated circuit lead areas welded by means of resistance welder.

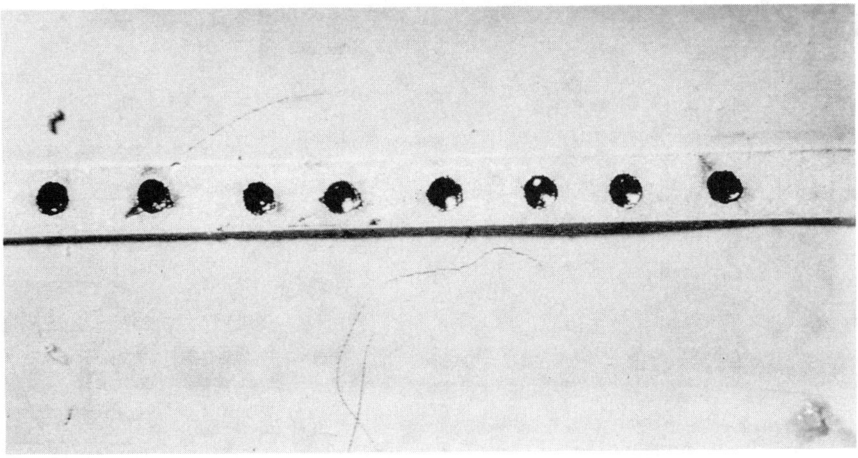

Figure 8-10 Gold-plated kovar lead laser-welded to gold-plated circuit board.

of this size will yield from 3500 to 4000 g, which is actually greater than the resistance weld of equal size.

In Fig. 8-11, the coaxial circular lines represent the mechanism for instantaneous melting and cooling of the kovar during welding. The pattern indicates how the thermal pulse energy propagates radially from the center outward during melting, and its path

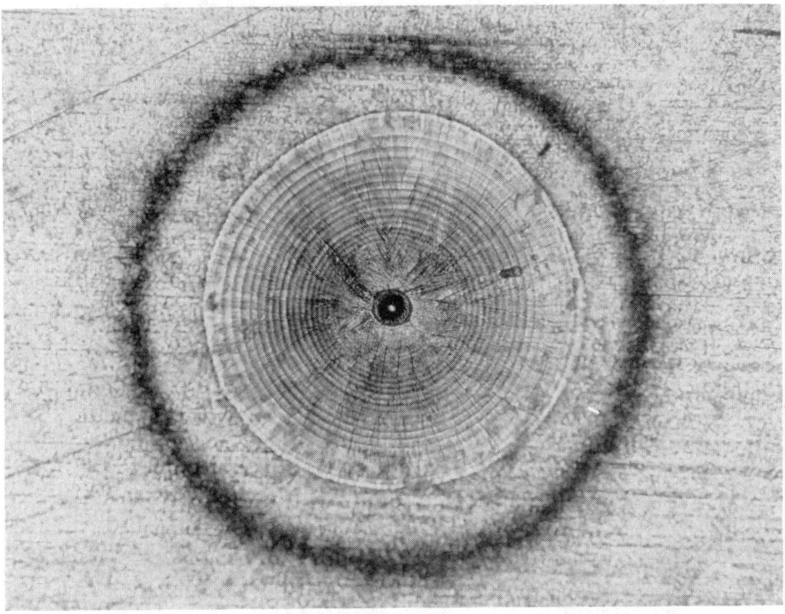

Figure 8-11 Top view of kovar weld, using a 10-mm focal length lens.

Figure 8-12 Cross-sectional view of the laser weld shown in Fig. 8-11 (nital etch).

solidifies as the cooling proceeds from the periphery to the center. The black outer circular area is the residue of the charred surface contaminants. The etched cross-sectional view of the weld in Fig. 8-12 shows the microstructural pattern after fusion of the metal and sudden solidification. Each sheet is 2 mils thick and good penetration is indicated.

Figure 8-13 illustrates the manner of welding integrated circuit leads to a circuit board, using Hughes Aircraft's first ruby laser system following the discovery of the ruby as a lasing element by that company. The view in Fig. 8-13 shows the laser head improvised by the author when employed at Rockwell International Corporation.

Figure 8-13 Relation of the laser head to the workpiece.

Although the equipment is the first crude model of a laser welder, thousands of welds and holes were made in microcircuits with this system rated at 1.4 J.

Pulsed and Continuous-Wave Laser Welding

With pulsed laser welding, generally the power density is reduced and the pulse duration is increased to allow sufficient time for the laser energy to be absorbed by the metal. In pulsed-type operation, usually a ruby, Nd-YAG, or Nd-glass rod is used, while with CW laser welding a carbon dioxide laser is preferred, although Nd-YAG lasers can also be used continuously. The laser power for welding metals of ordinary size ranges from 10^4 to 10^5 W/cm^2. Higher powers are employed for thicker and harder metals, such as titanium, stainless steel, and tungsten. Since there is no theoretically formulated equation for welding any given material, the preliminary work on welds will have to be performed empirically, and a graph can be prepared for each specific material under specific conditions. A welding schedule may be prepared by following the procedure given in Table 8-1. The data obtained for Table 8-1 can be translated into graphical form as shown in Fig. 8-14.

In continuous-wave laser welding, penetration of the laser beam into the material being welded is by thermal conduction, which limits the depth of penetration of the beam. Although solid-state lasing elements can be employed successfully for continuous-wave welding, for deep penetration a CO_2 laser is used. This molecular laser can yield multikilowatts of radiation, is capable of penetrating up to 2 cm of steel, and is particularly useful in seam welding. With pulsed laser beams, seam welding is accomplished by superimposing one weld partially over another, and the resultant weld is just as strong as that obtained with a CW laser beam. Advantages of pulsed-laser seam welding are the relatively low-cost operation, speed of performance, and easy controllability of the production work.

Theoretical Considerations

In welding two pieces of metal together, the temperature of the faying surfaces must be raised to the melting points of the two metals. The energy used to melt the metals must come from the laser beam. When the two metals are of the same type, the melting problems are reduced to the characteristics of one metal. However, if two different metals are being joined, the characteristics of each metal must be taken into consideration to establish a welding schedule. In addition, the surface characteristics of the metal facing the incident laser beam must be precisely considered because the important coupling property between the wavelength, surface condition, and reflectivity of the metal will determine the amount of thermal energy conducted to the underlying metal. The incident laser energy is dissipated by several inherent factors: (1) reflection from the surface of the metal, (2) conduction to contiguous areas, and (3) insufficient optical coupling of the radiation wavelength with the metal. The remaining energy is used in raising the temperature of the two metals to the fusion state. As stated earlier, the laser rod receives the energy for radiation from the flashlamp. But, in practice, only 1 to 2%

Table 8-1 Laser feasibility test firings for kovar (lens focal length 20 mm)

Thickness, mils					Voltage, kV				Capacitance, μF				Weld size, mils*	Energy input, joules†	Laser output, joules‡	Remarks	
1	2	3	4	5	1.1	1.2	1.5	1.8	200	400	500	600				Weld	Hole
X					X				X								
X					X					X							
X					X						X						
X					X							X					
	X					X			X								
	X					X				X							
	X					X					X						
	X					X						X					
		X					X		X								
		X					X			X							
		X					X				X						
		X					X					X					
			X					X	X								
			X					X		X							
			X					X			X						
			X					X				X					
				X			X				X						
				X				X				X					
				X			X					X					
				X				X			X						
X					X					X							
X						X				X							
X							X			X							
X								X		X							

* Measure weld size with microscope and place across appropriate X.
† Calculate input energy in joules and place across appropriate X.
‡ Calculate output laser energy using Eq. (3-1) and place across appropriate X.
Note: Any variation of this table may be tried for other electrical inputs and laser outputs.

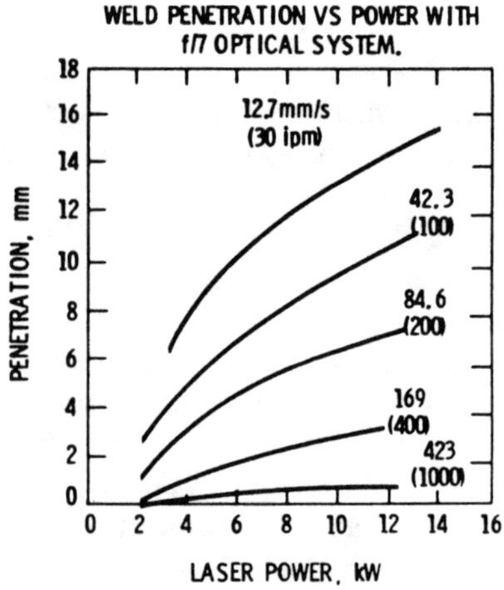

Figure 8-14 Laser power versus penetration in millimeters (Courtesy of AVCO Everett Research Laboratories, Inc.).

of the incident light energy is converted into laser radiation. This subject will be taken up further as our study progresses.

As an example, suppose 1 joule of laser-beam energy is incident on a 2-mil-thick kovar sheet and is focused on an area 2 mils in diameter. The following thermal conditions will exist:

The heat content of the incident 1-J laser beam will be

$$H = \frac{1}{4.186} = 0.24 \text{ calorie} \tag{8-1}$$

Let us assume that the radiant energy, when incident on the kovar material, has incurred no loss by reflection or by secondary radiation. The energy absorbed by kovar may be calculated as follows:

$$\text{Volume } V \text{ of kovar heated} = \pi\left(\frac{d^2}{4}\right)t \tag{8-2}$$

$$= \pi r^2 t \tag{8-3}$$

where d is the diameter of the weld in mils, r is the radius of the weld in mils, and t is the thickness in mils of kovar.

Substituting numerical values for these quantities, we have

$$V = 3.1416(0.001)^2(0.002)^2$$

Expressing this relation metrically, we obtain

$$\begin{aligned} V &= 3.1416(1 \times 10^{-6})(2 \times 10^{-3}) \times 16.387 \text{ cm}^3 \\ &= 102.96 \times 10^{-9} \\ &= 1.03 \times 10^{-7} \text{ cm}^3 \end{aligned}$$

The weight W of the kovar of density 8.35 g/cm^3 will be

$$\begin{aligned} W &= 1.03 \times 10^{-7} \times 8.35 \\ &= 8.6 \times 10^{-7} \text{ g} \end{aligned}$$

The heat H, in calories, required to raise the temperature of 8.6×10^{-7} g of kovar 1°C will be

$$\begin{aligned} H &= 8.6 \times 10^{-7} \times \text{specific heat of kovar} \\ &= 8.6 \times 10^{-7} \times 0.105 \text{ calorie/°C} \\ &= 0.903 \times 10^{-7} \text{ calories/°C} \end{aligned}$$

The melting point of kovar is 1450°C, and its heat of fusion (calories necessary to melt the kovar without change of temperature) is 64 cal/g. Therefore, to raise the temperature of 8.6×10^{-7} g of kovar to its melting point, the heat required will be

$$\begin{aligned} H &= 0.903 \times 10^{-7} + (8.6 \times 10^{-7}) \times 64 \\ &= 1309.35 \times 10^{-7} + 550.4 \times 10^{-7} \\ &= 1859.75 \times 10^{-7} \\ &= 1.86 \times 10^{-4} \text{ calorie} \end{aligned}$$

In order to heat the two layers of kovar that are to be welded together, the heat required will be doubled as follows:

$$H = 1.86 \times 10^{-4} \times 2$$
$$= 3.72 \times 10^{-4} \text{ calorie}$$
$$J = 0.156 \times 10^{-2} \text{ joule}$$

In the preceding problem, to convert calories into joules, the number of calories was multiplied by 4.186, the conversion factor.

On the basis of the preceding assumptions, which do not take into consideration the heat dissipation by conduction, reflection of energy from the surface, diffusion, and so forth, it is evident that 1 J of laser energy of 0.24 calorie equivalent is much greater than the required welding energy. Thus, instead of obtaining a weld, a hole will be produced through the two layers of kovar. This condition suggests that the laser energy should be attenuated, according to our calculations.

The laser energy from the laser head can be attenuated by several means. One of the most common procedures is to regulate the voltage, the capacitance, or both. This can be calculated from Eq. (1–18). Another method is to adjust the size of the focal spot incident on the workpiece; however, this latter scheme imposes restrictions on the wavelength of the laser beam and on the thickness of the material being welded. An equation to calculate the focal spot size is

$$A = 1.169\left(\frac{f}{a}\right)^2 \lambda \tag{8-4}$$

where A is the focal spot size, d is the diameter of the focal area, f is the focal length of the lens, a is the diameter of the laser rod, and λ is the wavelength of the radiant beam.

When the expected diameter of the microweld is known, the following relation may be used to determine the lens-to-workpiece distance S_t:

$$S_t = \frac{da}{2}\sqrt{\frac{\pi}{1.7\lambda}} \tag{8-5}$$

If the materials being welded are gold, silver, copper, and aluminum or are plated with any one of the first three metals, the diffusivity characteristics and the high conductivity of these metals play important roles in joining them together or to other metals. This relationship is given by the following expression:

$$\alpha = \frac{\zeta}{\rho H_e} \tag{8-6}$$

where α is the thermal diffusivity of the metal (cm^2/sec), ζ is the thermal conductivity (cal/cm-sec-°C), ρ is the density of the metal (g/cm^2), and H_e is the specific heat (cal/g-°C).

Steel, stainless steel, nickel, kovar, inconel, zinc, beryllium, chromium, and tantalum, all of which have low diffusivities and low thermal conductivities, are more readily welded than aluminum, magnesium, and the noble metals. The noble metals are difficult to weld unless their surfaces are treated or coated with a readily removable substance such as graphite. A laser beam ordinarily has no effect on glass because the latter is transparent to the beam. However, glass can be cut or drilled after the surface is coated with a low-thermal-conductivity material, such as a deposit of carbon or soot. Nontransparent plastic materials are also reasonably easy to weld when a large-diameter focal spot is used on the workpiece.

In seam welding any given specimens, the total energy used per linear unit would be the product of the laser energy employed for a single weld times the number of pulsed welds per unit length considered. A seam-welded kovar specimen is illustrated in Fig. 8-15.

Microcircuit Joining with Laser Beam

Since we will be concerned in this book with semiconductor laser devices, such as gallium arsenide laser diodes or their derivatives, it is appropriate to discuss the failure mechanism of the bond structure of the microcircuit leads and the manner by which the failure can be eliminated.

It has been determined by failure analysis of microcircuits that a large percentage of failures is due to conventional welding or thermocompression bonding problems. Almost all semiconductor microcircuits consist of a 2 × 2 mm to 10 × 250 mm area of silica or sapphire substrate (die), on which a silica epitaxy is vacuum-coated and microcircuits are deposited in this epitaxy. The deposited circuits terminate in aluminum bonding pads on the substrate. To these pads are bonded interconnection leads ranging in diameter from 0.5 to 1.0 mil (0.127 to 0.254 mm); the leads are made of either aluminum or gold. Other lead materials, such as molybdenum-gold, titanium-platinum-gold, or chromium-gold, have been tried with some success. However, in all these cases an ideal material has not been commercially adopted to avoid excessive failure

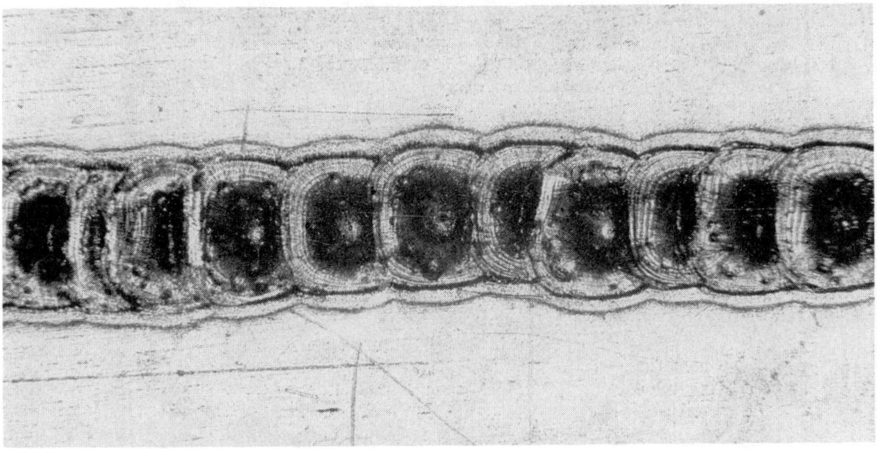

Figure 8-15 A seam-welded kovar-to-kovar specimen.

of the bond interface at the pads. Thus, it is readily seen that the reliability of microelectronic devices (microcircuits) is related to the conditions at their bond interfaces.

To form a junction interface, interaction of the atoms of the faying surfaces must occur. Such an action may be explained by considering an elemental system with variable electrical parameters in which a change in the electrical characteristics of the interacting elements would be a function of environmental stress. This system would be subject to changes in interface properties, which may give rise to various modes of failures. In such failures, isolated stresses intrinsic to the interface-forming media exceed the energy of the interface domain and are affected by time and temperature. The time-temperature behavior of the interface is measured by the changes in electrical parameters of the microcircuit device considered.

In a union of heterogeneous metallic phases, such as aluminum and gold bond phase, the character of intermetallic constituents thus formed depends on the electromotive potential difference between the metallic phases. This difference influences the interaction of interjoining elements that are far apart in the electromotive series and may advance into a degrading phase. The latter interaction is further favored by incompatible interfacial media, which may cause the interacting elements (metals) to react catastrophically. An example of the latter condition is the frequently encountered semiconductor device failure due to opening of the gold-aluminum bond because of weak and porous intermetallic phase formation at the bond interface, as illustrated in Fig. 8-16.

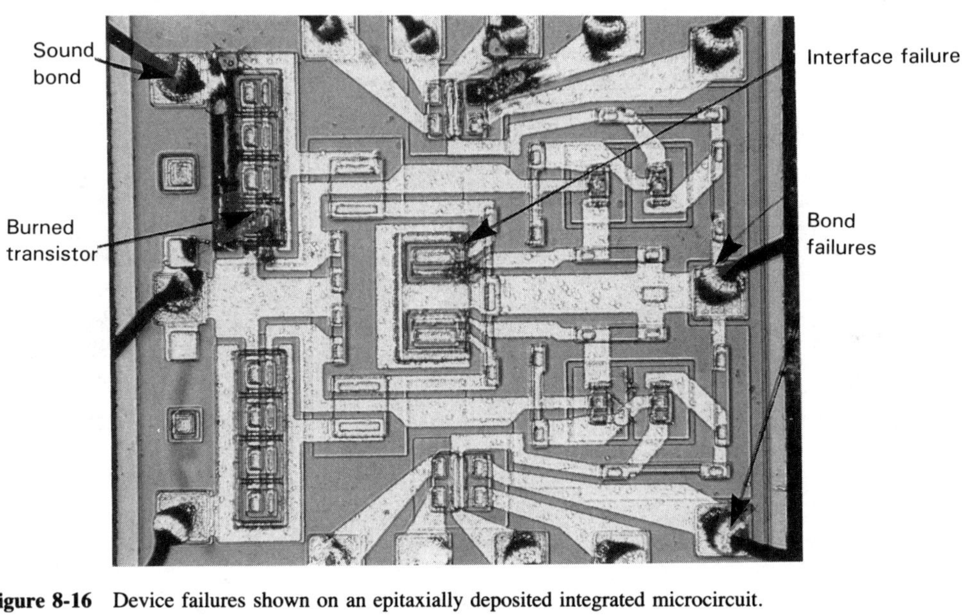

Figure 8-16 Device failures shown on an epitaxially deposited integrated microcircuit.

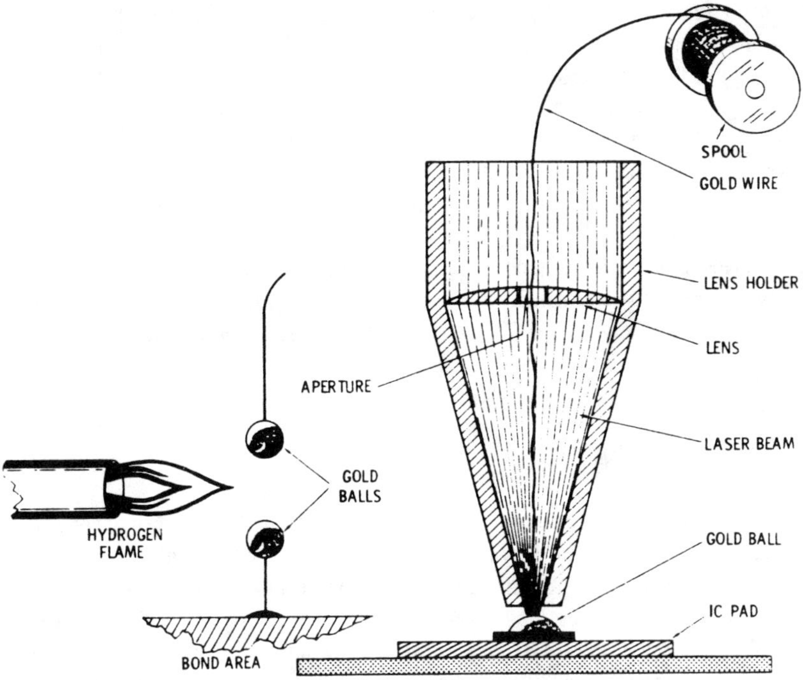

Figure 8-17 Laser bonding of a microcircuit lead of 1-mil-diameter gold wire.

When a gold lead is bonded to a porous aluminum pad, the combination can form microgalvanic cell sites under ambient conditions in a poorly sealed package. In the presence of silicon from the epitaxy and heat from thermocompression bonding, gold and aluminum may react to form $AuAl_2$ or Au_2Al, the silicon acting as a catalyst. In an advanced stage of this condition, the silicon may enter into the reaction to form an Au-Al-Si complex. This complex is porous and mechanically weak and brittle because it is a degraded form of a bimetal. Thus, the new constituent offers high electrical resistance and gives rise to an electrical parameter, $V_{BE(SAT)}$, in the semiconductor device. This condition results in opening of the gold-aluminum bond at the interface.

Accordingly, an effective method of eliminating bond-interface failure is to maintain the two bonding materials in an intermetallic alloy phase that will not be vulnerable to ambient environmental conditions within the device package. When the joining surfaces are fused to form an alloy, the tendency for the alloy to separate into its constituents would be practically nil. The application of a laser beam in the joining of the deposited circuit pad sites could form a cohesive bond, eliminating the degradation and opening of the bond interface. This is particularly true because of the instantaneous heating and cooling of the bond junction by the laser beam.

Laser welding of microcircuit leads is new to the laser industry, and a system of this type is not commercially available at present. A laser optical head designed and developed by the author is shown in Fig. 8-17. The device consists of a conical housing with a planoconvex lens at its base section, which can be mounted on the exit aperture of any commercial laser welder. The laser focusing lens is provided with a circular aperture at its center through which the gold or aluminum lead is led to the exit tip of the conical housing; this is the point that makes contact with the bond nugget.

Before starting the welding operation, the terminal point of the lead at the exit tip is heated with the usual hydrogen flame to form a ball integral with the lead. The conical laser beam surrounding the aperture area of the lens is converged to an arca equal to that of the microcircuit pad, at which the beam transfers pulses of thermal energy to the joining metals and melts them together to form an alloy, if the joining metals are of different materials. Since the pad material is ordinarily made of aluminum deposit, and aluminum has great affinity for combining with silicon, the resulting alloy bond is strong. The joining lead is fed from a spring-biased spool adjacent to the conical housing.

Dual Laser Beam Welding

The dual laser beam welding technique is used with laser welders whose generated laser beam is invisible, such as that from the fundamental wavelengths of a Nd-YAG, Nd-glass, or carbon dioxide laser. Starting with the visible laser beam, a ruby laser is capable of delivering high-average laser power, with pulse energies on the order of 5 to 50 J at rates of 1 pulse per second or less to prevent the laser rod from becoming too hot and losing emission efficiency with an increase in beam divergence. A warm-up pulsing time is necessary for continuous pulsing in production work with adequate cooling of the ruby rod. On 10- to 2-mil-thick kovar leads, moderate energies of 2 to 10 J allow continuous-pulsing operation to be performed, and seam welds of sheet material can be made by overlapping the spot welds, as illustrated in Fig. 8-15. An average pulse energy of 6 J at 4 to 5 msec, pulsing at 60 to 180 pulses per minute, can be employed with continuously water-cooled equipment.

A Nd-YAG laser rod is also applicable to continuous welding of metals and certain types of plastics, although an argon ion laser would be preferable for plastic welding. The Nd-YAG laser is particularly suitable for seam welding because energy levels of hundreds of watts can be obtained from the laser rod. Because Nd-YAG has high efficiency and can provide a small focal spot, it can be used for miniature circuits and for scribing semiconductors. It all these cases, if the equipment is used in a manual welding scheme, the Nd-YAG laser beam must be mixed with a visible laser beam, such as that from a He-Ne laser generator, as illustrated in Fig. 8-18.

Neodymium-glass or alexandrite rod is another candidate for laser-welding operations. It has laser output energies in excess of those from a ruby rod. Because of the low thermal conductivity of the Nd-glass material, the pulse repetition rate is restricted to about 10 to 15 pulses per minute. After a warm-up pulsing time, the laser rod can be pulsed continuously at higher repetition rates.

To weld the common bus terminals on multichannel crossbar switches, Western Electric has developed an edge detection optical system that triggers the dual laser beam system at exactly the instant when a terminal on a moving stage is detected. The

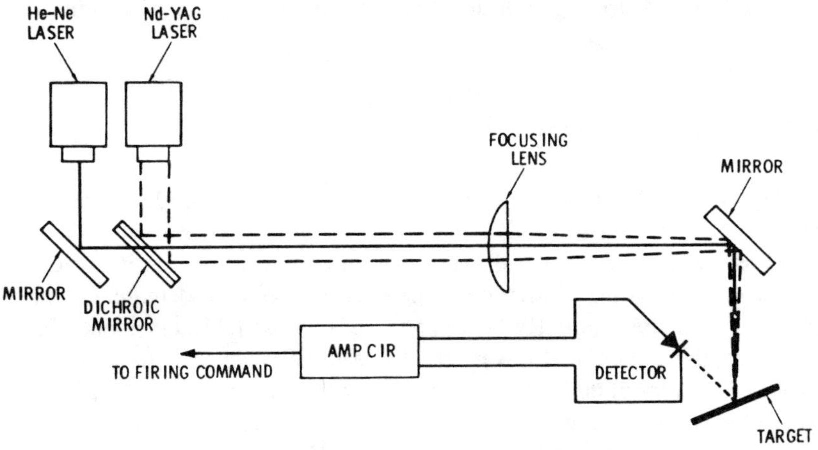

Figure 8-18 Two-beam laser welding scheme employed by Western Electric.

scheme of focusing the beam is shown diagrammatically in Fig. 8-18, in which two laser generators, a Nd-YAG and a He-Ne laser, project their beams at the target simultaneously. Since the Nd-YAG laser beam is invisible, with a wavelength of 10,640 Å, it is focused on the target by means of the red light from the He-Ne laser beam of 1-mW intensity. The two laser beams are then directed by means of mirrors to a common focusing lens, from which the beams are reflected onto a translating mirror and then to a target.

A portion of the laser beam from the target is sampled by a photodetector, which senses bus terminals of various sizes and actuates the Nd-YAG system to deliver the right amount of laser energy for welding. This procedure, which permits up to 200 welds to be made per minute, yields routinely two welds per second. This is faster than the rate at which the operator can handle the loading and unloading of the switch terminals. A transfer mechanism handles the conveyor-carried crossbar switches in a light-tight chamber to avoid any stray laser beam reflections to the operator.

The carbon dioxide laser operates with the highest efficiency and at high laser power output in a single mode. The laser may be operated continuously at high-energy pulses and at long wavelengths that are easily absorbed by low-conductivity metals and some nonmetals. By pulsing the laser energy at rates of 100 pulses or more per second, deep penetration in seam welds of heavy materials (e.g., stainless steel, nickel, kovar, and titanium), as well as rubber materials of certain types, can be performed because reflectivity losses become negligible at the infrared wavelength of 106,000 Å. However, the laser beam from carbon dioxide is highly absorbed by many optical systems. For this reason, the lenses must be made of relatively costly silica, germanium, gallium arsenide, quartz, and the like for nonobstructive beam transmission. Or, reflective optical systems must be employed for welding miniature microelectronic circuits or devices; however, the beam is suitable for scribing semiconductor wafers and trimming thin- and thick-film resistors. The carbon dioxide laser is found to be superior to all other lasers in cutting and drilling thin and thick metals and sheets of nonmetallic materials.

Microsoldering System Using a CO_2 Laser

A 20-W, flowing-gas carbon dioxide laser system operating at 100,600 Å has been developed by Apollo Lasers, Inc. (Fig. 8-19). The system is provided with a 1-mW He-Ne laser, whose beam is projected through the common focusing optics onto the workpiece for localization of the beam focus. The specimen to be soldered is mounted on an *X-Y* translator, and the visible He-Ne laser beam is focused through a lens of 2.5 inch focal length on the gold-plated kovar lead to be joined to the microcircuit pad. The solder used is a mixture of 60% tin and 40% lead with an organic flux. A specimen of a soldered lead is shown in Fig. 8-20.

The power of the CO_2 laser is adjusted to a nominal 18 W at a 200-msec pulsewidth. The workpiece under the laser beam can be moved in the *X* and *Y* directions manually or by means of a microprocessor programmed by a computer. This procedure is found to be 5 to 10 times faster than hand-soldering the leads. When it is automated, even higher rates of soldering can be obtained from the system. The equipment is energized from a 115-V, 60-Hz, 20-A power source equivalent to household power. The carbon dioxide mixture consists of 6 parts of CO_2, 18 parts of nitrogen, and 76 parts of helium, consumed at a rate of 1 cubic foot per hour. The system contains a television screen

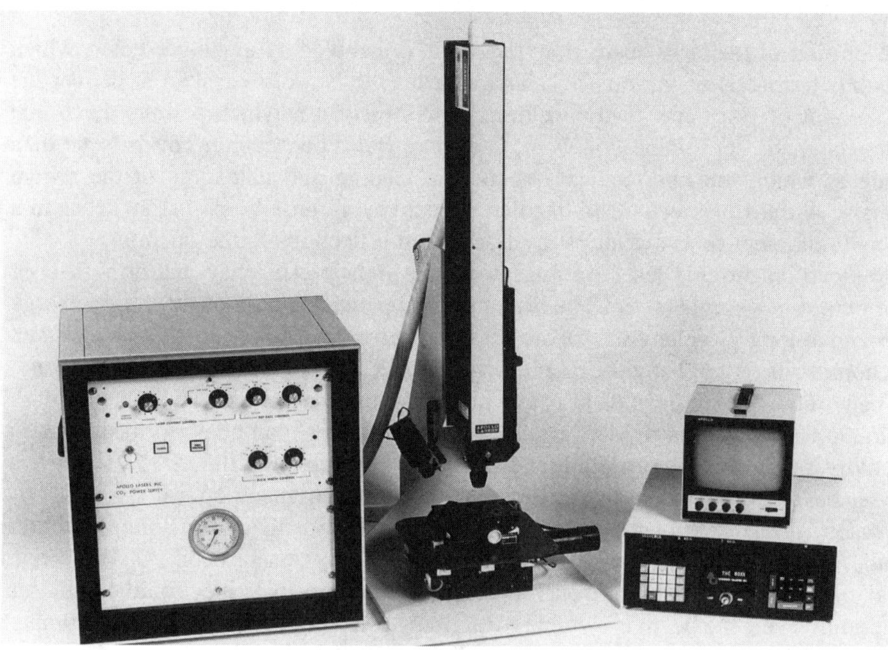

Figure 8-19 The Apollo Lasers microsoldering laser system, Model LMS-1 (Courtesy of Apollo Laser Systems, Inc., an Allied Company.)

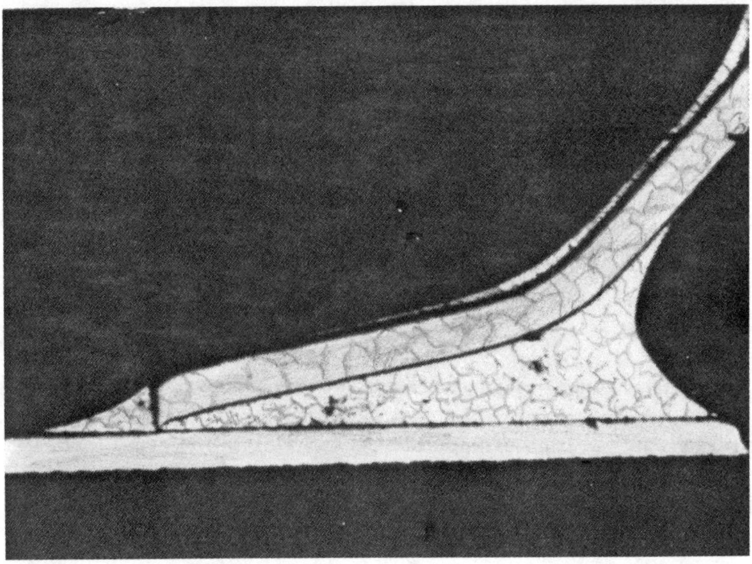

Figure 8-20 Cross-sectional view of a gold lead soldered to a microcircuit pad using laser-beam energy, 75×.

(located on the right-hand side of the illustration in Fig. 8-19) for viewing the soldering process in action.

8-3 CUTTING AND DRILLING WITH A LASER BEAM

A laser beam is also employed for cutting or drilling through metals, plastics, ceramics, rubber sheets, and even glass if its surface is properly prepared for beam absorption. For either cutting or drilling, a long-focal-length lens is used to reduce the tendency of the beam to form a conical hole or slanted cut as well as to produce a narrow kerf in cutting. The laser beam can also be jet-assisted by using either an inert gas, such as argon or helium, or an active gas, such as oxygen, depending on the type of application. This scheme is shown in Fig. 8-21.

The inert gas is generally employed to avoid oxidation of the materials during cutting or drilling, and oxygen is used to obtain exothermic reactions with metal processing. The latter scheme produces a clean cut and a high rate of cutting or drilling. All flammable materials, such as plastics, textiles, and rubber, are cut or drilled by enclosing the focal area with a peripheral jet of argon or helium gas to avoid burning the material beyond the area being cut or drilled. The inert gas prevents atmospheric oxygen from combining with the laser beam heat at the area being processed by the laser beam.

In drilling with a laser beam, a ruby, Nd-YAG, or carbon dioxide laser is used in the pulsing mode; a CO_2 laser can also be used in the continuous mode. Short, high-

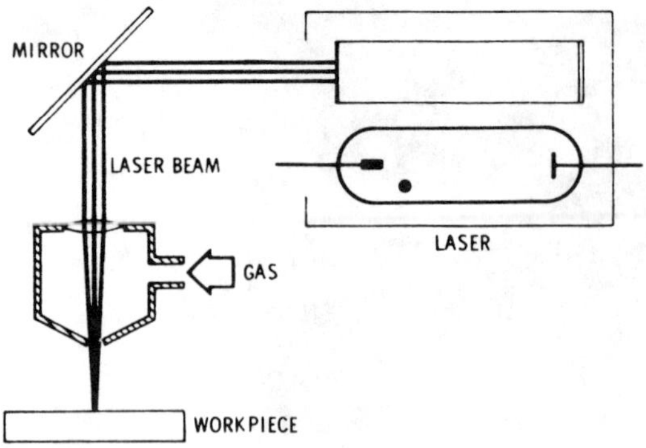

Figure 8-21 View illustrating the scheme for gas jet-assisted laser cutting of materials.

intensity pulses from the laser beam will vaporize the material to produce a hole, the size of which may be varied from 0.01 mm to several millimeters, depending on the size desired and the laser power available. A carbon dioxide laser yields the highest depth-to-diameter ratio when active jet assist is used. The jet assist increases the heat at the focus by oxidizing the material, and the kerf is narrow and smooth.

The mechanism of hole drilling may be explained as follows. The laser beam incident on the metal surface elevates the temperature at a rate approaching 10^{10} degrees per second, vaporizing the surface material to a depth of a few micrometers. The metal below the surface melts and, together with the vaporized metal, an enormously high pressure of several hundred atmospheres is formed in the hole. The pressure results in expulsion of the metal at high velocity from the hole in the form of a plume. This process continues until all the metal is removed from the hole area, which takes less than a second. The gas jet further aids in removing the vaporized metallic particles and thus cleaning the hole. With plastics, the material is removed by burning, and the clear-cut edges are formed by expulsion of the burned or charred material with an inert-gas jet assist.

As the material is removed from the area being worked on, the beam focus gradually moves inward until the beam cuts through the metal. When the laser beam is Q-switched, a depth-to-diameter of ratio of up to 250:1 can be obtained; the cutting rate is also increased with the jet assist, as illustrated in Fig. 8-22. Cutting plastics or metals requires the same technique, with differences in jet-assist gases. A Nd-YAG laser beam produces a narrower kerf than a carbon dioxide laser beam. However, with a Nd-YAG laser, the optical system must be brought closer to the workpiece by employing a short-focal-length lens.

The laser energy required to drill (or cut) a hole in a given metal may be given by the expression

$$H = \rho V\left[m(T_f - T_a) + L_h + L_v\right] \tag{8-7}$$

where H is the energy in calories required to remove the hole material, ρ is the density of the material (g/cm^3), V is the volume of the material (cm^3) and is equal to $\pi d^2 t/4$ (where t is the thickness and d is the diameter of the focus area), m is the thermal capacity or specific heat (cal/g-°C), T_f is the temperature of vaporization (°C), T_a is the ambient temperature taken as 20°C, L_h is the heat of fusion of the metal (cal/g), and L_v is the heat of vaporization of the molten metal (cal/g).

As an example, suppose a 1-mm-thick zinc plate is to be drilled with a hole 2 mm in diameter, using a Nd-YAG laser beam. How many calories will it take to complete this job, assuming a reflectivity of 65% for the zinc surface?

$$
\begin{aligned}
H &= 7.133 \times 3.1416 \times \frac{0.04}{4} \times 0.1[0.0925(906 - 20) + 24.1 + 425.6] \\
&= 7.133 \times 3.14 \times 0.01 \times 0.1[0.0925(886) + 449.6] \\
&= 7.133 \times 3.14 \times 10^{-3} \times 530.6 \\
&= 11.8 \text{ calories}
\end{aligned}
$$

where ρ is 7.133 g/cm^3, d is 2 mm, t is 1 mm, m is 0.0925 cal/g-°C, T_f is 906°C, T_a is 20°C, L_h is 24.1 cal/g, and L_v is 425.6 cal/g.

A recent candidate for microdrilling holes in metals and plastics is another solid-state laser rod made of alexandrite, which is beryllium-doped aluminum oxide (see Section 3-3). This material competes with Nd-YAG and in some cases is said to be superior to Nd-YAG. In a series of comparative tests made with alexandrite and Nd-YAG under the same environmental conditions, it was demonstrated that the alexandrite is capable of drilling holes with a better aspect ratio. (Aspect ratio is defined as the ratio of the depth of the hole to its midsection diameter. Thus, the higher the aspect ratio, the narrower the hole.) For example, when the same thickness of aluminum was drilled with alexandrite and with Nd-YAG, the alexandrite laser yielded one-half the

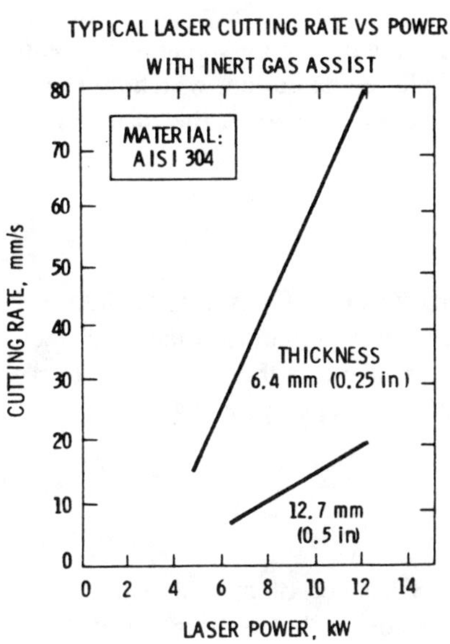

Figure 8-22 Typical laser cutting rates with an inert-gas jet assist (Courtesy of AVCO Everett Research Laboratories, Inc.).

aspect ratio produced by Nd-YAG; this property is usually desired in all laser drilling and can be obtained only by use of a jet-assist gas.

Furthermore, the holes drilled with alexandrite contain less solidified material at their upper edges than those drilled with Nd-YAG, without the use of any jet-assist gas. The advantage of alexandrite is its formation of spikes when operating in the pulsing mode, and the peak power is higher than that of Nd-YAG. This beryllium-doped material operates at a shorter wavelength (7500 Å) than Nd-YAG (10,640 Å); thus the former laser has more energy in its beam than the latter.

A recent development using quarter-wave phase retardation coatings on mirror surfaces from which the beam is directed to the material being processed has eliminated the roughness in the drilled or cut surfaces, producing a very clean and sharp surface. This is accomplished by converting a linearly polarized CO_2 laser beam that is projected parallel to the direction of laser head movement to a linearly polarized beam perpendicular to the direction of movement. Also, in cutting metals, when the plane of beam polarization is parallel to the cut surface, the kerf will be as narrow as the beam and will have parallel surfaces. Such a scheme increases the speed of cutting and the quality of the cut edges. The scheme further leads to a circularly polarized beam, which has no preferential direction of polarization, and thus eliminates any asymmetry of the drilled and cut surfaces.

The usual method of producing circular polarization is to convert the linearly polarized beam as it emerges from the laser head into a circularly polarized beam by first directing the laser beam onto a mirror surface coated with a circularly polarizing material having a 90° phase retardation property. Such a phase shift system must have the characteristics of low beam absorption and high reflectivity of the laser beam, usually a CO_2 laser beam that emanates at 106,000 Å. The coating of the mirror could be zinc sulfide or thorium fluoride or both, evaporated on the mirror surface in alternate layers over the silver coating of the mirror. The S and P polarization components of the beam can be controlled by the amounts of ZnS or ThF_4 coating. The phase retardation of the polarization components can be achieved by having the S component of the beam reflected before the P component, resulting in a 90° phase retardation, which gives rise to the circular polarization of the beam. The laser beam intensity at the focus and at the center of the focus with respect to the focal radius may be expressed as

$$I_r = I_c \left[\frac{2\, J(\pi dr/\lambda f)}{\pi dr/\lambda f} \right]^2 \tag{8-8}$$

where I_r is the focused intensity with respect to the radius of the focus, I_c is the intensity at the center of the beam focus, J is a Bessel function, r is the radius of the focal spot, d is the lens diameter, f is the focal length of the lens, and λ is the wavelength of the spatially filtered laser beam. In this equation, I_c is further equal to $\pi d^2/4\lambda^2 f^2 P$, where P is the laser beam power.

Resistor Trimming

Resistor trimming is the process of adjusting the resistance of thin- and thick-film resistors to permissible tolerances. During the manufacture of the resistors, it is difficult to control the film resistance to precise values. The manufacturer fabricates the resistors

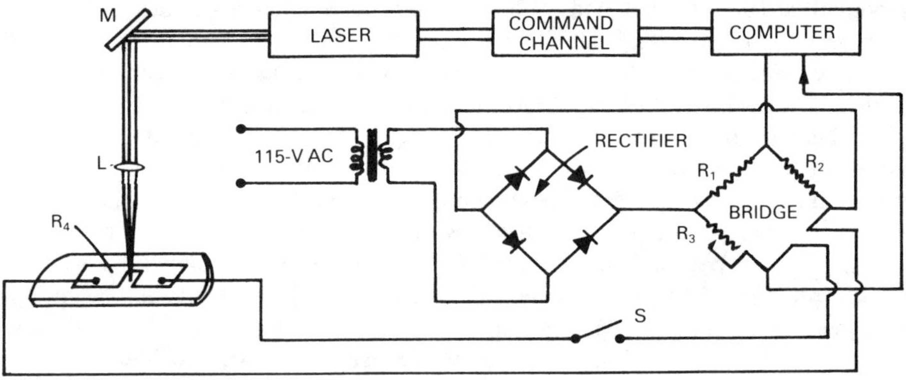

Figure 8-23 Schematic diagram of a simplified resistor-trimming technique.

with lower resistances than intended and then trims (to increase resistance) them to the required resistance, thus permitting close tolerances to be achieved in the device.

There are three types of resistor trimmers in general. The first type is the abrasive trimmer, which uses an abrasive-powder jet to abrade the metal; the procedure is closely watched to eliminate contaminants from the microcircuit. The second type is the capacitive trimmer, which projects an electrical pulse on the metal to melt and remove a portion of it; however, because of the difficulty of obtaining repeatable trimming performance, industry has not fully adopted this type. The third type is the laser trimmer (Fig. 8-23), one model of which is shown in Fig. 8-24. The laser trimmer has the following advantages: (1) the resistance value is precisely controlled to an exact tolerance; (2) a narrow kerf size can be made through the resistor; (3) only negligible heating of the material occurs, so that the characteristics of the contiguous structures in the

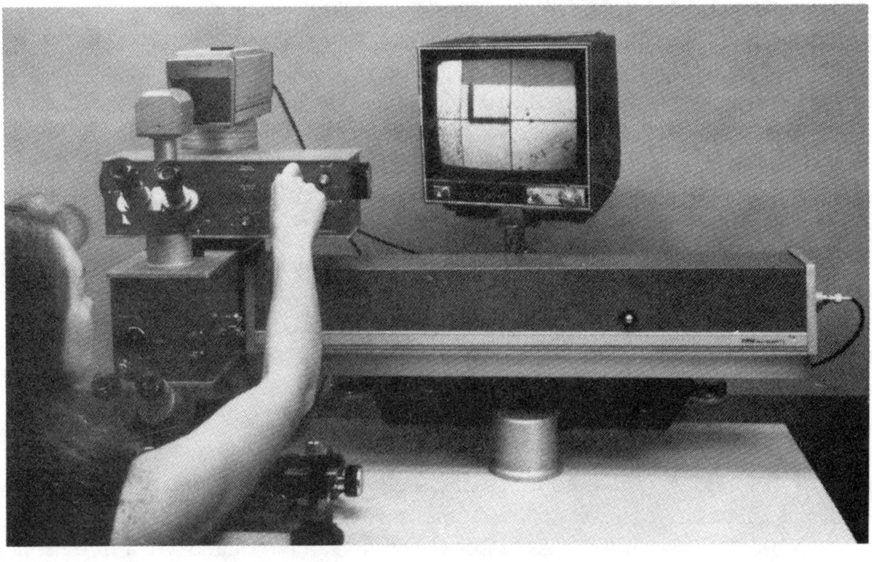

Figure 8-24 The Quantrad laser resistor trimmer. (Courtesy Quantrad Corp.)

microcircuit are not affected; and (4) little or no residue is produced during trimming. This process is cleaner than all other trimming methods. In addition to trimming resistors, a laser trimmer can also fabricate cylindrical resistors by cutting spirally through a nickel-chrome coating on a cylinder to form the resistor.

The laser trimmer shown in Fig. 8-24 is manufactured by Quantrad Corporation. The system employs a pulsed xenon laser, which is ideal for trimming thin-film hybrid microcircuits to adjust their resistors, for ferrite micromachining, for ROM encoding, and for removing photographic emulsion without damaging the film substrate. The radiation from the laser head is green (5000-Å wavelength) and is suitable for photographic error correction. The radiation also can be employed for trimming nickel-chrome resistor films on silicon substrates that require narrow kerf, about 3 to 100 μm. The unit can also trim thick-film hybrid circuits using a Q-switched YAG laser. In addition, integrated circuit silicon wafers can be scribed with a kerf of 10 to 20 μm. The Nd-YAG laser is optically pumped by means of a krypton flashlamp.

Although ruby, Nd-YAG, argon, alexandrite, and carbon dioxide lasers can be employed for trimming resistors, only two types—Q-switched Nd-YAG and Q-switched carbon dioxide lasers—have been chosen by the industry for their practicability and fast repetition rates in trimming. Ruby is used for manual and slow operations where the need is selective rather than routine production. Both Nd-YAG and carbon dioxide lasers are repetitively Q-switched during trimming. The Q-switched Nd-YAG produces laser pulses of nanosecond duration and eliminates substrate heating. The carbon dioxide laser produces a relatively larger spot or kerf line because of its longer wavelength of 106,000 Å; it also requires an expensive optical system with germanium or silicon lenses. However, the CO_2 laser operational cost is low. Both lasers are adaptable to automatic trimming of resistors in a continuous production format, limited only by the *X-Y* table speed and the bridge response time.

A laser trimmer unit typically consists of a laser generator, a Nd-YAG or carbon dioxide laser, standard focusing optics, a workbench with a manual or motorized *X-Y* translator, a microcircuit substrate holder, a binocular microscope for manual trimming operation, a set of resistor probes for sensing the resistance during trimming, and a Wheatstone bridge. The probes are attached to the Wheatstone bridge and, together with the resistor being trimmed, make up two legs of the bridge, as shown in Fig. 8-23. The resistor to be trimmed, R_4, is made a leg of the bridge and is located on the opposite side of a variable resistor R_3 to which resistor R_4 is to be adjusted. Resistor R_4 is positioned at the focus of the laser beam converged by means of lens L, which is preferably a long focal length type. Resistor R_3 is adjusted to the exact resistance required, and when the laser beam is turned on, the *X-Y* translator begins to move the resistor until it has exactly the same resistance value as R_3, whereupon the bridge is balanced and the movement of the translator as well as the emission of the laser beam stops immediately. All adjustments and operations are performed by computer command in an automatic trimmer.

The use of carbon dioxide laser trimming has one significant advantage: the reflections of 106,000-Å radiation are practically harmless to the human retina when accidentally incident on it. However, this author recommends the use of safety goggles at all times. Of course, with Nd-YAG laser operation safety goggles are a must at all times, even though its radiation is also in the infrared. With computerized trimming systems equipped with a television viewer, safety goggles may be dispensed with. Both Nd-YAG and CO_2 lasers pulse with high peak power, but the Nd-YAG has the added advantage that its radiation can be focused to a spot 10 times smaller than that of the CO_2 laser. Also, Nd-YAG radiation is more readily absorbed by the resistor material, which may be nickel-chrome, palladium, or a mixture of silver-gold and other metallic powders formed into a cementing ink that makes up the resistor material.

Semiconductor Scribing

Laser scribing is performed on semiconductor wafers, which are usually made of silicon dioxide, ceramic, sapphire, or similar crystalline substances. Ceramic and sapphire wafers 1.25 to 5 inches in diameter are processed on one of their surfaces by coating them with a layer or layers of *n*-type and *p*-type active silicon material from which miniature electronic circuits are made. Leads of aluminum or gold, which are vapor-deposited integrally with the circuits, are externally bonded to kovar leads for joining to printed circuit boards. From 250 to 400 or more individual circuits are processed on each wafer, each circuit having a square or oblong configuration. The processed wafer is mounted on a scriber, and thin scribe lines are cut in the wafer between the circuits. The square or oblong sections are broken with tweezers or a knife edge from the back side of the wafer to obtain the individual microcircuits, which may contain transistors, diodes, resistors, or integrated circuits of these components.

Ordinarily, the scribing is performed with a diamond scriber. However, the diamond scriber sprays fine particles of the silicon substrate on each side of the kerf. These particles sometimes are difficult to remove by conventional methods. When they are contained in the fabricated circuit package, they may contaminate the circuit components by reacting with the package atmosphere to form electrically conducting particles. These particles, in turn, may cling between any two circuit conductors and cause destruction of that section by short-circuiting.

Consequently, laser scribing is a distinct and advantageous improvement over diamond scribing. For laser scribing, a fine-focus lens of long focal length is used to make a straight line or a series of fine holes in the wafer on the peripheries of the circuits. About 10 holes per millimeter length are made between the microcircuits. Each hole extends to a depth of about three-quarters of the thickness of the ceramic wafer. In either method, no contaminants or crystalline particles are thrown over the microcircuit surfaces; no cleaning of the surface is necessary. The laser-scribed wafer is broken into square chips (dies) in the same manner as the wafer scribed with a diamond point. A laser scriber is similar to a laser cutter, except that the scriber does not have to be provided with a gaseous jet.

8-4 THERMAL TREATMENT OF METALS AND SEMICONDUCTORS

The process involving the heating and cooling of materials to regulate and control their physical properties is known as thermal treatment or simply heat treatment. Thermal treatment serves principally to harden the metal (alloy) to increase its tensile strength, or to soften it so that it can be machined easily; after machining, the metal may be hardened if necessary. In the case of castings, the process relieves the internal stresses created as a result of unequal cooling rates at various areas of the casting. In this process, known as annealing, the crystalline structure of the metal becomes homogeneous.

Thermal Treatment of Metals

Although there are many methods of bringing about the desired degree of crystalline transformation in a metal, some metallic structures, due to their design configurations, cannot be properly heat-treated by conventional methods. Laser thermal treatment plays an important role here because the beam can be directed selectively to any single spot or area.

The treatment consists of heating the metal to its metallurgical transformation temperature, sustaining this temperature for a predetermined period of time, and then rapidly cooling the metal by bringing it in contact with a cold gas, liquid, or solid surface to retain its internal crystal structure at the transformation phase. The question then may arise, "How can a laser beam perform all these operations now performed commercially on a routine low-cost basis?" The answer to this is simple. Not all metals, shapes, configurations, and physical states permit heating and cooling under the commercially specified process conditions. Distortion is a principal element in commercial heat treatment, and it can be eliminated in laser thermal treatment. Also, with laser heat treatment, a section of a part can be hardened or softened without affecting the remaining sections, or the surface of a part can be hardened with no effect on the core of the part.

Unlike laser beam welding or drilling, in which a focused laser beam is used, in laser thermal treatment the temperature of the focus would be too high for some work, so a defocused laser beam may be used. Another reason for the use of a defocused laser beam is that the incident beam can cover a large surface area. Furthermore, the types of materials and parts that are to be heat-treated are not huge railroad beams or building structures, which can be more effectively and economically heat-treated by commercial methods. However, all types of miniature electronic parts—including certain turbine blade surfaces for aircraft, gear-tooth surfaces, and the like—that cannot be subjected to the conventional methods can be heat-treated for hardening or stress relief by laser thermal treatment.

There are several methods of using a laser beam for thermal treatment. One method uses a carbon dioxide laser beam operating in the continuous-wave or pulsed mode. The laser beam energy is conducted through a flexible fiberoptic conduit, such as a bundle of quartz fibers, and focused with a long-focal-length lens to a point in air; then the divergent beam is utilized in an area ranging from several square millimeters up to 100 cm^2, depending on the power level used. A workpiece area equal to the divergent beam area should be treated at a time, and is cooled by means of an inert gas, such as argon or helium, injected concentrically through the laser conduit at a rapid rate to the area of treatment. The laser beam energy varies from 500 W to 10 kW, depending on the size of the surface area treated. In some cases, the focused laser beam is oscillated back and forth to cover the area to be treated, at a rate of 2 cm/sec. Figure 8-25 presents curves plotted for the material AISI 1045 with 6 to 15 kW of laser power used to case-harden the surface.

No standard equation displaying the energy profile of the thermal treatment can be given because of the numerous variables involved, such as the material's composition, surface shape, texture, and density; the ambient temperature; and the permissible angle of incidence of the laser beam. However, an approximate estimate of the beam energy in joules or watts can be made with the following simplified relation:

$$J = (At\rho H\ dT\ s)2.093 \tag{8-9}$$

where J is the beam energy (joules), A is the area of the metal surface to be treated (cm^2), t is the thickness of the surface treated (cm), ρ is the density of the material (g/cm^3), H is the specific heat of the material (cal/°C-g), dT is the change in temperature above the ambient (°C), s is the time of treatment (seconds), and the factor 2.093 is a conversion figure.

To convert the heat energy in joules into the energy flow rates in watts, which is the usual measurement unit in work of this type, we may write

$$\text{watts (W)} = \frac{\text{joules}}{\text{seconds}} = \frac{I^2Rs}{s} = \frac{IVs}{s} = IV \tag{8-10}$$

where I is the current (amperes), V is the potential (volts), R is the resistance (ohms), and s is the time (seconds). Since R in this case can be neglected, the value IV should be used for calculating the watts.

From Eq. (8-10) we may obtain an expression for the energy in watts:

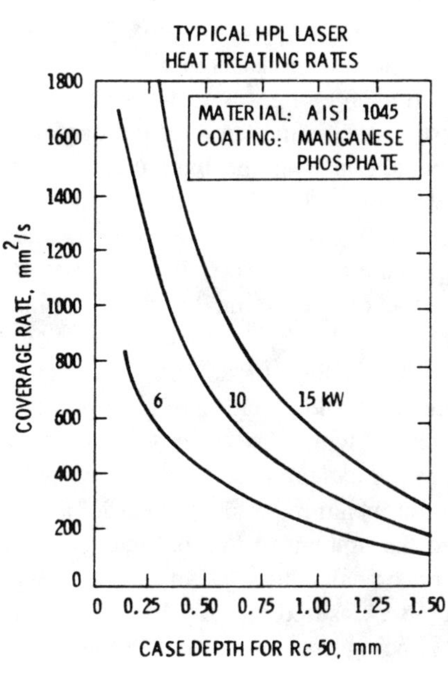

Figure 8-25 Typical laser heat-treating rates (Courtesy of AVCO Everett Research Laboratories, Inc.).

$$W = 2.093(At\rho H\, dT) \tag{8-11}$$

As pointed out above, Eqs. (8-9) and (8-11) are only empirical; actual tests should be made on a given material under constant working conditions and graphs should be prepared before a mass production treatment is undertaken. The test may follow the similar techniques given for welding schedule preparation, shown in Table 8-1.

Thermal Treatment of Semiconductors

Annealing semiconductor devices has become a routine procedure, using the laser beam from either ruby, Nd-YAG, or argon. The process is used to restore a semiconductor lattice structure to its natural crystalline form by removing stress profiles and damage due to dopant impurities. Either a pulsed or a continuous-wave laser beam can be employed for recrystallization. While it takes a pulse energy of 1 to 2 J with ruby at 0.05 μsec to anneal a given spot on a semiconductor device, it takes a 0.005-J pulse to perform the same restoration at 0.1 second with a Nd-YAG laser. For the same operation, an argon beam power of 10 W at a spot size of 0.03 mm and pulse duration of 10^3 μsec is employed. Since the ultraviolet spectrum from an excimer is strongly coupled with silicon, KrF or XeF has also been used to anneal semiconductor devices. Electron beams have also been employed for annealing, but tests have shown that laser annealing of semiconductors is superior to either furnace annealing or electron beam annealing.

8-5 ALLOYING WITH LASER BEAMS

Laser beams from high-power lasers can alloy metals in cases where ordinary methods cannot be used effectively and conveniently. One procedure consists of alloying a hard metal on the wear surfaces of a softer metal, such as alloying a tungsten steel as the cutting edge on a hardened steel tool. The tungsten steel is melted over the hardened steel edge, which becomes partially molten and alloys with the base metal. After cooling, the tool is ground down to the required cutting edge. Metal tubings that are subject to elevated temperatures or to the firing temperatures of furnaces are impregnated with carburizing material, which is then reduced to the metallic form by laser thermal energy. The reduced metal alloys with the surface of the tubing to create a high-temperature-resistant facing.

An ultimate application of laser alloying is in the field of dentistry. Nickel or stainless steel base metal is alloyed on its surface with gold. Melting of gold or silver fillings for dental restoration can be readily performed with a laser beam in the dental laboratory before mounting the prepared filling in the patient's tooth. Again, in order to obtain the full effectiveness of the laser heat, the beam must be applied defocused to the metal to cover the entire workpiece. Here the amount of thermal energy to be conducted to the workpiece does not have to be precalculated by any laser energy equation. The energy delivered to the workpiece is measured visually by the melting of the alloying material. Currently, a gas torch is being used in dental laboratories to perform this operation. One disadvantage of gas-torch melting is that alloying metal is lost as the torch blows on the molten metal or on the metallic flakes. Laser alloying or melting could be economically performed by proper manipulative beam techniques.

8-6 GARMENT-CUTTING LASER SYSTEMS

The rapidly advancing laser application field is a very important factor in everyday economy. New applications of laser radiation have been responsible for increasing the production of consumer products.

One of these applications involved the development of a garment-cutting laser system by Hughes Aircraft Company in 1971. As a unique application of the laser beam, this development has revolutionized the garment-cutting industry. The first laser garment cutter was installed by Genesco, Inc., one of the largest garment-cutting plants in the United States. The laser system, programmed by a computer, cuts garments one at a time at lightning speed. It has reduced the garment-cutting time to one-fifth of that necessary with conventional mechanical cutters. The benefits claimed for the laser cutter are low cost, fast delivery of the garment to the consumer, quick response to fashion changes, reduction of inventories, and a kerf width tolerance of a single thread of the garment. The system can also be used to cut shoe leather, plastics, and rubber sheets of all types and configurations.

The garment cutter developed by Hughes Aircraft Company is shown in Fig. 8-26. The system uses a carbon dioxide laser manufactured by Coherent Radiation Corporation. The movement of the laser beam projected on the garment material is controlled by a programmed computer, which directs the beam by means of programmed instructions for cutting any pattern in mixed batches. Merely by pressing a button for the selection, the system can be programmed to cut any pattern selected from a number of patterns in the computer memory storage section. The clothing material rolls off from a bolt and is moved by a large conveyor belt toward the laser head. Although the garments are cut in a single layer, the speed of the laser cutting procedure more than compensates for this disadvantage; 15 to 20 suits can be cut per hour.

The laser cutting system employs silicon mirrors in a pentaprism arrangement to direct the carbon dioxide laser beam to the clothing material. The laser beam gyrates as it follows the outline of the computer-selected pattern. The laser beam is focused on the clothing material by means of gold-plated aluminum mirrors. The cut is made by burning through the material, and the kerf is claimed to be narrow, sharp, and clean. The temperature of the cutting area (under the garment) is controlled by a cooling system.

Another versatile laser garment-cutting machine developed by Messer Griesheim, Frankfurt, Germany, employs a carbon dioxide gas plasma excited by 5000-V current and operates with an efficiency of 30%, as reported by the company. The system, shown in Fig. 8-27, is capable of cutting textiles, plastics, leather, rubber, ceramics, wood, and glass as well as mild and alloy steels at a cutting rate in excess of 250 inches per minute. The laser beam seals the edges of polyester materials as it cuts. The system operates with an output power of up to 250 W multimode when operating continuously,

Figure 8-26 The Hughes garment-cutting laser system (Courtesy of Hughes Aircraft Co.).

or 1 to 300 Hz in the pulsed mode. The beam diameter at the exit aperture is 13 mm and is reduced to 1 mm with a focusing lens. The power input to the unit is supplied by a three-phase, 220-V or 440-V source with a power of 5.5 kVA.

The carbon dioxide plasma consists of a mixture of 4.5% carbon dioxide, 82% helium, and 13.5% nitrogen. The mixture flows through a discharge cavity, shown in the schematic diagram of Fig. 8-28, at a rate of 10.6 cubic feet per hour and at a pressure of 13 to 40 millibars. Tank T supplies the mixture into the plasma tube P, from which the mixture flows into the atmosphere. As the mixture travels through the optical resonator (cavity), a voltage is impressed to produce the electric discharge and thereby the laser radiation, which oscillates between mirrors M_1 and M_4 for amplification. Mirror M_1 is 100% reflective and mirror M_4 is semitransparent with a dielectric reflective surface. The emergent laser beam is focused with a planoconvex gallium arsenide lens, L, with its convex surface against the emergent beam. For cutting textiles, plastic, nylon, and polyester materials, the laser beam is focused to a focal point of 50 μm or less. For cutting various patterns, the beam is guidance-controlled by means of a computer, punched-tape numerical control, or photoelectric tracing of a drawing layout.

Several laser firms have been considering the development of small laser systems suitable for use by small-business garment-cutting firms. At present, the high cost of the laser cutter would make it unaffordable by small garment-cutting plants and the cost would not be competitive with that of the currently used mechanical garment cutters.

Figure 8-27 Messer Griesheim all-purpose laser cutting system (Courtesy of MG Cutting Systems).

8-7 MODE LOCKING OF A LASER BEAM

Mode-locking is amplitude modulation of a laser beam in the optical cavity at the Fabry-Perot frequency, which is equal to the laser resonator mode spacing. At this state, each mode contains symmetrical sidebands that interact with contiguous modes. Because of the constant relationship between a mode and its individual sidebands, this effect produces phase and amplitude locking of the mode and results in a series of short pulses at the chosen frequency of modulation. In an intracavity system of this type, the mode-locking and wavelength selection prisms are usually combined to reduce the intracavity losses to a minimum. The index of refraction of the prism is modulated by an ultrasonic transducer bonded to the prism; the transducer operates in the radio-frequency range. Pulsewidths of up to several hundred microseconds have been obtained with this technique. By employing an ion laser for synchronous pumping of a dye laser, tunable pulses of high peak power are obtained. (See Chapter 14.)

8-8 ACOUSTO-OPTIC CAVITY DUMPING

The technique of cavity dumping is somewhat analogous to Q-switching, whereby the stored energy within the optical cavity is accessed. When the technique is used with mode-locked lasers the energy per pulse can be increased as much as 30 times, and when it is used with continuous-wave lasers the scheme can generate shaped pulses

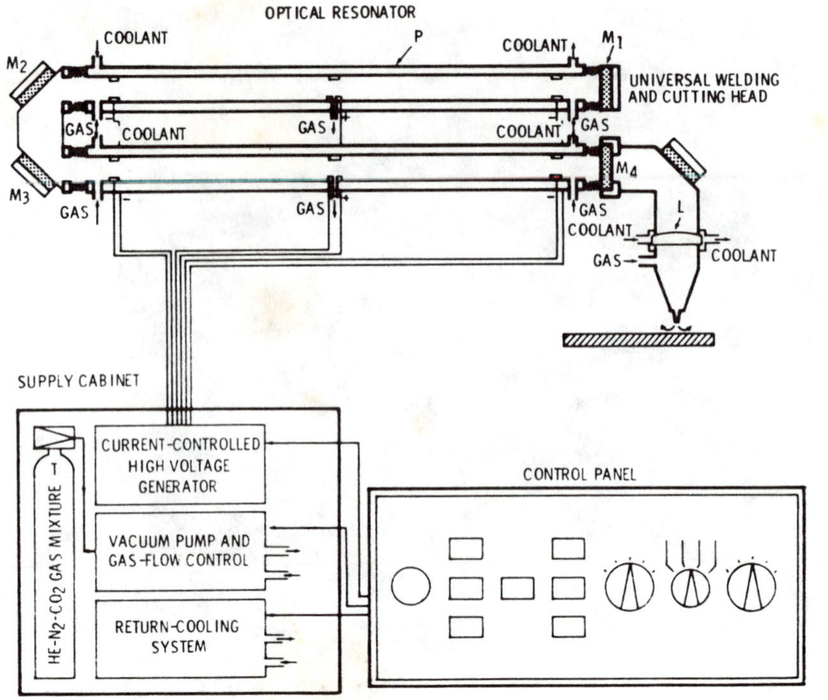

Figure 8-28 Diagram illustrating the principle of operation of the Messer Griesheim plasma system.

having coupling efficiencies up to 50%. The technique will be explained by means of the illustration shown in Fig. 8-29.

In a Brewster-angle-window plasma tube with external mirrors M_1 and M_2, where M_2 is the output mirror, a third mirror M_3 is positioned parallel to the laser beam axis. In this position, the system is normally operative as an ion laser. When mirror M_3 is tilted upward so that it cuts off the laser beam from mirror M_2, the intracavity energy will be deflected out of the cavity and mirror M_2 will receive no radiation. This action is achieved by means of acoustic waves from a radio-frequency generator, with pulse duration equal to the cavity round-trip time. In actuality, the radio-frequency generator is an acoustic-optic modulator driver at high frequency. When the acoustic wave is motivating the mirror, the laser light is deflected; when the light is continuous-wave,

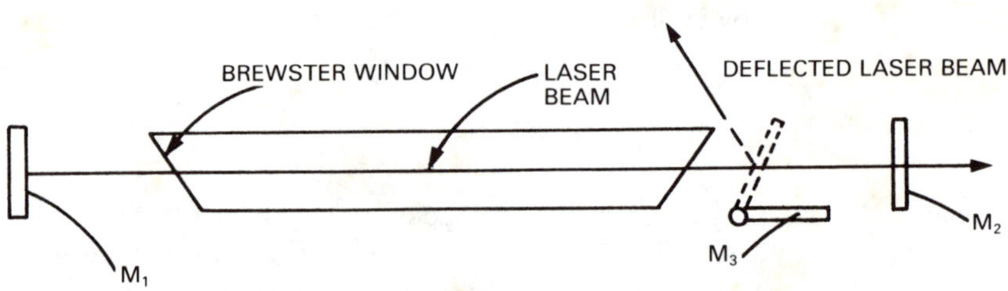

Figure 8-29 Diagram illustrating the cavity dumping scheme.

part of the beam is diffracted by the modulator, resulting in a partial frequency shift, and the remainder of the beam passes through the modulator unaffected. As the beam passes a second time through the modulator, both the diffracted and undiffracted beams are directed to the rear mirror, and a portion of the undiffracted beam becomes diffracted. The two diffracted laser beams optically interfere and produce an amplitude-modulated output that contains a beat frequency of the same magnitude as the acoustic frequency.

When this technique is used with a mode-locked ion or liquid-dye laser, the pulsewidth or shape of the pulse is unaffected because the entire radiation is within an ultrashort optical pulse. In this state, a cavity dumper can generate trains of ultrashort optical pulses at variable repetition rates. Thus, the mode-locked, cavity-dumped pulses can provide higher power than a mode-locked system does alone. In addition to amplitude modulation, the system tends to couple a larger portion of the plasma light for projection out of the cavity.

Assuming that the mode-locked pulse is within the Gaussian envelope, the resident intensity may be approximated by the relation

$$I_m = I_c \exp\left\{-4 \ln\left[2\left(\frac{\pi\omega_e}{\Delta\omega}\right)^2\right]\right\} \tag{8-12}$$

where I_m is the mode number intensity, π is the mode number, ω_e is the width of the center lines, and $\Delta\omega$ is the bandwidth of modes larger than the central mode amplitude by one-half magnitude, the central modes being greater than those at the leading and terminating edges when the laser emission is much higher than the threshold level.

QUESTIONS AND PROBLEMS

8-1 What is materials processing with laser beams? What are the various requirements in materials processing? What types of laser equipment can be used for materials processing? Should laser beams be visible for processing materials? Name the general types of materials processing performed with laser beams.

8-2 What does a laser welding machine consist of? Do all laser welders require cooling? Can ionic lasers weld? Which is the best welding laser generator? Can welding be continuous-wave, pulsed, or both? Describe a typical laser welding system, giving each of its parts.

8-3 For what type of welding is a long-focal-length lens used? For what type of welding is a short-focal-length lens used? Should all laser welders be provided with a binocular microscope? Do laser welders have a television screen? Which is more accurate for aligning the workpiece with respect to the laser beam—a microscope or a television screen? Which is safer to use from a radiation safety standpoint? If the laser beam welds a workpiece without the laser head contacting the workpiece, what assurance has one that the faying surfaces are in intimate contact?

8-4 At what average power does the laser welder manufactured by GTE Sylvania operate? What laser generator does it use? What pulse energies does it generate? At what pulsewidth is it used? What is its pulse rate for the highest energy? In what modes can it be operated? Can the system weld microcircuit leads? At what spot diameter? How is the laser beam focused in this system?

8-5 How does Control Laser Corporation's laser welder operate? What type of laser does it use? At what laser power? What is its beam divergence? Up to how many laser heads can this welder have when cascaded?

Can each unit be operated separately in the cascaded arrangement? Why is the laser beam upcollimated in this system? Up to what power output is generated by this unit?

8-6 How are the welding parameters determined? If a wavelength in the infrared range is used in welding, how is the beam focused? Explain. How does the distance between the lens and the workpiece influence the weld quality? Explain. How are electrical parameters chosen for a given workpiece to be welded? If the welding parameters for a material are not given, how does one establish these parameters?

8-7 How does a self-focusing laser welder focus the beam on the workpiece? Does such a welder eliminate the use of safety goggles? Who manufactures such a welder? Does such a welder ensure positive welds over other noncontact-type welding machines? What thicknesses of metals can be welded with this machine? What are the significant advantages of such a machine over conventional laser welders? Can semiconductor devices be welded with this laser system? What defects are eliminated when semiconductor leads are welded with this machine? Discuss.

8-8 How many laser pulses will be required for proper welding of microcircuit leads? What would a typical welding schedule be for this type of welding? How many grams of force would a pull test on this type of weld yield? What are the coaxial circular lines on a laser weld? What etchant would you use for showing the grain structure of a kovar weld?

8-9 With pulsed laser welding, why should the pulse duration be increased and power density decreased? What types of lasers are generally used for welding metals? What power density is used? Discuss how a welding schedule can be prepared. What type of laser beam is used for deep penetration into the metal? How large a power density can be expected from the laser source you gave in the last question? What is the advantage of pulsed seam welding? Of continuous-wave seam welding?

8-10 When two different types of metals are to be welded together, what characteristics are to be considered? State the factors by which the incident laser energy is dissipated during welding. How does welding power vary with density of a metal? How does heat of fusion vary with conductivity of the metal? How does the melting point of a metal vary with density?

8-11 Two sheets of kovar, each 3 mils thick, are to be welded together, employing a 2-mm focal spot. If the density of kovar is 8.35 g/cm^3, its specific heat is 0.105 cal/°C, and its melting point is 1450°C, how many joules will it take to weld the two pieces together? If the heat dissipation from all causes is 40%, how many joules would be required to weld the same metals (the first question neglects any heat dissipation).

8-12 How can one attenuate the laser energy from its source? How does the thermal conductivity of a metal affect the thermal diffusivity? State the equation for thermal conductivity. What metals have relatively low conductivity? How do their conductivities affect their diffusivities? Can glass be cut or drilled with a laser? How?

8-13 What is epitaxy in microcircuits? Of what material are microcircuits that are deposited in the epitaxy? What metals are used for microcircuit leads? What causes failures in microcircuits? How does a laser beam prevent lead-bond failure in microcircuits? What complex substance in the epitaxy becomes the root of failure in microcircuits? Why? What electrical parameter does this complex material develop in the semiconductor device? Explain.

8-14 Why should one resort to a dual laser beam welding technique? Does Nd-YAG require this technique? Does a frequency-doubled Nd-YAG laser require this technique? Why should the lasing element be given an initial warm-up time? What type of gas is used for welding a plastic sheet? Why is this gas chosen? What laser element is used in welding thin materials? What laser element is used in welding thick materials?

8-15 Which laser operates with the highest power and efficiency? Why can a CO_2 laser beam be used continuously in high-energy pulses? Which mode, pulsing or continuous-wave, is preferable for deep penetration? For what materials should one employ this type of operational mode? What one laser beam is highly absorbed by heavy metals? What type of lens material should be used for the type of laser beam you gave in the last question? What type of laser beam should not be employed for welding thin- and thick-film resistors?

8-16 For cutting or drilling materials of all types, what type of lens should one use? When does one employ a short-focal-length lens? Why? What lasers can be used for drilling or cutting in general? State the physical mechanism involved in drilling a metal. What additional characteristic of the metal should be considered (over welding) in drilling it? State the equation for drilling a metal.

8-17 In drilling a zinc sheet material, it is required to remove a volume of the material equal to 2 mm^3. Using the physical characteristic given in the example in the text, calculate the number of joules required to remove this metal by laser drilling.

8-18 What is resistor trimming? Why does one laser-trim a resistor? Give the advantages of laser trimming. What types of lasers are chosen for resistor trimming? Why? What does a laser trimming system consist of? How is the resistor adjusted to exact tolerance by using the trimmer? Which laser beam is more readily absorbed?

8-19 What is laser scribing of semiconductor wafers? What are its advantages over diamond scribing? What type of lens is used in this operation? How are metals and semiconductors thermally treated with a laser beam? Describe. What type of laser is used in the thermal treatment of semiconductor devices? State Eq. (8-8).

8-20 How are semiconductors annealed with a laser beam? Describe. What types of laser are suitable for this work? How is alloying of metals with a laser beam performed? What applications does the alloying technique have? How is the laser energy delivered to the workpiece measured? How can the laser beam be used most effectively?

8-21 Describe the technique of cutting garments with a Hughes garment cutter. Give the component parts of the cutting system and explain the function of each. What other materials can this system cut? Differentiate the techniques used by the Hughes and Messer Griesheim garment cutters. What does the carbon dioxide plasma mixture contain? Give the individual percentages of the components.

8-22 What is mode locking of a laser beam? What does the Fabry-Perot frequency equal? How are phase and amplitude locking accomplished? What provision is made to reduce intracavity losses in such a system? What is the function of an ultrasonic wave in this technique? Up to how many microseconds can pulsewidths be in this technique?

8-23 What is cavity dumping as related to stored laser energy? What is the importance of cavity dumping? How much coupling efficiency can be obtained from such a system? Describe the technique of cavity dumping of a laser beam in a plasma tube. What is the function of an acoustic-optic modulator in this system? How is the undiffracted beam diffracted during the dumping process? With what types of laser system can mode-locking be performed? When a mode-locked laser system is cavity dumped, what becomes of the resultant energy of the radiation in the tube?

SELECTED SOURCES OF RELATED DATA

Affolter, P. K. 1984. Drilling and cutting with pulsed Nd-YAG Lasers. *Lasers and Applications* March, pp. 81–85.

American Society for Metals. 1981. *Sourcebook on Electron-Beam and Laser Welding*. Metals Park, Ohio.

Austin, D. H. 1979. Dual-wave annealing with lasers. *Appl. Phys. Lett.* 34(9):558.

Basov, N. G. 1982. *Materials and Apparatus in Quantum Radiophysics*. New York: Consultants Bureau.

Belforte, D. A. 1984. Materials processing laser systems. *Lasers and Applications* August, pp. 55–62.

Branst, L. 1985. Machining center allows 3-D part processing. *Lasers and Applications* September, pp. 48–50.

Control Laser Corp. 1984. Laser Metalworking System, technical brochure. Orlando, Fla.

Coherent, Inc. 1984. Coherent Laser Systems, technical publications. Palo Alto, Calif.

Darchuk, J. M., and L. R. Migliore. 1985. Guidelines for laser cutting. *Lasers and Applications* September, pp. 91–97.

Elza, D. 1985. Lasers take to the factory floor. *Photonics Spectra* March, pp. 75–78.

Forrest, G. 1986. New industrial applications of lasers. *Laser Focus* March, pp. 20–44.

Herrmann, J., and B. Wilhelmi. 1987. *Laser for Ultrashort Pulses*. New York: North Holland.

Hunsberger, R. G. 1982. *Integrated Optics: Theory and Technology*. New York: Springer-Verlag.

L&A Staff. 1983. Absorptive coatings could boost metalworking efficiency. *Lasers and Applications* January, pp. 34–36.

L&A Staff. 1984. Laser sputtering moves into ceramics. *Lasers and Applications* January, pp. 38–40.
Moretti, M. 1987. Lasers in manufacturing. *Laser Focus* February, pp. 26–30.
United Technologies. 1980. High-Power Welding, technical bulletin. Los Angeles.
Viswanadham, R. K., et al. 1983. *Science of Hard Materials*. New York: Plenum.
Witteman, W. J. 1987. *The Carbon Dioxide Laser*. Optical Sciences Series, Vol. 53. New York: Springer-Verlag.

CHAPTER

NINE

SYSTEM CHARACTERIZATION OF LASERS

9-1 INTRODUCTION

We shall now emphasize the system integration and characteristics of various types of lasers, together with their performance and design where applicable. The types of lasers that are of practical importance will be taken up here, covering their construction, operation, performance quality, and physical parameters. We have already learned that the electron transition between two energy levels that are widely separated results in the emission of photons of higher energy than those that are closely located in the same atomic lattice. We have further learned that emissions from molecular lasing materials are confined to the vibrational and rotational transitions.

Among the monolithic laser types, the solid-state lasers contain a crystalline host with lattice sites compatible with retaining dopant ions of similar magnitudes and amenable to the formation of a lasing source by substitution ions within a limited domain of the lattice structure. For this reason, we have found that the substitution of an ion for another in the crystalline lattice would not exceed more than a few percent of the total host material, whether such substituted ions are trivalent, divalent, or monovalent. There will exist several energy levels that will give rise to more than one radiant frequency; however, this will rarely be commercially practical for our purposes. Furthermore, the host crystal should be optically transparent to permit the radiant energies to exit from the host. The dopant in the host crystal must also be soluble in the crystal at both the melting temperature and the solid-state temperature. The valence of the anion dopant should also be equivalent to that of the host cation, and when they differ a charge compensation should be included.

The inception of vibronic lasers by Bell Laboratories and later the development of the crystal alexandrite by Allied Corporation showed that the emission of photons is intimately related to the emission of phonons (thermal quanta), which are vibrational energies dominant in the crystal lattice of the chromium-doped beryllium aluminate ($BeAl_2O_4$). The Allied material behaves both as a four-level vibronic laser and as a three-level ruby laser. In the first mode of behavior, the alexandrite is broadly tunable from 7960 up to 8260 Å as the temperature is increased respectively from room temperature up to 80°C. Since in alexandrite the chromium ions (Cr^+) have replaced some of the aluminum ions, the concentration of chromium ions would be dependent on the number of aluminum ions replaced; this amounts to a chromium ion concentration of 0.1%, equivalent to 3.5×10^{19} ions/cm^3, to a maximum content of 0.4%. The higher this concentration is, the higher the kinetics and efficiency of the dopant. The crystal can operate in both pulsed and continuous-wave modes.

For an ultimate tunable laser source, a dye laser and notably a ring dye laser capable of producing several watts at one or more megahertz is available. For generation of ultrashort optical pulses, synchronously pumped dye lasers operating at up to 100 MHz frequency and picosecond pulse durations can be cavity-dumped acousto-optically and tuned continuously over a wide wavelength range. Operating at low average and high peak power, this permits the frequency-doubling of crystalline lasers exempt from thermal effects and selection of the desired dye media tunable to varied wavelengths. The output from the system has high stability for long periods of time, and its reliability makes it suitable for time domain spectroscopy and phenomena relevant to chemical and physical studies. The critical parameters consist of matching the mode locker to the long-term stability of the dye laser.

There are other lasers, some of which we have introduced with preliminary discussions, such as excimer lasers derived from inert gases combined with halides in their excited states, helium-cadmium lasers, copper vapor lasers, a beryllium vapor laser (a toxic vapor), an electron beam–excited laser, and the like, all of which we shall study with special concern for their generation, characteristics, and ultimate uses.

9-2 GAS LASER SYSTEMS

The gas laser systems comprise the He-Ne lasers, inert-gas lasers, excimers, ionic lasers, and molecular lasers. All gas lasers, with the exception of a few, are pumped either electrically or by microwave radiation. The category of lasers that we shall study here consists of those that are excited electrically.

He-Ne Lasers

The He-Ne laser radiation has as many as 40 possible transitions between the *s* and *p* quantum states. Our study will concern the 6328-Å wavelength radiation, which is the most prominent and strongly radiative in the red spectrum. This radiation has low gain and its power is limited to several watts; this output cannot be increased by an increase in tube length, and its gain cannot be realized without the presence of helium buffer in the gas mixture. Although very high outputs can be achieved with the longer wavelengths, such as the 115,000 to 339,000 Å lines, these lines have small gain and do not have particular industrial importance. The metastable energy level of helium in a collision of the second kind with neon atoms principally populates the upper-energy neon levels. The lifetimes of these energy states result from entrapment of the radiation in the excitation levels. Since no transition can occur to nonexistent low quantum states of helium, the metastable helium levels have long lifetimes and thus are capable of transferring their energies to the neon energy levels, populating them to lasing status. Applied power can also increase the population, but to a limited extent, until saturation sets in.

The He-Ne laser emits very intense coherent light, which can be collimated, after spatially filtering it, to any beam diameter desired; it is then focused to any focal spot size by prudent lens selection. The beam can be polarized either randomly or linearly. Polarization of the beam is necessary to avoid beam output fluctuation, in addition to permitting control of the beam intensity by means of an external polarizer and providing for beam modulation. The linear polarization of the beam may vary from 100:1 to 1000:1, depending on the laser application.

In the excitation of the He-Ne laser in a discharge tube, the helium atoms initially absorb the excitation energy and transfer it to the neon atoms by collision, resulting in transition of the neon atoms to a higher-energy quantum level, from which they return to the ground state while emitting red radiation. If these photons encounter other atoms in excited states, the excited atoms emit additional photons with the same frequency as the photons encountered; this interaction is known as stimulated emission. The emitted photons oscillate between the feedback mirrors within or external to the resonant cavity and yield additional photons from still other excited atoms.

Some of the radiation passes out from the output mirror, because of its partial transmissivity, to the exterior of the tube, and the process repeats itself as long as the applied voltage is sustained on the plasma tube. The useful lifetime of a He-Ne laser ranges from 15,000 to 25,000 hours and may be longer with further advances in the design of the plasma tube. One of the recently developed He-Ne plasma tube designs is illustrated in Fig. 9-1.

An important consideration in selecting a He-Ne laser unit is the power output density and the signal-to-noise ratio required in the system. This often requires experimentation for the particular application. If atmospheric attenuation is not of significance, a small-power laser may be selected with little or no concern about the signal-to-noise ratio. Manufacturers with extensive experience in such considerations should be consulted to save considerable time and expense in determining the proper gas laser system.

A readily usable equation, offered by Hughes Aircraft Company Laser Products Division, for determining the diameter of a He-Ne laser beam at a distant target is as follows:

$$d = d_0\left(1 + 6.492 \times 10^{-9}\frac{D^2}{d_0^4}\right)^{1/2} \tag{9-1}$$

where d is the laser beam diameter in centimeters at the target, d_0 is the beam diameter in centimeters at the exit port of the optical resonant cavity, D is the distance in

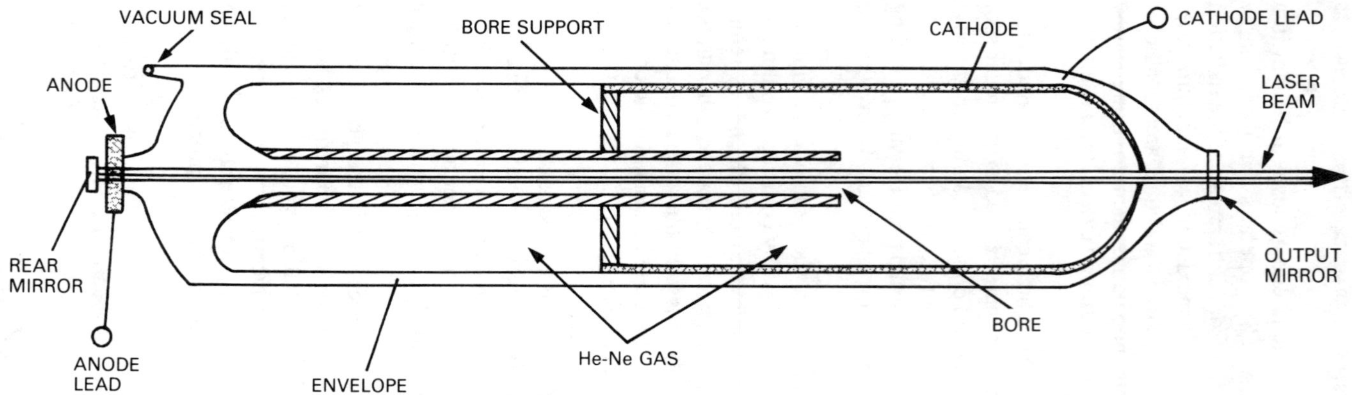

Figure 9-1 Plasma tube design of a recent He-Ne laser.

centimeters between the laser system and the target, and 6.492×10^{-9} is a constant characteristic of the helium-neon laser beam at 6328-Å wavelength.

Since the carbon dioxide laser diverges much more than the He-Ne laser in traveling through atmosphere, an equation that expresses the focal spot size at a distant target may be given as

$$d = d_0\left(1 + 1.82 \times 10^{-6}\frac{D^2}{d_0^4}\right)^{1/2} \tag{9-2}$$

If a laser beam is collimated at an expansion of, for instance, 10 times, then the beam diverges 10 times less and therefore the beam focus at the target for distances beyond 60 feet will be significantly smaller. Curves representing distance versus beam diameter for He-Ne lasers are shown in Fig. 9-2.

An ultimate design for a He-Ne laser head manufactured by Hughes Aircraft Company is shown in Fig. 9-3. This laser head contains a beam attenuator located on the top section near the beam output mirror. This attenuator variably reduces the beam intensity as well as cutting it off to prevent exposure when the laser head is temporarily not in use. It has a key-activated input switch and a remote-control connector to shut it off or to start it remotely. While the unit is in operation, an indicator light integral with the unit gives a warning that the laser beam is in operation, to prevent the operator or a visitor in the area from coming in contact with the beam inadvertently. A number of accessories can be attached to the laser head for any special requirements.

Specifications for a modern He-Ne plasma unit of typical construction are listed in Table 9-1.

Optical Cavity and Mirror Profile

Optical cavities are constructed (1) much larger than the wavelength and (2) so that the electromagnetic field is not completely confined by the feedback mirrors. The feedback mirrors are either flat or spherical; the latter type represent a confocal cavity profile in which the resonant frequencies f_r are given by the expression

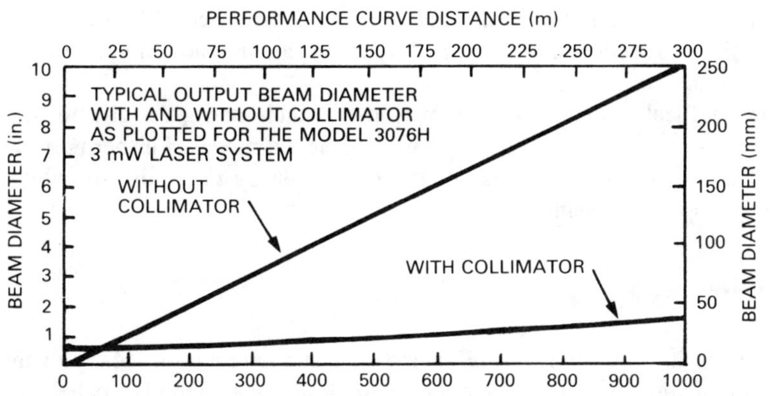

Figure 9-2 Performance curves for He-Ne laser beam focus at target with respect to distance.

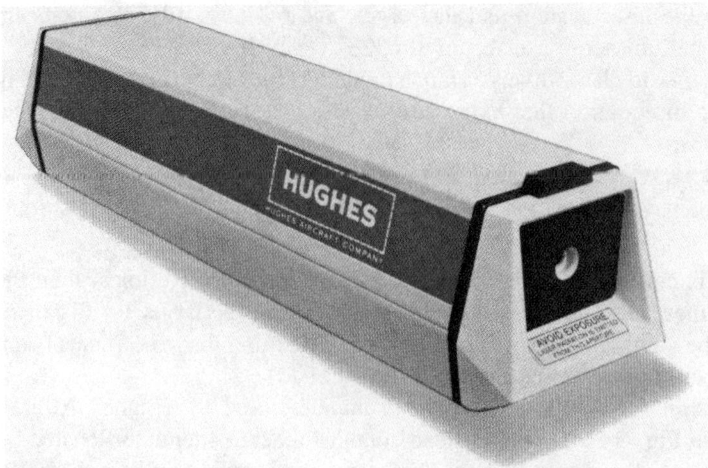

Figure 9-3 Huges Aircraft He-Ne laser head (3000 series).

$$f_r = \frac{c}{2L}\left[n + \frac{1}{2}(L_1 + 2p + 1)\right] \tag{9-3}$$

where f_r is the resonant frequency in the confocal cavity, c is the speed of light, L is the distance between the two mirrors, n represents the number of half-wavelengths between the two mirrors, L_1 stands for half the number of angular zeros, p is the number of radial zeros, and the entire expression promulgates on axial modes.

An approximation for optical cavities that are not associated with confocal beam structures is

$$f_u = \frac{c}{2L}\left[n + \frac{1}{\pi}(L_1 + 2p + 1)\cos^{-1}\sqrt{\left(1 - \frac{L}{R_1}\right)\left(1 - \frac{L}{R_2}\right)}\right] \tag{9-4}$$

where f_u is the frequency of a nonfocal optical cavity, c is the speed of light, L is the distance between the two feedback mirrors, and R_1 and R_2 are the radii of curvature of the two feedback, confocal mirrors.

Because the confocal optical cavity is marginally stable and gives rise to high losses, flat mirrors are usually preferred to avoid degenerate modes. Furthermore, as the electric field intensity rises in excess of the normal decay rates, the stimulated emission rate may improve considerably.

Ion Laser System

An ion laser operates in a multiline mode in its simplest configuration. Many of the commercially manufactured ion lasers are either argon lasers or krypton lasers because of their very high power yield at spectral ranges from the ultraviolet through the visible

Table 9-1 Specifications for a typical He-Ne laser unit

Description	Characteristics
Tube construction	Hardseal, coaxial
Lasing element	He-Ne gas
Wavelength	6328 angstroms
Power output, TEM_{00}	0.4 to 8 milliwatts
Beam diameter at $1/e^2$	0.50 to 1.0 millimeter
Beam divergence	1.0 to 3.0 milliradians
Longitudinal mode spacing ($c/2L$)	400 to 700 megahertz
Maximum warm-up time	About 15 minutes
Input voltage	115/230 VAC, 50/60 Hz
Power consumption	20 to 40 watts
Cooling	Usually air-cooled

to the near-infrared region. Argon ion lasers produce the highest visible power outputs with wavelengths in the blue and green spectra. As a typical high-power argon laser, Lexel Corporation's argon ion laser is chosen for discussion here because of the high-quality laser characteristics of this unit. Up to 4 W can be obtained from Lexel argon ion lasers in multiline operation at a spectral range between 5145 and 4579 Å, with beam diameters at $1/e^2$ points from 0.9 to 1.3 mm, operating in a TEM_{00} mode. The divergence of the beam is 0.8 mrad for the Model 75 series and 0.6 mrad for the Model 95 series. The cavity length for the Model 75 series is 64 cm and for the Model 95 series is 96 cm. Each of the plasma tubes is provided with an internal prism wavelength selector for single-line operation and an external prism for the multiline plasma tubes. A Lexel plasma tube with multiline operation is shown in Fig. 9-4, and a single-line plasma tube is shown in Fig. 9-6.

In Fig. 9-4, the rear and output mirrors are aligned on the plasma tube axis. The output from this tube is multiline with a Gaussian power distribution at TEM_{00} fundamental mode. The beam is capable of being focused to a small focal spot size. The Gaussian laser beam diameter is generally measured at the $1/e^2$ power point equivalent to 86.5% of the total laser power impinging on the output mirror. A graph of the Gaussian beam power distribution is shown in Fig. 9-5. The multiline laser beam

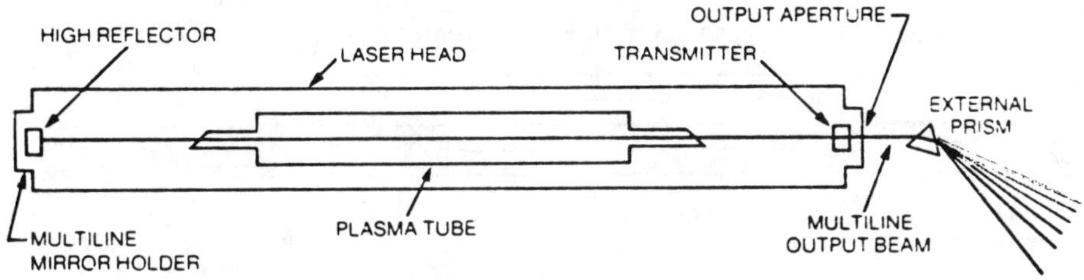

Figure 9-4 A multiline-operated Lexel plasma tube.

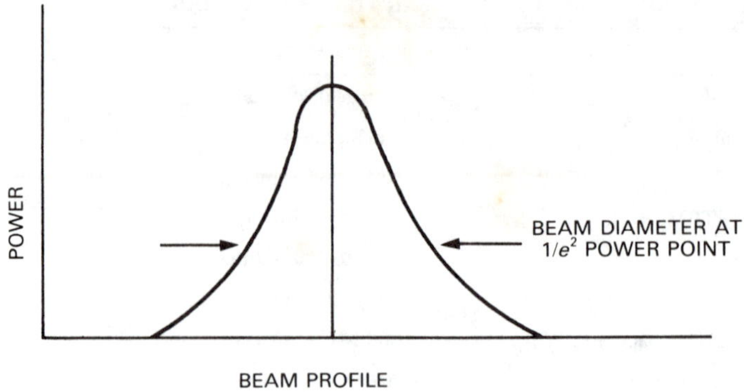

Figure 9-5 Gaussian beam power distribution in the fundamental mode.

can be separated into its component six wavelengths by use of an external prism, as shown in Fig. 9-4.

Krypton ion lasers also operate similarly to argon ion lasers; they can be made to emit wavelengths through the entire visible spectrum by use of optical elements. The most prominent krypton lines are in the red spectrum between 6471 and 6764 Å. The power rating of the krypton lasers is measured at 6471 Å, since the maximum output of the gas is at that wavelength.

When it is desired to select a single monochromatic line from the argon ion laser, a prism is interposed between the rear mirror and the Brewster window of the tube, as illustrated in Fig. 9-6. Only one wavelength can be obtained from the tube by rotating the prism to the particular wavelength. The single-wavelength line is then amplified in the tube prior to its emergence from the output mirror; this single wavelength has higher energy than a similar wavelength obtained from a multiline plasma tube.

Both the argon and krypton laser systems manufactured by Lexel Corporation operate at the fundamental mode of TEM_{00} with Gaussian distribution as shown in Fig. 9-5. Their nominal beam diameters vary from 0.9 to 1.3 mm, with beam divergence from 0.6 to 1.2 mrad. The polarization ratio of the beam is 100:1 in the vertical vector. Both of the plasma tubes operate from a 120-V AC, 50/60-Hz, and three-phase power source.

Although a single-line, single-frequency wavelength can be obtained from a multi-

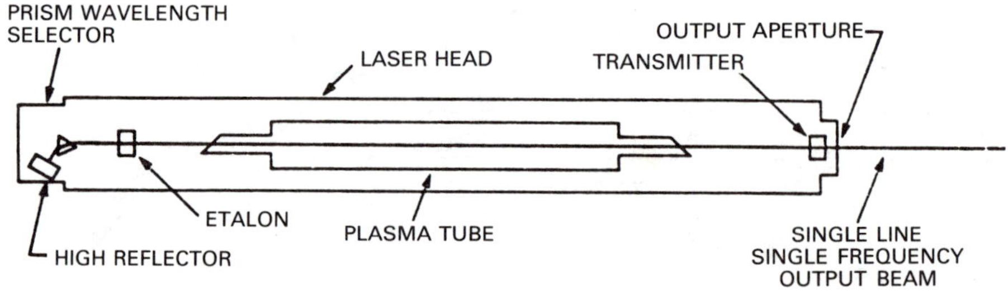

Figure 9-6 A single-line ion laser tube manufactured by Lexel Corporation.

line ion tube, the power of such a wavelength will be less than that of a single-line wavelength selected internally by means of the prism and the etalon interposed between the rear reflective mirror and the rear Brewster window, as shown in Fig. 9-6. The single line, separated by the prism acting as the wavelength tuner, will be amplified within the plasma tube before it is projected to the exterior of the tube. Either the argon or the krypton laser can be manufactured as a single-line or a multiline plasma tube.

The small-diameter beam from the tube, measured at the $1/e^2$ power point, projects from the output window with 86.5% of the plasma power, which corresponds to a shutter diameter that will allow the beam to be transmitted through the output aperture.

A typical performance specification for an argon plasma tube output is given in Table 9-2.

For a simplified electric circuit in which such a system may operate, the reader is referred to Fig. 1-20, with the exception that the capacitor network and the flashlamp should be removed and the ion tube should be substituted. The terminals of the plasma tube should be connected to the power supply leads furnishing voltage to the flashlamp. The reader should now be able to draw a circuit diagram for operating an ion tube, from knowledge gained from the text and from suggestions given respecting the alteration of the diagram shown in Fig. 1-20.

Mode-Locked Ion Laser

An acousto-optic mode-locked ion laser manufactured by Spectra-Physics is illustrated in a simplified form in Fig. 9-7. The output of this laser is an 82-MHz pulse train, each pulse duration being less than 200 picoseconds. The system can be coupled to an

Table 9-2 Specifications for an argon plasma tube

Description	Type and units
Lasing element	Argon
Wavelength	
Multiline	100 milliwatts to 4.0 watts
Single line	10 milliwatts to 1.4 watts
Beam diameter	About 1 mm
Beam divergence	About 0.7 milliradians
Beam polarization ratio	100:1, *E* vector vertical
Cavity length	0.6 to 1.0 meter
Longitudinal mode spacing	150 to 250 megahertz
Amplitude power stability	±0.2 to ±3.0%
Optical noise	0.5 to 1.5 RMS
Plasma tube bore material	Beryllium oxide
Input voltage	190 to 245 volts
Cooling type	Air, or water at 1.5 gallons per minute or more

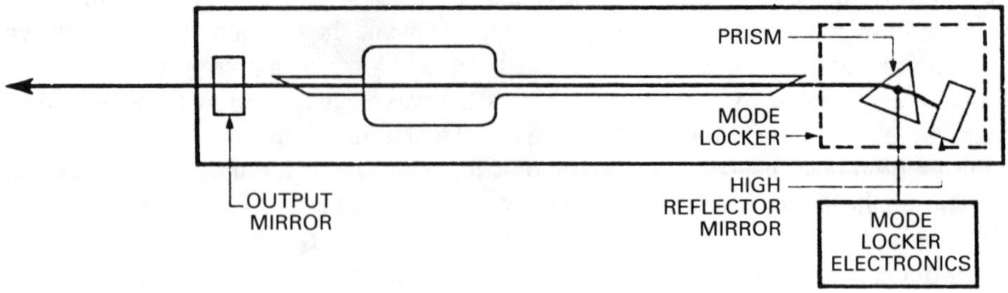

Figure 9-7 Spectra-Physics mode-locked ion-laser scheme.

acousto-optic cavity-dumping circuit to generate pulsed outputs at variable repetition rates; this output can be either discretely or continuously tuned over a broad-wavelength spectral region. These pulses are ideal for use in conjunction with signal-averaging techniques. Because of the high-peak pulse trains of the system, efficient generation of second-harmonic frequency doubling can be achieved without inducing extraneous thermal effects. Also, time domain spectroscopy as well as studies of radiative and nonradiative relaxation phenomena can be conducted because of the long-term stability and discrete or continuous tunability of these outputs. The acousto-optic modulator of the mode-locked ion laser is located within the optical cavity. The modulator is in the form of a prism, to one surface of which a piezoelectric transducer is attached. The transducer is driven at a high frequency to generate a standing acoustic wave within the modulator. This wave simulates periodic variations in the index of refraction of the prism along its axis and normal to its two polished surfaces.

As the beam enters the prism, it is partially diffracted and partially shifted in its frequency equivalent to the acoustic frequency. After these two beams pass through the modulator, they are reflected by a mirror with a dielectric coating into the modulator and then into the optical cavity. The frequency of the diffracted portion of the beam becomes upshifted, or downshifted to its original mode frequency. Thus, the modes that match the mode-locker frequency remain in the cavity and are mode-locked into a fixed frequency, as shown in Fig. 9-8.

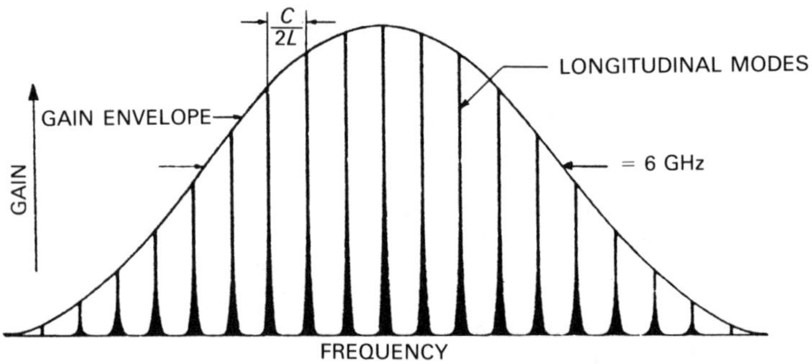

Figure 9-8 Amplitude, frequency, and phase relationship in a mode-locked ion-laser system (after Spectra-Physics).

The condition illustrated in Fig. 9-8 results in a train of pulses of 200-psec durations with periodicity precisely consistent with the intracavity mode spacing. The output pulses of such a mode-locked ion laser may be employed for optical pumping of a dye laser in a synchronous profile.

Energy to Achieve Oscillation

When the gain in an optical cavity becomes equal to loss, the oscillation just begins to occur for some particular mode. As the atomic line approaches the span of the cavity resonance, the radiation frequency will be almost equal to this resonance. To obtain this condition, we may express the function by an equation,

$$N_e = 4\pi \sqrt{\frac{\pi}{\ln 2}} \cdot \frac{f_c^2 \Delta_u t(1-\eta)}{c^2 L} e^r \qquad (9\text{-}5)$$

where N_e is the number of atoms in the excited state, f is the center frequency, Δ_u is the width of the Doppler line, c is the speed of light, t is the transition lifetime of the radiation, L is the distance between the centers of the feedback mirrors, η is the mirror reflectivity, e is the Napierian base, and r may be given by the expression $2\sqrt{\ln 2}(f-f_0)\Delta_u$.

From Eq. (9-5) it may be inferred that as the plasma intensity increases in the cavity, the stimulated emission rate rises to the extent of saturating the gain, and the Doppler shift width associated with the locked oscillation motivates the density of the excited atoms to increase. Thus, with continued optical pumping, the frequency as well as the radiant density increases until a steady-state condition is attained, whereupon the beam projects out of the output mirror.

9-3 TUNABLE DYE LASERS

Dye lasers have already been discussed in a previous chapter. However, in this section we shall take up the more specialized performance of dye lasers and the techniques followed in accomplishing these functions. The pump irradiation of the dye laser can be either pulsed or continuous. When the pumping source is pulsed or mode-locked, tunable high-peak power pulses are produced, and when the source is continuous the laser output is continuous and tunable. To produce a tunable output from the ultraviolet through the visible to the near-infrared spectrum, several readily changeable dye lasers are required, and a number of laser companies have already made such systems available. The pumping source can be one of the ion lasers or a frequency-multiplied (doubled or tripled) Nd-YAG laser. The durations of the pulses in the pulsed-mode operation vary from microseconds to picoseconds, depending on the power capability required. In the Q-switched operational mode, the pulses have high peak power and short durations. Also, the power output from a dye laser using a coaxial flashlamp is greater per pulse (up to 20 joules) at a low repetition rate than that from a dye laser employing linear flashlamp pumping.

When additional narrowing of the spectral bandwidth is needed, a diffraction grating of proper grazing angle is inserted at the output end of the resonant cavity and an expanded laser beam is projected on the grating, which is rotated at its optical axis for tuning. To further narrow the linewidth (bandwidth) of the emergent spectrum, an etalon, similar to that shown in Fig. 9-9, is introduced between the expanded beam and the grating. The combination of the grating and the etalon provides a still further improvement in the narrowing of the output bandwidth down to about 0.025 Å. If polarizers are inserted between the grating and the advancing laser wave front, the extinction ratio can be increased.

The temperature of the dye solution affects the wavelength and efficiency of emission. In an elevated-temperature solution the fluorescence becomes broadened and lasing is shifted toward the red. Overcooling the dye solution decreases its efficiency. If the dye cavity is long, light scattering increases, reducing the efficiency of emission. The bubbles formed during mechanical pumping of the solution can be reduced by the use of textile filters. Also, the spectral characteristics are somewhat different for different optical pumping schemes. Optically pumping the dye with a laser does not usually dissociate the dye as quickly as flashlamp pumping does; dissociation by radiant energy is a chemical process. The operating life of a dye laser is measured by its half-life—the time it takes for the energy per dye volume to decrease to one-half the strength of the original laser output. Emission is also volume dependent, and increasing the volume of the dye flowing per second will directly increase the efficiency of emission; the highest efficiency of emission achieved to date is in the neighborhood of 24%.

If the dye is synchronously pumped with a mode-locked ion laser, output pulses ranging down to 0.6-psec durations can be obtained. For the latter type of pulsed laser, the choice would be the Spectra-Physics Model 375 dye laser, which can be used in time domain experiments, Raman spectroscopy with fluorescence suppression, gain profiling, and the like. Using a synchronously mode-locked and cavity-dumped dye laser, power peaks at kilowatt levels have been achieved. Most dye lasers provided

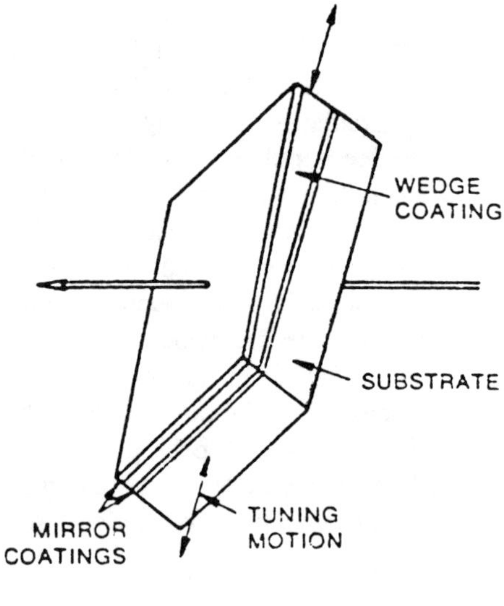

Figure 9-9 The Spectra-Physics coated tuning wedge.

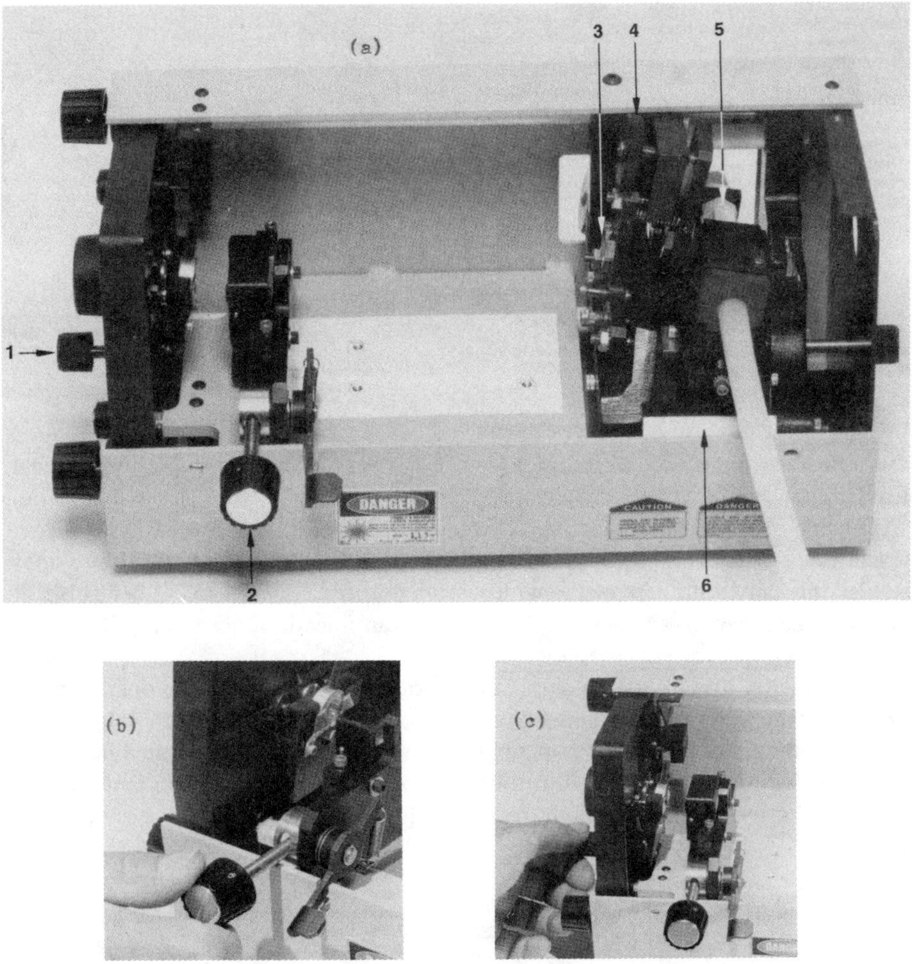

Figure 9-10 The Spectra-Physics Model 375 general-purpose dye laser.

with etalons as tuning elements exhibit sidebands; Model 375, which employs only a wedge interference filter for tuning, is exempt from such sidebands because of its unique system design. The wedge, shown in Fig. 9-9, is a highly selective Fabry-Perot tuning element and provides broadband, narrow-line operation from 4600 to 9500 Å with the standard wedge and from 4200 to 5000 Å with the blue wedge. This wedge further increases the tuning resolution of the laser; it is also provided with adjustable start and stop wavelength limits. The Model 375 dye laser is shown in Fig. 9-10.

The versatile, high-intensity Model 375 dye laser is adaptable for use with commercially available dyes and contains optimized optics and tuning elements to provide a broad selection of tuning range. In the figure, (*a*) is an internal view of the system showing, among other things, the fine-tuning adjustment in (*b*) and the tunable filter in (*c*). Also shown are (3) a snap-in reflector to accommodate different wavelength ranges,

Figure 9-11 Phase-R 7–10 J, 5-mW, tunable dye laser, Model DL-3100B.

(4) the adjustable pump-focus reflector, and (5) a nozzle that projects a thin optical quality laminar foil of any dye of desired wavelength range.

A tunable dye laser, manufactured by Phase-R Corporation, Model DL-3100B, is provided with a flat-flat reflector optical cavity with two SP-10 Schott prisms. It provides high pulse intensity with narrow linewidths from 0.1 to 1.0 Å at 4 to 7 J per pulse. It has a beam diameter of 22 mm and functions as an injection-locked amplifier. The system is shown in Fig. 9-11.

The Molecron dye lasers, Design DL-II series, on the other hand, operate most efficiently if they are optically pumped and excited to emission along the polarization axis that is common to both the pump and the dye laser beams, as illustrated in Fig. 9-12. A half-wave plate is employed to rotate the polarization axis (which is normal to the plane of the two beams) through the dye cell to the vertical. The dye is optically

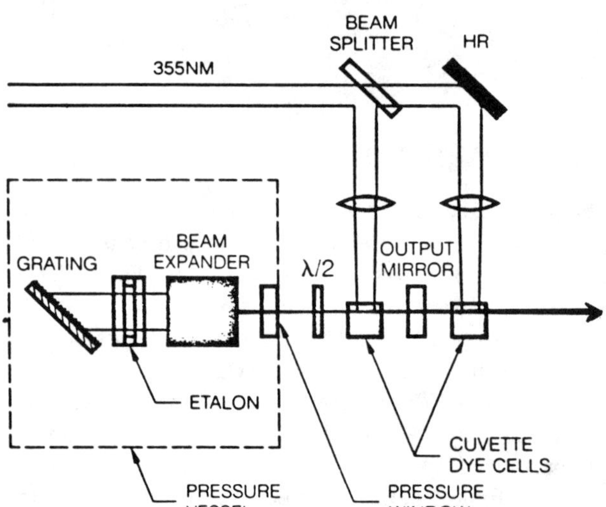

Figure 9-12 Dye optical pumping by means of third-harmonic wavelengths from Nd-YAG laser.

pumped by means of a Nd-YAG third harmonic, and the wave plate shown in the illustration is sandwiched between crystalline quartz and magnesium fluoride in the airspace, simulating an achromatic optical element.

The dye can be optically pumped by second, third, and fourth harmonics of a Nd-YAG laser. Although the second and third harmonics can effectively pump the visible wavelengths, as shown in Table 9-3, the fourth harmonic, the 2660-Å wavelength, is not usually employed because it is not very stable and the dye has a short lifetime. Furthermore, the last wavelength can be effectively produced by a harmonic-fundamental mixing technique. The cuvette cells provided in the circuit illustrated in Fig. 9-12 achieve adequate stability for repetition rates up to 10 pps, and when a good polarizer is used as an analyzer, the Molecron system produces dye polarization ratios between 50:1 and 2000:1, which aids in frustrating the background fluorescence.

In Fig. 9-12 the two cuvette cells contain the dye solution at a concentration of about 10^{-4} molar; the solution is transversely pumped by the beam of a second-harmonic Nd-YAG laser. The resultant dye laser beam passes through a half-wave plate into the pressure chamber containing a beam expander and a double-wall etalon, and the expanded beam impinges on a diffraction grating. The selection of a wavelength is made by the grating, which projects the beam back into the cuvettes, where the beam is amplified and projected to the exterior of the system. The externally projected beam may be mixed with a harmonic wavelength from the Nd-YAG laser after the diameters of the two beams have been matched at the beam expander. In another dye laser system, the Molecron design permits optical pumping of the dye solution axially, which is claimed to yield the best combination of efficiency, beam quality, and low background fluorescence. The specifications of a randomly chosen Molecron dye laser Model DL-18 are listed in Table 9-4.

WARNING. Since most of the dyes used in laser work are toxic, the solution or solvents that contain the dye material should not be spilled on the hand and the vapor phase that may be evolved during handling should not be inhaled or be absorbed in the mucous membranes, to avoid serious hazard to the body. See "Dyes can be deadly," *Laser Focus,* May 1975.

9-4 DIODE LASER SYSTEMS

A diode laser is a semiconductor device having p-type and n-type active layers disposed face to face about 1 μm apart to form a p-n junction, which acts as an electronic discharge cavity, as stated elsewhere in this book. When a negative bias is applied to

Table 9-3 Dye tuning ranges pumped by Nd-YAG harmonics

Nd-YAG pumped wavelengths	Dye tuning range
Second harmonic, 5320 Å	550–7600 Å
Third harmonic, 3550 Å	3800–5500 Å
Fourth harmonic, 2660 Å	3250–3800 Å

Table 9-4 Specifications of Molecron dye laser DL-18

Description	Parameters
Fundamental tuning range	
5320 Å pump	5500–7600 Å
3550 Å pump	3600–5500 Å
Frequency doubling/mixing range	2170–4200 Å
Wavelength accuracy—absolute	0.2 nanometer
Linewidth (5000 Å)	0.01 nanometer
Wavelength reproducibility	0.02 nanometer
Wavelength stability	0.01 nanometer/°C
Energy conversion efficiency	
5600 Å (5320 Å pump)	30%
5000 Å (3550 Å pump)	10/20%
Pulse duration	
5600 Å	4/12 nanoseconds
5000 Å	4/12 nanoseconds
Beam dimension (nominal)	8 millimeters
Beam divergence	0.3 milliradian
Polarization ratio	100:1

the *n*-type active side, a current flows from the *n*-type to the *p*-type active laser. The *n*-side consists of electrons and the *p*-side consists of holes into which the electrons are injected, causing recombination of the two entities with resulting emission of a radiant photon. If the energy involved in the process is equal to or in excess of the diode threshold energy, coherent radiation is emitted from the *p-n* junction. If the energy impressed on the diode is below the threshold energy, incoherent light is emitted and the diode is known as a light-emitting diode (LED).

The *p*-type active material is doped, in the case of a gallium arsenide laser diode, with zinc, and the *n*-type active material is doped with tellurium, selenium, silicon, or other dopant. There are numerous types of diode lasers that have other intrinsic structures and are manufactured by different manufacturers. A few of the most common diode lasers have structural constituents such as GaAs, GaAlAs, GaInAlAs, GaAlP, and many others. The diode laser generally operates in the infrared spectrum, emitting between 7800 and 9050 Å, the continuous-wave emission being from the shorter-wavelength region and the pulsed-mode emission from the longer-wavelength region of the range.

In a semiconductor diode laser, current conduction occurs only in the presence of free electrons. Such a diode can be considered as a four-level laser device having two sets of energy levels. One of these sets contains discrete levels of orbital electrons broadened into energy levels known as conduction bands. The other set is similarly associated with atomic shells with positive holes dispersed in the acceptor sites known as the valence bands, as illustrated in Fig. 9-13. Between these two bands is a Fermi level, also known as a forbidden band. The narrower the Fermi band, the more conductive is the device. Metals have the smallest or no Fermi band, and dielectric materials with no conduction electrons have very wide forbidden gaps.

The semiconductor devices have a smaller Fermi gap than the dielectric materials. Accordingly, by doping the conduction band and the valence band with impurities, one can produce a device in which the electrical conductivity can be controlled by the concentration of the dopants. In a device of this type, the *n*-doped levels are negatively charged and the *p*-doped sites, having fewer electrons, are charged positively in the lattice structures. The Fermi band denotes the reference level and contains no electrons. The transition region between the *p*- and *n*-type active sites is known as the *p*-*n* junction. The density or concentration of free electrons in the conduction band is a function of energy, and the probability that a given energy level will be occupied may be expressed by the relation

$$F = \frac{1}{1 + \exp[(E - E^*)/KT]} \tag{9-6}$$

where F is the Fermi probability function, E is the electron energy in a given level, E^* is the energy of the Fermi level, K is Boltzmann's constant (1.38×10^{-16} erg/°K), and T is the absolute temperature. When E is equal to E^*, the numerical value of E^* is 50%.

Diode lasers are excited to radiation by injecting a negatively biased current across the *p*-*n* junction. However, a new type of diode has been developed that is excited to emission by optical pumping. The diode is fabricated with multiple stripes associated with the active layer of GaAlAs, and when coupled optically it produces a CW power of 1.2 W. As the temperature is elevated to 100°C or above, the diode emits 200 mW in the CW mode. When operated in the pulsing mode, the device emits at 8320 Å. With a higher temperature, the diode emits 50 mW at 8360 Å in the CW mode, indicating that it is temperature-sensitive. Provision of a well-designed heat sink enables

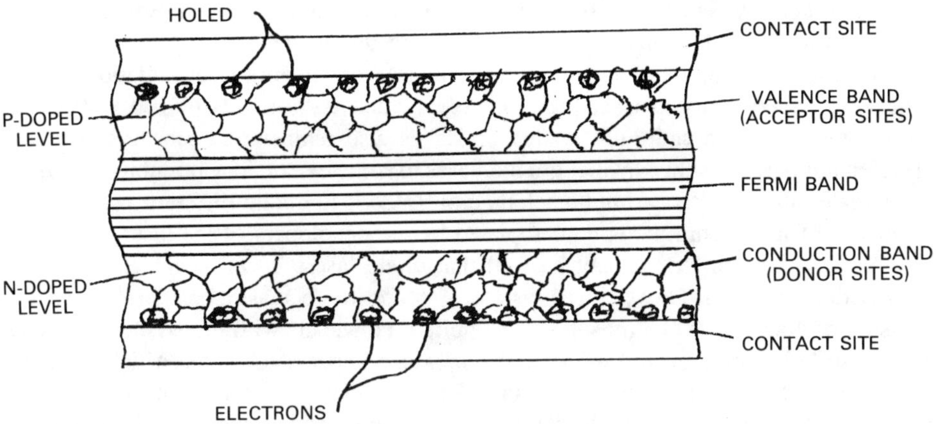

Figure 9-13 Conduction and valence bands of a semiconductor diode.

the diode to emit the high power stated above, according to the developers at Xerox Research Center, Palo Alto, California.

In describing the population of electrons and holes in the respective sites, we may express the electrons per volume per energy interval in the conduction band as

$$E_c = \frac{1}{2\pi^2}\left(\frac{2M_i}{h^2}\right)^{3/2}\left(E - E_c\right)^{1/2}\frac{1}{\exp[(E - F_n)/KT] + 1} \tag{9-7}$$

and for the number of holes contained in the valence band we have

$$E_v = \frac{1}{2\pi^2}\left(\frac{2M_2}{h^2}\right)^{3/2}\left(E_v - E\right)^{1/2}\frac{1}{\exp[(F_p - E)/KT] + 1} \tag{9-8}$$

where E_c is the number of electrons per volume per energy interval, M_1 is the effective mass of charges in the conduction band, M_2 is the effective mass of holes in the valence band, E is the energy level in the conduction band with E_c energy, F_n is the functional factor in the conduction band, and F_p is the same in the valence band, k is equal $2\pi/\lambda$, and T is the electron energy functional constant.

General Optronics is manufacturing InAsGaP lasers emitting at 130,000 Å with a power of 3 mW. These diodes are used in fiberoptic data transmission because of their efficiency as long-wavelength laser emitters. In contrast, Hitachi produces shorter-wavelength laser diodes emitting at 7800 Å for use in videodisks; they are reported to have a lifetime of 100,000 hours operating at room temperature.

The MIT researchers have developed a tunable laser platelet of cadmium sulfide that is optically pumped by a 4760-Å argon laser longitudinally to the crystal; this diode has achieved 9 mW of laser radiation with a 0.1-nanometer bandwidth. The crystal is tunable by means of a prism between the wavelengths of 4950 and 6010 Å at 95°K. It is thought by the developers that such a crystal emitter may eventually replace dye lasers because the crystal does not perform capriciously, as the dye does by fluctuating during its flow through the dye cell. The 5-μm-thick crystal functions as an external Fabry-Perot cavity because the beam quality is distinctly improved and further insertion of external optical elements is possible for tuning the crystal.

It is also reported that the cadmium sulfide crystal can emit laser radiation with as little as 25 mW of argon laser pump energy, which corresponds to 100 kW/cm^2 of argon radiation. The cadmium sulfide crystal is also temperature-sensitive, since at elevated temperatures the wavelength increases to some extent. The possibility of using other materials, such as cadmium selenide or ZnCdS, to obtain different wavelengths by means of this technique is contemplated by the developers of the laser.

As a generalization, it may be said that diode lasers have the lowest threshold current density. With GaAs, the threshold is found to decrease considerably when a pressure of 4000 to 5000 atmospheres is applied. However, with mixed elements such as GaAs-GaP, InSb, or InAs-GaAs, the pressure appears to increase the threshold current density, possibly due to inhomogeneities in the crystal entities. It is also interesting to note that magnetic fields affect the conduction band in InSb, increasing to some extent the stimulated emission from it. Furthermore, the direction of application of the magnetic field and its density level also affect whether the emission is spontaneous or stimulated. The internal quantum efficiency of the diode is higher than that of some of the more prominent laser elements. This efficiency is the ratio of the quantity of photons to the quantity of electrons crossing the junction per second, each electron being expected to produce one photon, although not all the photons leave the diode. The energy relationship of the electron-to-photon emission is given in Eq. (3-5).

9-5 EXCIMER LASER SYSTEMS

One of the latest developments in gaseous laser materials is the excimer laser, which lases with reasonably high laser output, generally in the ultraviolet spectrum. As described earlier, an excimer consists of a rare-gas atom and a halogen atom bound together in their excited state, but chemically they are unstable and dissociate when the exciting field is removed. Examples of excimer lasers are the rare-gas monohalides such as ArF, KrF, XeF, ArCl, KrCl, XeCl, and XeBr. Although rare-gas iodide has not as yet been reported, an iodine laser emitting at 1.3-μm wavelength has been produced with a 5145-Å-wavelength argon laser pump. The wavelengths of excimer lasers range from 3510 Å for XeF down to the long-wavelength x-ray region. The excimer radiation occurs in an electron discharge cavity, and helium or argon gas is used to initiate the excitation of the lasing elements. Excimer lasers exhibit high order of beam quality, pulse reproducibility, and tunability under laboratory conditions. Pulsewidths up to 40 nsec and pulse energies up to 100 J have been observed.

For a high-energy excimer to become a commercially valuable tool, it must operate in a constant-flow condition through the optical cavity. This condition requires that the excimer gases, subsequent to lasing, be expelled into the atmosphere, since at present there is no provision to recover them. This makes the operation of the excimer laser economically costly. Consequently, development effort is being directed toward trapping the flowing gas in an inert container for reuse or for recirculation; in the latter case, the optical cavity walls must be passivated against reaction with the halogen gas, such as fluorine, by using monel, Teflon, or the like. Another alternative would be to separate the inert gas from the halogen for reuse and to introduce them into the discharge cavity with just enough halogen to make the chemical combination with the inert gas. Then the gas flow system must be completely sealed off to eliminate the hazards of the evolved halogen gases. The halogen, fluorine, is commercially available in a 1:20 mixture with helium (F:He); this mixture can be used diluted so that the fluorine content is about 0.2% in the excimer mixture.

An excimer laser system developed by GTE Sylvania has achieved a record 40 W of ultraviolet radiation from KrF operating at a pulse repetition rate of 1 kHz. The intracavity gas transport provides a pressure of 15 psi (gauge), and the system attains an efficiency of 0.4% in performance. It is a sealed-off closed-cycle system, the krypton fluoride gas flowing rapidly between the transversely located electrodes. Pulse formation is accomplished by means of a hydrogen thyratron that fires only at the leading end of the pulse, thus achieving the full energy of the pulse at high repetition rates.

Lumonics, Inc., manufactures a number of excimer lasers, one of which, chosen at random, is the Model TE-430-2, a compact, high-pulse-rate excimer laser. Although this system is designed primarily to operate with KrF at 2490 Å, any of the other excimer gases may be used to perform the same function. For operation of the system, the laser head is evacuated to a few millitorrs by a vacuum pump, which is then disconnected. The laser head is backfilled in a partial pressure with a mixture of 1 to 5% rare gas, 0.1 to 0.5% halogen, and the balance helium. The gases are dispensed either from the three external cylinders or from a single cylinder with premixed excimer gas. The system then can be operated at pulse rates from a single shot up to 100 pps. The pulses are triggered from an internal pulse generator, and the half-life of the gas mixture, for example, an XeCl mixture, is about 5×10^6 shots at a wavelength of 3080 Å in the ultraviolet spectrum. The unit generates an average power of 2.5 W at a pulse repetition rate of 70 pps, with a pulsewidth of 4 to 7 nsec. Its applications include functions as a laser optical pump; in ultraviolet photochemistry, ultraviolet spectroscopy, and photochemistry; as an injection locking source for larger lasers; and in remote sensing. A cutout view of the excimer system is illustrated in Fig. 9-14.

Figure 9-14 shows (5) the advanced control panel from which all parameters of the system are adjusted, (2) the capacitor network for generating laser pulses, and (1) the inert laser vessel. The system receives its power from an integrally contained source (4) and is provided with a lasing gas through a solenoid valve (3). The thyratron (6) triggers the main power supply to produce the excimer laser beam, which is projected from the aperture (7) to the exterior of the unit.

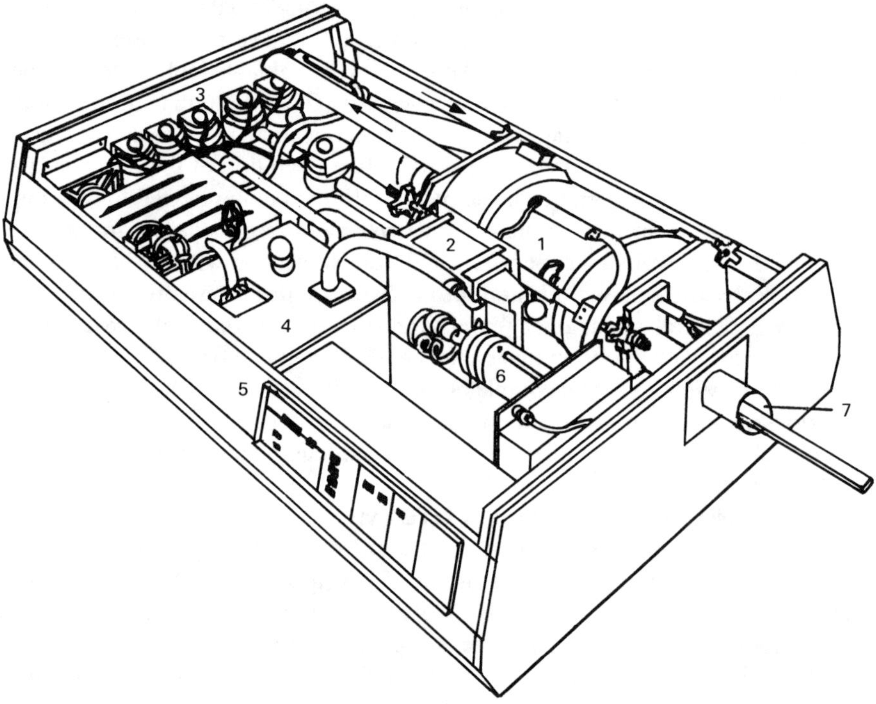

Figure 9-14 Lumonics Model TE-430-2 series excimer laser.

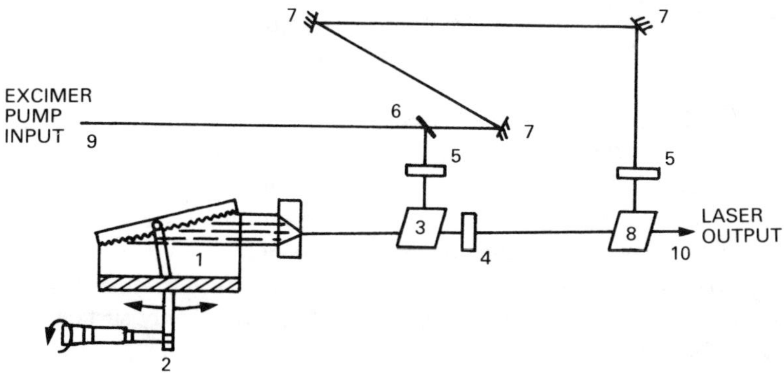

Figure 9-15 Schematic of the optical system of Lumonics Model EPD-330 series excimer-pumped pulsed dye laser.

A well-devised system for optically pumping a dye laser by means of an excimer optical pump is illustrated in Fig. 9-15, in which a Lumonics excimer optical pump (9) excites the oscillator dye cell (3) and the amplifier dye cell (8) amplifies the output beam as at (10). The oscillator cavity (1) employs a diffraction grating positioned at near-grazing incidence to the beam, in conjunction with four 15× prism expanders to yield a high conversion efficiency. A beam emitter (6) partially steers the laser beam to the dye cell (3) and partially transmits it through the optical delay line to the final amplifier dye cell (8) prior to emergence of the beam from the system. The expanded laser beam from the dye cell (3) is tuned by means of the tuning mechanism (1) and is projected to the output coupler (4). The cylindrical lenses (5) direct the input and delayed pumped beams, respectively, to the oscillator dye cell (3) and the amplifier dye cell (8).

Other applications of excimer lasers include isotope separation (isotopes are forms of an element with different atomic weights but with the same atomic number), enrichment of uranium by irradiating it, production of deuterium, photolithography, chemical separation of nuclear wastes, optical pumping of dye lasers as shown in Fig. 9-15, and purification of silane (SiH_4) used in the production of organic chemicals that are impossible to produce by other methods. The excimer laser is also being investigated for target implosion in the nuclear fusion effort.

9-6 SOLID-STATE LASER SYSTEMS

Solid-state lasers are those that produce radiation from ruby, neodymium-YAG, neodymium-glass, alexandrite, and other synthetic glass or plastic rods. These are ion-doped crystalline or glass rods and do not include the semiconductor diode lasers. All solid-state laser rods are optically pumped, as we have had occasion to refer to earlier. That is, they receive their excitation energy from the illumination of a high-intensity flashlamp or a tungsten arc lamp. Therefore, energy transition is achieved by optical coupling of the photons from the flashlamp with the dopant ions of the host materials, such as aluminum oxide (sapphire), calcium tungstate, glass, yttrium-aluminum-garnet, or $BaAl_2O_4$ in alexandrite. The dopant content of these hosts ranges from less than 1% to about 3%. A typical solid-state YAG laser head manufactured by Lasermetrics, Inc. is shown in Fig. 9-16.

The flashlamp for almost all solid-state lasers consists of a quartz tube containing xenon gas and having electrodes made of stainless steel, nickel, aluminum, tungsten, or any other suitable material. The flashlamp is either helical or linear in configuration, and its length is usually equal to or slightly greater than that of the laser rod in order to cover the rod completely with its illumination. A few turns of high-voltage triggering wire are sometimes looped around the flashlamp to start the electrical discharge in the tube. Once the discharge begins, the high voltage from the power supply sustains the potential across the tube. The flashlamp radiation may be continuous or pulsed. The flashlamp generally starts with a higher voltage and settles down to a sustained lower voltage within a fraction of a second. This occurs because the internal resistance R across the tube is directly proportional to the electrical discharge length L and inversely proportional to the cross-sectional area A of the discharge column. Thus, we may write

$$R \cong \frac{L}{A} \tag{9-9}$$

In a solid-state laser rod, the ionization occurs in the dopant rather than in the host material. In ruby the dopant, or the active element, consists of chromium ions; in neodymium-YAG the dopant is the neodymium ions; and in lanthanate glass the dopant is the lanthanate ions. The flashlamp photons possess high optical energy and are capable of transferring their energy to a dopant atom in an amount equal to or greater than the energy required for excitation of the atom, whether chromium or neodymium. Subsequent to the excitation, the dopant ions behave in the same manner as the gaseous ions in an argon or He-Ne discharge cavity. The excited atoms are raised to a quantum state higher than their original quantum state, and as they return to their ground or original state they emit photons of energy equal to the exciting energy of the radiation field.

When the energy transition from the flashlamp photons to the dopant ions is not sufficient to raise the atoms to an excited state, no radiation is emitted from the atoms; the energy transition may dissipate in the form of heat from the laser rod. This heat, together with the heat radiated from the flashlamp, elevates the temperature of the laser rod. The increase in temperature causes loss in efficiency of emission. Therefore, in such an event the laser rod has to be cooled by circulation of air or pure deionized water through the laser head to remove the detrimental heat from the rod.

Most of the heat is concentrated in the central axis of a circular laser rod, where the heat acts as a multiple lens system, decreasing the intensity of the radiation from the rod. For this reason, the author has tried cylindrical rods with central bores and found that the thermal stresses within the rod are alleviated and that the emission from the rod is almost unaffected for short periods of operation. Naturally, the resultant laser beam is tubular. However, for practical purposes, this configuration does not seem to have any effect on the laser energy incident on the workpiece. The author has also used two flat-sided quadrangular bars cemented together. The cement behaves as a reflector

Figure 9-16 A 9300 series Nd-YAG laser by Lasermetrics, Inc.

of heat directed from the flashlamp. Both of these methods will reduce the thermal effect on the laser rod, and both techniques will retain the normal efficiency of the rod for intermittent pulsing operations without external cooling. The design characteristics of a nominal-size Nd-YAG laser rod are listed in Table 9-5.

It should now be clear why a laser beam is said to be spatially coherent; that is, any two selected photonic wave fronts execute a harmonic, unidirectional motion in the resonant cavity and are of the same phase, frequency, amplitude, and direction. This is because the lasing photons are produced by equal-energy photons, and since any increase or decrease in the transition energy will correspondingly increase or decrease the frequency and hence the photon energy hf, the resultant radiation will be correspondingly of different wavelengths. Such a radiation wave lacks the capability of being in phase and ceases to be coherent. While perfect coherence has not yet been obtained, laser radiation is the closest to perfection. When waves of equal frequency and amplitude are not in phase, an interference phenomenon takes place in which the waves partially or wholly cancel each other and thus produce interference fringes or complete darkness, depending on the degree of interference.

Structures and Characteristics of Solid-State Lasers

As we have already studied, any solid-state laser rod consists of a host and a dopant, which is responsible for the emission of stimulated radiation. The ruby laser rod consists of aluminum oxide (Al_2O_3) as a host material containing about 5.0% chromium oxide (Cr_2O_3) as the dopant. A Nd-YAG laser rod consists of yttrium-aluminum-garnet as a host with about 3% neodymium ions. Other solid-state lasers are similarly structured with a host and a dopant. Since space does not permit us to take up all the solid-state lasers, we shall discuss only a few representative laser rod materials.

Table 9-5 Specifications for a typical Nd-YAG laser

Description	Nature or parameters
Material	Nd-YAG laser rod
Laser beam wavelength	10,640 angstroms
Nominal power output	60 watts up, TEM_{00}
Laser beam diameter	2 to 10 millimeters
Laser beam divergence	0.2 to 3.0 milliradians
Energy stability, pulse to pulse	±2 to ±8%, long to short wavelengths
Stability at peak power	±5%
Optical pump	Xenon, nitrogen, krypton arc, etc.
Flashlamp trigger (nominal)	2 to 15 volts, 50 ohms, rise time variable from 3 to several hundred nanoseconds
Power supply, nominal	208 VAC, three phase, 50/60 Hz, 5 kW
Laser head dimension	50 cm wide, 50 cm high, and 60 cm deep
Cooling, when used	2 gallons per minute, clear water at 15° to 35°C

In order that a host material can contain a dopant for lasing, the host and the dopant ions must have inherently the same size lattice sites. The difference between the host lattice size and the dopant ion size should not exceed about 15%. In the case of the ruby laser, the density of chromium ions in aluminum oxide is approximately 1.6×10^{19} ions per cubic centimeter; this number is equivalent to the number of electrons in a conductor of unit current. A stimulated transition occurs when negative frequency absorption by the ionic dopant in the host takes place. This condition can be attained when the frequencies of the stimulating photonic energies are in coherence with those of the absorbed energies of the stimulated chromium atoms. In such a transition, both the phase and the amplitude profiles are in coincidence, respectively, with those of the energy-transmitting photons. Stimulated emission occurs at a rate in direct proportion to the intensity of the exciting radiation and in consequence of the excess of stimulating illumination over the absorbed radiant energy. Thus the resonant cavity acts as a feedback amplifier, producing a net closed-loop gain by the coherent oscillation between the two end mirrors of the lasing element. With high-reflectivity mirrors, the threshold of emission is low, with the result that a high density in the radiation can be attained. The efficiency of laser emission is dependent on the characteristics of the flashlamp containing the gaseous radiant element. For example, most flashlamps contain xenon at a pressure of about 150 mm of mercury. If an exciting flashlamp energy of about 800 joules is used to stimulate a solid-state rod, the resultant laser energy will amount to from 0.1 to slightly over 1%.

Energy-Level Characteristics

We have already stated that a ruby laser has three energy levels (Fig. 3-8) and a Nd-YAG laser has four energy levels (Fig. 3-9) in which stimulation occurs. Laser action can occur in these levels only when the lasing element undergoes negative absorption from the illuminating source. The transition rate of energy must also be fast to the terminal level at which the stimulated radiation will settle. In the case of four-level excitation, such as in Nd-YAG laser stimulation, the transitions terminate at an intermediate level, while in ruby the transition occurs between the upper excitation level and the ground-state level. Also, the transition from the upper terminal level to the ground state should be at such a rapid rate as to avoid the accumulation of thermally ejected electrons in the ground state, to aid the stimulated quantum energy in its transition to the ground-state level.

In defining the radiative processes, we arrive at two functional forms where we encounter spontaneous radiation and stimulated radiation in the interaction of the cavity waves with the atoms. A third relationship is implied as the absorption of photonic energy to attain the two radiative states. Referring to Fig. 3-8, it is seen that the excitation of atoms from the ground quantum state to quantum state 3 required absorption of photonic energy at a rate in proportion to the number of atoms that absorbed this energy, designated as dN_3/dt, and the constant to indicate this absorption would be $B_{1,3}$ to energize N_1 atoms from the ground state. To relate these functions differentially, we may write

$$\frac{dN_3}{dt} = B_{1,3} N_1 \rho_f \quad (9\text{-}10)$$

and the spontaneous radiative emission may be given as

$$\frac{dN_3}{dt} = -A_{3,1} N_3 \quad (9\text{-}11)$$

or the stimulated radiative emission may be given as

$$\frac{dN_3}{dt} = -B_{3,1} \rho_f \quad (9\text{-}12)$$

where ρ_f is the energy involved per unit of atomic population volume at the radiative frequency f. The quantities A and B are known as the Einstein coefficients of radiative functions.

The threshold energy of a lasing element excited by the pulsing mode depends on the pulse duration of the illumination from the flashlamp; the pulse duration is further dependent on the efficiency of the electronic energy converted into radiation for optical pumping. The Nd^{+3} ion is one that can be stimulated to emission of a number of wavelengths in different host materials with suitable lattice sites. Accordingly, the neodymium ion can give rise to wavelengths characteristic of lattice sites where it becomes lodged; this action is influenced by the presence of a charge compensation or other lines from other elements. Also, using different techniques of charge compensation gives rise to relatively different spectral lines.

9-7 METAL VAPOR LASERS

As progress continues in laser materials development, new types of lasers are emerging from the laboratory for commercial use. Among these materials are several recent vapor-phase rare-earth lasers. A potentially important laser projected for fusion by the Energy Research and Development Administration (ERDA) is the terbium vapor (Tb^{+3}) laser. This material in its triply ionized state lases at 5450 Å in the visible spectrum, with an optimum gain of 0.5% per centimeter in the active cavity. Rare-earth lasers are destined to yield high pulse rates with high efficiency, and the $TbCl_3$-$AlCl_3$ vapor complex is considered to be one of the best potential materials for fusion work by Argonne National Laboratory.

Copper vapor lasers also have high optical conversion efficiency for optically pumping tunable dye lasers. Lawrence Livermore National Laboratory has developed copper vapor lasers with efficiencies as high as 40%. While conventional copper vapor lasers exhibit high beam divergence using stable resonators, the Livermore investigators have used unstable resonators by oscillating the emission in the transverse mode and thus have reduced the beam divergence several orders of magnitude. The laser system has been used as an amplifier as well as for optical pumping of a dye, such as rhodamine 6G, in the pulsing mode and has achieved the highest optical conversion efficiency obtainable at present. The copper vapor laser produces pulses of 20- to 40-nsec durations at repetition rates of 1 to 15 kHz; its output wavelengths vary from 5110 to 5780 Å. For adequate operation, the copper vapor must be kept at a temperature of 1500°C, but when copper ions are dissociated from the compound, such as copper chloride or copper bromide, the system can be operated at temperatures from 400° to 600°C, making its operation compatible with conventional tubes and sealing schemes. Either type, though, can achieve efficiencies as high as 2% with special techniques. Because of its blue-green color and high penetration into deep water, this laser has become very promising for hydrographic instrumentation purposes. Measurements of depth are made from low-flying aircraft, and the determination of depth is based on the difference in beam transport times between the bottom and the surface of the water, as detailed in Chapter 5.

Xerox Corporation in Pasadena, California, manufactures 15-mW helium-cadmium (He-Cd) lasers for use with the company's high-speed printers. It is reported that this system uses less power per emission rate than He-Ne lasers generally used for this purpose. Spectra-Physics also manufactures He-Cd lasers for the printing industry. Liconix produces 10-mW He-Cd lasers emitting at 3250 Å and 40-mW lasers operating at 4420 Å, and Omnichrome manufactures a unit that can produce as high as 50 mW at 4420 Å. The potential usage of He-Cd lasers will be in the reprographic and medical fields. For photographic work and printed circuit board inspection, the He-Cd laser appears to be excellent, as already demonstrated in industry.

Iodine vapor lasers with moderately high powers have been produced in West Germany and in the Soviet Union. In the United States, efforts at Sandia National Laboratories have been confined to increasing the emission efficiency of the iodine laser, which is currently about 1% at 13,000 Å. Max Planck Institute in Germany is reported to have produced iodine laser systems that can develop more than 500 J of radiation per pulse of less than 500 psec. Although Sandia's iodine laser produces a little over 10 J/psec, the system is said to have better control over the pulse duration with precise focusing on the fusion pellet. To further improve the iodine laser's efficiency, doped flashlamps are being tried; in this case, the emission characteristic of the flashlamp must correspond to that of the iodine laser. Further effort is concentrated on the production of atomic iodine from the diatomic molecule, which would be required in the excitation of the iodine to laser radiation.

9-8 FREE-ELECTRON LASER

Among the emerging technologies is the development of a free-electron laser generating technique by which the generated laser beam is tunable from the ultraviolet through the visible and infrared to the microwave region of the electromagnetic spectrum. With this technique, powers in the range of gigawatts can be produced more efficiently than when conventional methods are used. In the conventional procedure, the emission of photons results from the transitions of atom-bound electrons from a higher to a lower quantum state. In the free-electron technique, the electrons that produce photons are not bound to any atom but are free; they are generated in an electron accelerator and ejected at high speeds into a magnetic field, where they become stimulated to emission of coherent radiation.

The free-electron laser generator illustrated in Fig. 9-17 must be considered as a composite of two different techniques, each being separate from the other. For instance, in one technique the system comprises the free-electron generator, the magnetic field generator known as a "wiggler" magnetic system, and the mirrors M_1 and M_2 used as beam oscillators. In the second technique, mirror M_2 is deleted and a laser source with two beam-bending mirrors, M_3 and M_4, are inserted to project a beam from a high-energy laser, such as carbon dioxide, through the magnetic field for amplification of the carbon dioxide laser beam.

Starting with the first technique, the high-speed electrons discharged from the electron accelerator enter the magnetic field, in which the electrons deflect periodically between the wiggler magnets. The undulatory oscillation decelerates the electrons, giving rise to emission of coherent radiation, which oscillates between mirrors M_1 and M_2 for further amplification. Following the emission, the electron beam emerges from the magnetic field and is caught in an electron trap, from which it can be diverted to any measuring instrument or to a detector.

In the second technique, in which the free-electron beam acts as an amplifier for a second laser beam, mirror M_2, shown in the figure, does not exist; in its place a laser source and two bending mirrors, M_3 and M_4, are positioned as shown with respect to the electron beam. A laser beam of, for instance, carbon dioxide radiation or frequency-doubled Nd-YAG radiation is projected through the wiggler magnets and, after oscillating between the magnetic poles, emerges to the exterior of the system through output mirror M_1. Output mirror M_1 is partially reflective and partially transmissive and acts as a feedback mirror for the laser beam. In this operational mode, a free-electron beam is ejected from the electron accelerator into the wiggler magnetic field, where the electrons execute undulatory motion between the unlike poles of the wiggler magnets. As the electrons decelerate during the undulatory motion, they radiate their energies

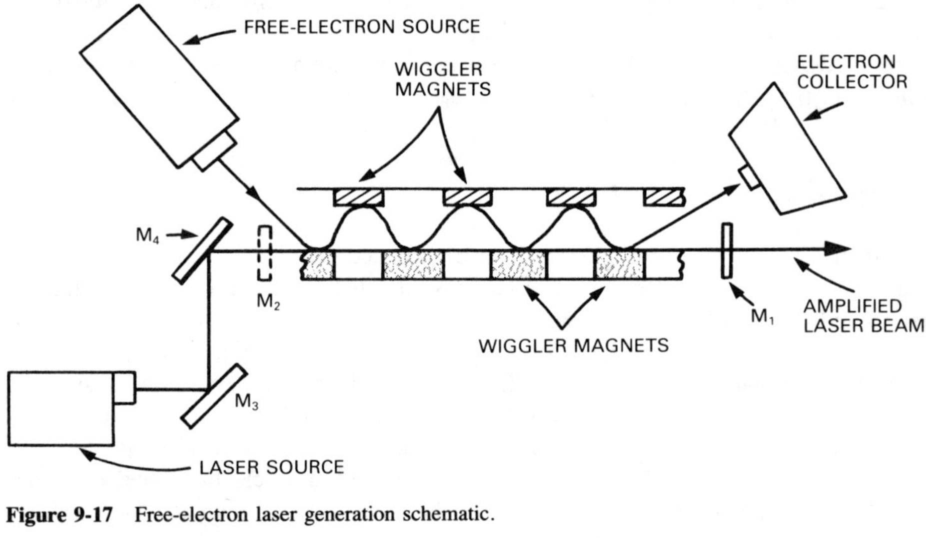

Figure 9-17 Free-electron laser generation schematic.

into several higher-order modes of the optical resonator. These energies are extracted by the traversing laser beam in the resonator, increasing the laser beam intensity.

The wavelengths emitted from the free electrons can be varied by varying the magnetic intensity of the wigglers or by varying the speed of the electrons ejected from the electron accelerator. Tuning of the different wavelengths from the system is achieved by increasing or decreasing the free-electron energies by the two methods stated earlier. Such a scheme makes the self-starting system function as an oscillator. The efficiency attained in this system is about 3% at present, but further improvements of the technique are expected to the extent that laser beams thus produced are contemplated for use in weapons systems on satellites. The system operating as an oscillator has attained up to 180 kW, extracting only about 0.2% of the electron energy of 43 million electron volts at 33,000 Å. Such a system, when perfected, will have civilian applications in solid-state and molecular spectroscopes.

QUESTIONS AND PROBLEMS

9-1 In solid-state laser rods, what is the relation of the lattice site to dopant ion size? What characteristic should host crystals have to permit the radiant energy to exit from them? Should the dopant be soluble in the host crystals? What is the relationship of the anion dopant valence to the host cation valence?

9-2 What are phonons? What is the relationship of a phonon to the emission of a photon? What is vibronic laser? What is alexandrite? In what range of wavelengths is it tunable? What is the dopant ion in alexandrite? What is the dopant concentration in alexandrite? In what mode or modes can the alexandrite crystal operate? What other laser is tunable at, for instance, 1 megahertz?

9-3 Give five types of gas lasers. How many possible transitions are there in a He-Ne laser? Which of the two gases in the He-Ne laser is responsible for the gain? What is the relationship of metastable helium levels to neon energy levels? Discuss. What is the range of linear polarization of the He-Ne laser? What is the function of He ions in the excitation of He-Ne gas? What is the useful lifetime of a He-Ne laser? What is the function of an attenuator in the Hughes He-Ne laser head? How is the indicator light connected in the circuit?

9-4 What is an ion laser? Name several ion lasers. Which ion laser produces the highest visible light? Which ion laser can emit the highest ultraviolet output power? What is a multiline mode of operation? Can more than one wavelength be obtained from a multiline ion laser? Where is the prism placed in a multiline ion laser? What is meant by Gaussian power distribution at TEM_{00}? Explain. What relation does fundamental mode operation have to the focus size of the beam? What is the function of a Brewster window?

9-5 What is a mode-locked ion laser? Can this system be coupled to an acousto-optic cavity-dumping processor? What does the processor do? What effect do high-peak pulse trains have on the second-harmonic frequency doubling? Why could time domain spectroscopy be conducted with this type of radiation? If a piezoelectric transducer is attached to a prism, what is the function of such a structure? In the transducer-prism combination, how does the refractive index of the prism vary? What is the output of a mode-locked ion laser used for?

9-6 What is a tunable dye laser? When could tunable high-peak power pulses be produced? What type of pumping (irradiating) source is used in this type of tunable laser? How is the narrowing of the spectral bandwidth accomplished? What is the function of an etalon? What is the effect of the temperature of the dye solution on the emitted wavelength?

9-7 Explain the effect of overcooling a dye solution on its efficiency of emission. Give two other factors that reduce the efficiency of emission of a dye solution. When does a dye solution dissociate faster, with a flashlamp pump or a laser beam pump? What is the highest efficiency obtained with a dye laser to date? What is the function of a half-wave plate in the polarization of a laser beam?

9-8 What harmonic wavelength of a Nd-YAG laser can pump the dye laser most efficiently? What one factor frustrates the background fluorescence? What scheme can make possible the mixing of a Nd-YAG beam with a dye laser beam? Why should one be cautious in working with a dye laser? Is a dye laser pumped axially or transversely? Explain.

9-9 Describe the construction of a diode laser. Which active layer is known as the donor? In which direction does the current flow in a diode laser? Explain why. When does a diode emit coherent radiation? When does a diode emit incoherent radiation? Which one is visible? How are diode lasers pumped? Explain. Is a diode laser temperature-sensitive? How does diode laser temperature affect the wavelength? Can a diode laser be optically pumped? If so, which type can thus be pumped? Is such a diode tunable? Is it temperature-sensitive?

9-10 What is an excimer laser? Give examples of excimer lasers. Is an excimer laser optically pumped? Why? What characteristics are exhibited by excimer lasers? What maximum energy pulses have been obtained from an excimer laser? Are excimer lasers tunable? Why is the operation of an excimer laser costly at the present time? Why should the discharge bore of an excimer laser be passivated? What material can be used as the bore for an excimer laser without passivating the bore? Can an excimer be operated in a sealed-off system? Explain how. How is the Lumonics excimer laser constructed? What is its half-life? How many shots can it yield at a wavelength of 3080 Å? Give one special characteristic of the Lumonics excimer.

9-11 What are solid-state laser systems? Name them. Can an optical pump be used with solid-state lasers? What types of optical pumps can be used to excite solid-state lasers? What is the general structure of a solid-state laser rod? Does the emission occur in the host or in the dopant? Can a wavelength of 5000 Å optically pump to produce 3000-Å wavelength radiation? Explain why. Where does thermal concentration occur in a circular, cylindrical solid-state laser rod? How does this heat affect the emission from the rod?

9-12 What is meant by the words "spatially coherent"? Explain. Can a laser beam be incoherent? Explain. What significant phenomenon occurs when two beams of equal frequency and amplitude are not in phase? When does complete darkness occur when two beams are not in phase? Explain. What are interference fringes?

9-13 What does a ruby laser rod have for a host material? What does it have for a dopant material? What percentage is the dopant in a ruby laser rod? What is the dopant in a Nd-YAG laser? What percentage is the dopant in the Nd-YAG laser rod? What occurs when negative frequency absorption by the ionic dopant takes place? How many chromium ions are there in 1 cm^3 of aluminum oxide host in ruby laser rod? In what relation does stimulated emission occur with respect to the exciting radiation? Explain.

9-14 Explain the transitions that occur between three-level energy states and four-level energy states. What does the threshold energy of a lasing element excited by a pulsing mode depend on? What does the pulse duration of an optical pump depend on? Why can Nd ions give rise to lattice-site characteristic wavelengths? What is charge compensation?

9-15 Name several examples of metal vapor lasers. How does a metal vapor laser operate? Can rare-earth materials produce high-power lasers? Where can the latter type of lasers be employed? Do copper vapor lasers have high conversion efficiency? What percent efficiency can copper vapor lasers have? What are the applications of copper vapor lasers? What are the wavelengths a copper laser can produce? At what temperature does a copper laser cavity operate? When copper ions are dissociated from copper bromide, at what temperature can such a system operate? Why? What laser wavelengths have become very promising for hydrographic instrumentation?

9-16 Why does Xerox Corporation use He-Cd lasers with its high-speed printers? Explain. What are the wavelengths that can be emitted from He-Cd lasers of different manufacture? Can iodine, a solid halide, produce laser action? If so, at what wavelength does it operate? What is the application of an iodine laser?

9-17 What is a free-electron laser? Describe it. Is the laser beam produced by an electron-aided method tunable? In what general optical range does a free-electron laser emit? Draw a diagram showing the mechanism of producing a laser beam with the aid of free electrons. How can the emitted wavelengths from a free-electron laser system be varied? Will varying the speed of the free electrons ejected from the accelerator increase the frequency of the emitted laser radiation? What is the system efficiency of such a scheme? What applications does free-election laser radiation have? Explain.

SELECTED SOURCES OF RELATED DATA

Agrawal, G. P. 1986. *Long-Wavelength Semiconductor Lasers*. New York: Van Nostrand Reinhold.

Baer, T. M. 1986. Diode laser pumping of solid-state lasers. *Laser Focus* June, pp. 82–92.

Brau, C. 1987. Recent developments in free-electron lasers. *Laser Focus* February, pp. 40–46.

Capasso, F. 1984. Multilayer avalanche photodiodes and solid-state photomultipliers. *Laser Focus* July, pp. 84–101.

DeShazer, L. G. 1987. Advances in tunable solid-state lasers. *Laser Focus* February, pp. 54–58.

Forrest, S. R. 1982. Photodiodes for long-wavelength communication systems. *Laser Focus* December, pp. 81–90.

Giuliano, C. N., et al. 1983. Can phase conjugate resonators enhance laser performance? *Laser Focus* February, pp. 55–64.

Holmes, L. 1986. Solid-state lasers: New technology and higher performance standards. *Laser Focus* November, pp. 58–68.

Holmes, L. 1986. Metal-vapor lasers: Special capabilities for special applications. *Laser Focus* December, pp. 76–80.

Hora, H. 1981. *Physics of Laser-Driven Plasmas*. New York: Wiley.

Hunsberger, R. G. 1982. *Integrated Optics: Theory and Technology*. New York: Springer-Verlag.

Jaeger, R. E., et al. 1987. Sensing Magnetic Fields, Using an Optical Fiber. U.S. Patent No. 4,650,281.

Jet Propulsion Laboratory. 1988. Axial pumping of Ho-YLF laser by laser diodes increases efficiency. NASA Technol. Briefs, Vol. 12, No. 6 (June).

Kunimoto, W. 1984. Cryogenics for infrared laser detectors. *Lasers and Applications* April, pp. 71–76.

Letokhov, V. S. 1983. *Nonlinear Laser Chemistry*. New York: Springer-Verlag.

Levanson, M. D. 1982. *Introduction to Nonlinear Laser Spectroscopy*. New York: Academic Press.

Marshall, T. C. 1985. *Free Electron Lasers*. New York: Macmillan.

Penilloux, B. E. 1988. Phase retarders to achieve optimum polarization. Presented at SPIE O-E/LASE Convention, Los Angeles.

Power, B. 1985. Vacuum instruments for measuring and controlling laser gases. *Laser Focus* November, pp. 86–91.

Pummer, H. 1985. The excimer laser: Its fast growth and improvements. *Photonics Spectra* May, pp. 73–82.

Schoen, N. C. 1982. Free electron laser for megawatts of microwaves. *IEEE J. Quantum Electron.* QE-18(9):1318–1321.

Van der Ziel, J. P., et al. 1981. Subpicosecond diode laser. *Appl. Phys. Lett.* 39(7):525.

Walter, R., et al. 1982. Laser-stimulated fluorescence of submicrometer regions within cell mitochondria. *Proc. Natl. Acad. Sci. U.S.A.* 79:466.

Yamamoto, S., et al. 1982. Room-temperature CW diode laser emits below 700 nm. *Appl. Phys. Lett.* 41(9):796–798.

CHAPTER

TEN

LASER INSTRUMENTATION IN MEDICAL SURGERY

10-1 INTRODUCTION

During the past decade, scientists have been intensely interested in the biological effects of laser beams on the human body, and considerable study and investigation have been conducted to determine the behavior of different types of laser radiations in this respect. Since the development in 1959 of the first helium-neon maser activated by microwave excitation, many materials have been tried as laser sources. Those found effective as lasing materials are ruby, neodymium-YAG, neodymium-glass, carbon dioxide, carbon monoxide, nitrogen, argon, xenon, krypton, germanate glass, lanthanate glass, sulfur glasses, gold and copper vapors, various organic dyes, erbium, excimers, and a gamut of other laser sources. All of these have been investigated for application in the biomedical sciences. One of the applications that is of particular interest is in bloodless surgery in medicine.

Although the study and investigation of laser beams for medical use have been confined to various research centers and hospitals throughout the United States and abroad, other groups and private individuals in the medical profession are beginning to apply laser beams for treatment of patients. The studies and the experimental work have been directed to determining the characteristics of various types of laser beams and to the varied techniques of their manipulation, in an attempt to create specialized applications for each particular type of laser radiation. The findings have disclosed that laser beams of different wavelengths have significantly different effects on different body tissues, as well as on the elemental substances inherent in body tissues.

These findings have further led to the realization that the type of laser used on a tissue structure must be selectively chosen. That is, a laser beam from a ruby rod is readily absorbed by retinal tissue and therefore is effective for attaching a detached retina (retinopathy). At this writing, several hundred thousand patients have been treated by laser for retinal attachment; this condition is common in diabetics. Because the ruby laser is visible, it can be focused through the pupil of the eye. However, the beam is absorbed only weakly by the blood cells. The visible beam from an argon laser is highly absorbed by the visceral tissues and is very effective for photocoagulation. Since the argon laser beam is compatible with fiberoptic materials it can be used continuously or in long pulses as a surgical scalpel.

The Nd-YAG laser is also a powerful surgical tool when the beam is conducted through a properly selected waveguide or fiberoptic bundle. The Nd-YAG laser beam is used either Q-switched or mode locked. The advantage of the Q-switched or mode-locked laser is that the beam can be precisely focused on the treatment site without significant effect on the contiguous tissues. Another advantage of this beam is that it can form an ionized plasma and thus can cut any unpigmented tissue in the body.

Although carbon dioxide is the most powerful lasing material for use in medical work for cutting and cauterizing small blood vessels in the liver, kidney, stomach, and pancreas, its principal drawback is that the long infrared wavelengths are highly absorbed by the focusing optics of the laser unit and are not compatible with ordinary fiberoptic waveguides. Accordingly, expensive lens materials or less costly but bulkier reflective optics must be used with it. Only recently, lenses and fiberoptic material have been fabricated for infrared ray transmission on a small scale from fluorite glass, gallium arsenide, zinc selenide, germanium, and the like, but they are costly to manufacture in mass production quantities at present. Table 10-1 illustrates the use of different types of laser radiation in the surgery and treatment of body tissues.

In medical surgery, various types of laser beams have already been employed in the fields of dermatology, oncology, hematology, and ophthalmology; in the surgery of brain tissues, neural nodes and trunks, gastric ulcers, and renal disorders; and lately in the treatment of genital herpes and in gynecology (diseases and disorders of female organs). Polyps and nodules in the esophageal tract, cancerous growths in body cavities, and carcinoma of the vocal cords and breast have been treated with laser beams. Cholesterol deposits in the cardiac blood vessels have also been "blown off" by means of a laser beam introduced into the blood vessel with a fiberoptic waveguide. Surgery of superficial and visceral cancer and benign tumors in various sites, coagulation of blood in surgical incisions, and photocoagulation in ophthalmic afflictions are among the applications of laser beams. Vitreous strands in the eye that may cause retinal detachment can be severed by means of a laser beam focused through the pupil of the eye, and many other delicate surgical operations can be achieved with laser beams and will be studied in the following sections.

10-2 COMPARATIVE LASER BEAM TREATMENTS

At the University of Cincinnati, use of a carbon dioxide laser beam has become routine in the removal of cancerous tissues, growths, and nodules. For black-colored malignant growths, it was found that laser beams from ruby, Nd-YAG, argon, and carbon dioxide were equally effective, with the exception of healing times. Malignant melanomas

Table 10-1 Randomly selected surgical applications of lasers*

Laser type	Spectral region, Å	Body tissue	Treatment
He-Ne	Visible, 6328	Dermal	Photoradiation, wrinkle removal
Argon	Visible, 4500	Otological, cholesterol clot, fallopian tube, gastrointestinal tract	Photocoagulation, skin blemishes, skin therapy, incision, ulcer
Argon	Ultraviolet, 3500	Nasal mucous membranes, dermal tissue	Ultraviolet therapy, germicidal agent, retinopathy
Nd-YAG	Infrared, 10,640	Stomach, liver, lungs, heart, kidneys, pancreas, brain, skin	Incision, control of hemophilia, cancer
CO_2	Infrared, 106,000	Stomach, liver, pancreas, skin, profuse bleeding, syphilitic tissue, herpes sores, melanoma, vocal cords, tonsils	Removal of tissue, sprained joint, cancerous tissue, fallopian tube
Erbium-YLF	Infrared, 12,280	Eye tissue	Ophthalmology
Ruby	Visible, 6943	Retinopathy, melanoma, skin blemishes	Eye disorders, cancerous tissue
Copper vapor	Visible, 5105	Herpes sores	Excision

* This listing is only exemplary and should not be considered complete. Wavelengths and applications are typical.

(tumors) of the skin were treated with a pulsed ruby laser and also with a continuous-wave carbon dioxide laser beam. The areas treated with the carbon dioxide laser healed more slowly than those treated with a pulsed ruby laser used at 1500 J/cm^2, a Nd-YAG laser at 380 J/cm^2, or an argon laser at 5 W. The patients had previously been treated with conventional methods, and no significant healing of the lesions was apparent. Therefore, they voluntarily submitted to the laser radiation treatment, which was effective and painless and resulted in healing of the tissues. During treatment, when pain was sensed by the patient, local anesthesia by means of a nerve block was administered.

Comparative experiments were also conducted using conventional scalpel, electrosurgery, electroexcision biopsy, excessive exposure to the sun, and x-ray therapy. It was concluded that the laser beam treatment was superior, since it was practically bloodless, rapid, and effective. A carbon dioxide laser beam was used in thermal coagulation necrosis for spreading melanoma and found to have no particular advantage over the other laser beams. However, the carbon dioxide laser was very effective and superior to its rival, electrosurgery, for profusely bleeding areas. These experiments have also shown that the use of laser radiation for cancer surgery in profusely bleeding organs, such as the liver, lungs, heart, and stomach, is necessary; accordingly, the work is continuing in this direction.

A team of scientists from Stanford Research Institute and Coherent, Inc., has developed an argon photocoagulator for the treatment of retinopathy, a pathological deterioration of the retina due to advanced diabetes. In the treatment, the Coherent Model 800 argon laser is used. The laser output of the system can be varied continuously from 0 to 1 W directed at the region of the cornea. The beam is transmitted through an articulate arm having reflective optics into a microscope positioned adjacent to the patient's eye. With the microscope, the physician adjusts the focus size of the laser beam on the retina through a contact lens over the cornea, as illustrated in Fig. 10-1. The focus size can be varied from 1 mm to 50 μm in diameter. Many of the laser treatments require no anesthesia, since the treatment is painless.

It is reported that at the University of Illinois Medical Center hundreds of diabetic patients are being treated annually. More than 90% of the approximately 6 million diabetics in the United States are said to require photocoagulation treatment ultimately. Although early treatment of retinal disease is important, a physician using laser photocoagulation must learn and know the exact dosage to be administered to any particular patient.

An erbium-doped YLF laser developed by Laser Surgical Systems, which produces 5 to 15 mJ of laser pulses radiating at 12,280 Å, is being used for ophthalmologic work such as cataract surgery. In this wavelength range, the system is an improvement over the Nd-YAG at a wavelength of 10,640 Å. This system is provided with a He-Ne laser for pointing the invisible wavelengths from the erbium laser, and when pointed at the eye the beam performs with a minimum of self-focusing. Auxiliary devices, such as a microprocessor and a photodetector, aid in focusing and in monitoring the optimum focal spot size and pulse energy delivered by the surgical operation. Experiments performed by physicians at the Chicago Eye Institute have demonstrated the precise

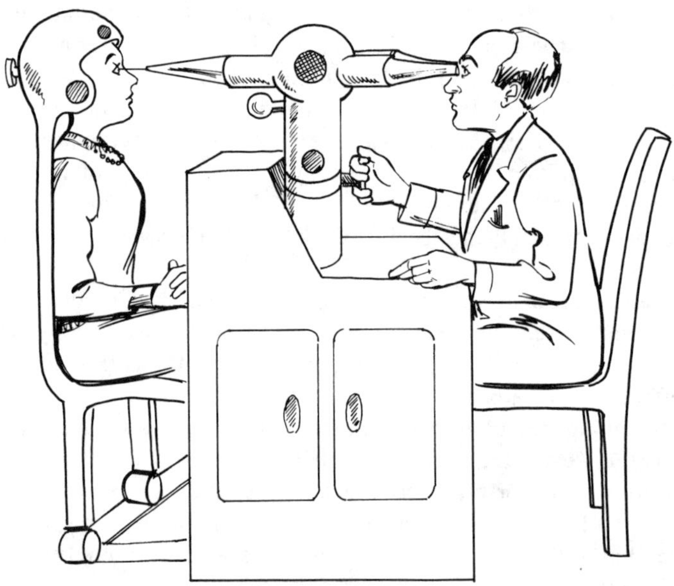

Figure 10-1 Scheme showing the manner of directing a laser beam at the retina during retinopathy.

and clean cutting capabilities of the laser beam with minimum capsular ablation from the ocular site.

10-3 SURGICAL LASER TREATMENTS

Treatment of Skin Disorders

In cosmetic surgery, the laser beam has been routinely employed in removing tattoos and skin blemishes from the human body. At the University of Cincinnati Medical Center, the procedure has become routine on patients with skin disorders. The mechanism of laser interaction with a tattooed site consists of injecting a beam from a ruby laser (for its better absorption quality) into the tattooed area. The beam penetrates the translucent skin with little or no harm, is absorbed by the dark dye pigment, and vaporizes the pigment. It is reported that at the Medical Center the carbon dioxide laser has become a standard device for removing skin blemishes, warts, and the like from the legs, chest, arms, or any other cutaneous part of the body. In some cases, the laser beam has been found to be a substitute for high-voltage x-rays, radium, or radioactive cobalt in the treatment of deep-seated cancer of the skin.

The procedure for removing skin blemishes, shown in Fig. 10-2, merely illustrates a typical technique for delivering the laser beam to the affected site; other techniques are also available. When using the invisible carbon dioxide laser beam, the practitioner should wear a pair of safety goggles in case of an accidental encounter with the primary beam from the laser generator, since the beam may also contain a He-Ne laser beam for visualization of the area being treated. The procedure is painless except a pricking feeling on the skin. The unwanted mark or blemish is removed by vaporization of the pigment or blood cells by the laser beam, which may take considerable time when large areas are to be treated.

Laser Treatment of Birthmarks

Port-wine birthmarks, which appear on the skin as a result of concentration of small blood vessels in a patch of the skin, are being treated by specialists at Beth Israel Hospital in Boston. The laser system consists of a 4-W blue-green argon laser beam, which is transmitted through one version of the Laserkinetics variable-focus, hand-held laser surgical stylus shown in Fig. 10-14. This surgical device uses a fiberoptic waveguide that delivers the laser beam to a collimator and focusing system; the focal point, as well as the distance form the exit port of the laser head to the work area, is variable by means of a thumb button on the stylus. Defocusing of the beam to cover an area a hundred times larger than the focus size is achieved by axial movement of the focusing optics to and from the work area. The blue-green argon ion laser beam is easily absorbed by the blood vessels, which shut off the blood flow to the birthmark area; the blood cells are vaporized from the port-wine birthmark site.

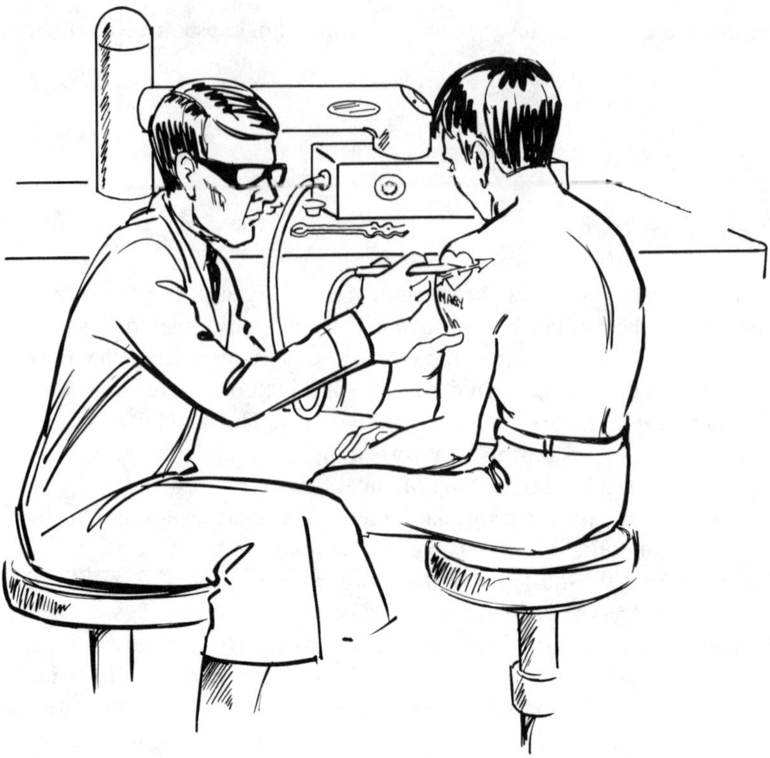

Figure 10-2 Simplified illustration of the removal of tattoos with a laser beam.

Vocal Cord Surgery

Vocal cords are fibrous elastic folds at the edge of larynx and produce the voice. In Fig. 10-3A the vocal folds are squeezed together in singing a high note, and in Fig. 10-3B the folds are in a position for quiet breathing. At times, abnormal horny growths,

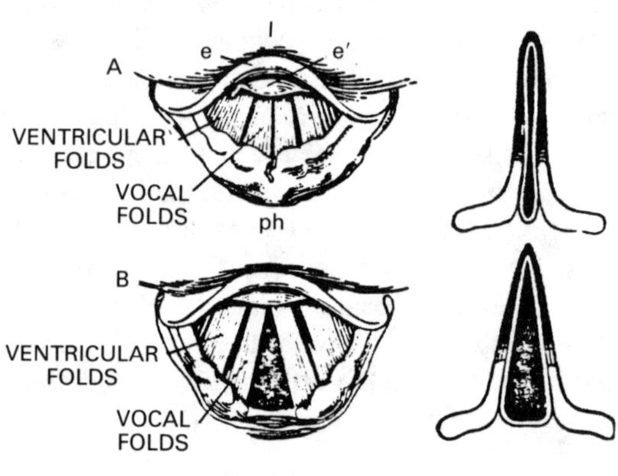

Figure 10-3 Vocal folds of human larynx shown from the top.

polyps, and nodules are formed on the folds and should be removed. A team of surgical specialists at Boston University has reported their effective use of a carbon dioxide laser beam in the removal of horny growths, polyps, and nodules from human vocal cords. The surgery has been highly precise and bloodless; healing has been prompt and successfully achieved without damage to the vocal cords or to the voice quality. These specialist also removed a huge tumor from the vocal cords of a 14-year-old boy who had undergone 24 previous unsuccessful, conventional operations. It was further reported that the boy acquired a normal voice for the first time in his life subsequent to the laser treatment.

Laser Microsurgery

Researchers at Uppsala University in Sweden have developed a microsurgical laser device that corrects faults in nerves and intricate body organs. Experiments have disclosed that such microsurgery is feasible with improved techniques even on the labyrinth of the ear, which could easily be damaged inadvertently, resulting in deafness. One problem was that the laser beam they used produced a mechanical shock through induced acoustic cell oscillation. The degree of shock was sufficient to cause the cells to explode and damage the contiguous structures. Apparently, U.S. researchers have not encountered this phenomenon in their applications of lasers to live biological specimens, such as the human nervous center, the brain.

Otological Surgical Instrument

Nasal sinus inflammation due to allergy or rhinal infection causes the eustachian tube to constrict or close up to the extent of interfering with the passage of the air to and from the inner ear (Fig. 10-4). This condition causes distention of the tympanum and ensuing pain and pressure in the ear. If the eustachian tube cannot be pneumatized, the otologist punctures a hole in the tympanum and inserts a grommet to enable free air passage into the inner ear of the patient. This is generally performed in a hospital to restore normal hearing.

A clinical argon laser instrument, with or without a He-Ne laser for focusing the beam through a long-focal-length lens and producing a single pulse of 5 to 10 J, has been developed by medical professionals at Stanford University and in Munich and used successfully to puncture a minute hole in the tympanic membrane. This technique was advanced by the author in his book *Laser Technology* published in the early part of 1979. The technique requires no grommet, since the laser beam cauterizes the periphery of the hole, and the treatment can be carried out in the otologist's office without the use of anesthesia.

In Fig. 10-4, M is the external ear, G the external acoustic meatus or external ear canal, T the tympanum or the ear drum, and P the middle ear cavity. The laser beam is introduced through the ear canal to a distance equal to the focal length of the optical lens from the tympanic membrane. A single pulse from an argon laser produces a fine hole in the tympanum, permitting passage of air in and out of the inner ear P and thus relieving the pressure inside the ear. The internal pressure may be either positive or

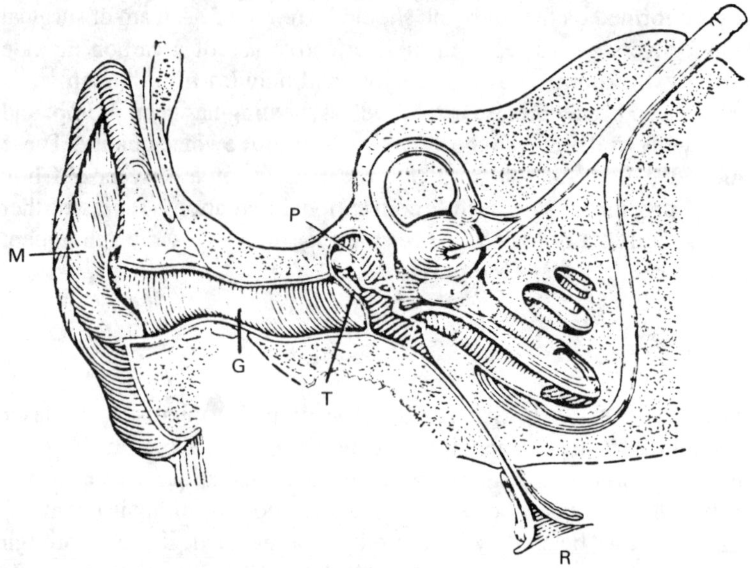

Figure 10-4 Sectional drawing of the human right ear.

negative; it is positive when exudates due to infection are accumulated within the inner cavity, and it is negative when external air is prevented from circulating in and out of the inner ear P.

Laser Tonsillectomy

A 30-W carbon dioxide laser system for surgery of inflamed tonsils of patients from 3 years of age to adults is being used routinely at the Rose Medical Center in Denver, Colorado. The laser beam is transmitted through a seven-mirror articulated optical system to a micromanipulator and a microscope, which focuses the beam on the periphery of the tonsil to incise it circumferentially. The blood vessels and lymphatic vessels are cauterized during the treatment so that no blood issues into the throat. The patient can eat and drink immediately after the operation, when desired. The operation takes about 10 to 15 minutes and there is little or no pain. A view of the tonsils is shown in Fig. 10-5.

Treatment of Sprained Joints

At times, the joints in the human body become severely twisted, spraining the disk between the bones of the joint. The disk, medically known as meniscus, becomes torn by the twist, which causes severe pain at the joint. The usual surgical procedure is to take the joint apart to cut away the torn membrane, which, when left, may tear away more of the membrane. The recovery takes several months. Recently, this orthopedic procedure has been carried out with a carbon dioxide laser beam at Sinai Hospital in Detroit. The laser beam is delivered to the patient's joint through a fiberoptic waveguide, called an arthroscope, which is provided with tiny biopsy channels to remove the affected cartilage. Because the 10-μm-diameter beam is directed to the meniscus without opening the joint, the risk of formation of arthritis, which exists in conventional methods, is eliminated. The patient can resume use of the joint in a few weeks.

Coronary Bypass Laser Technique

A 14-W argon laser is being used with an imaging fiberoptic waveguide to vaporize cholesterol clots deposited in the coronary arteries; if permitted to advance, this condition may be fatal to the patient. Researchers at the University of California at Davis have developed a technique in which a fiberoptic strand with a tube having a small balloon at its terminal portion is inserted into the artery, advancing the waveguide to the clot; at the clot, the balloon is inflated by blowing gas through the tube to block off the blood flow. The laser beam is then triggered to vaporize the cholesterol clot. With conventional procedures, the operation is more difficult and time-consuming. It is reported that about 100,000 patients with this disorder are being treated annually by conventional surgery.

Surgeons at the University of California have performed the world's first surgery to clear a partially plaque-blocked carotid artery (neck artery) using an argon ion laser beam. This operation reduces the complications that might be encountered in conventional stroke-prevention surgery, in which the neck is cut open, the artery is exposed, and a surgical knife is used to scrape off and remove the plaque. Following this procedure, the artery and the contiguous tissues are sewed up to close the wound.

The patient, a 72-year-old man, had suffered a dozen or more transient ischemic

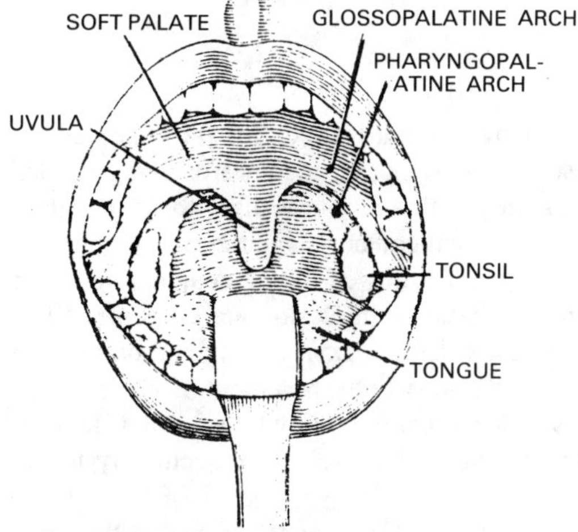

Figure 10-5 Front view of the tonsils, with open mouth.

attacks before the surgery because of the blocking of his left carotid artery, which supplies the left side of the brain. The plaque is a fatty deposit, and in this case it had blocked 80 to 90% of the blood passageway in the artery. The laser beam was introduced into the artery through a fiberoptic light guide and directed to the plaque growth on the internal periphery of the artery to vaporize the plaque. The vaporized debris was removed by suction peripherally to the light guide.

Brain Surgery with Laser

At the University of Pittsburgh, a team of medical researchers has successfully used a laser beam to remove a malignant growth in the brain tissue. In this operation, the laser beam was used to destroy the cancerous tissue and then the mass was removed by conventional methods. Many patients with Parkinson's disease (shaking palsy), which affects the nerve ganglia at the base of the brain, are reported to be aided by treatment with a laser beam. The laser beam treatment precludes the necessity of extensive surgery or probing the brain with an electrode. The laser beam is selected from several wavelengths that will be compatible with the tissue being removed.

Fallopian Tube Surgery

A carbon dioxide laser waveguide developed at the New Orleans Laser Institute is being used to restore the fertility of women whose fallopian tubes are blocked (Fig. 10-6), preventing insemination. The waveguide is introduced through the vagina and advanced to the blockage to vaporize it at a distance equal to the focal length of the focusing lens. It is reported that the ova of women treated in this manner are capable of being inseminated. Only one or two pulses of the laser beam are required to vaporize and open the blockage.

Gastrointestinal Surgery

An endoscopic laser coagulator introduced by Spectra-Physics for the treatment of bleeding in internal organs, such as the esophagus, stomach, and intestines, is provided with an argon-ion laser for coagulation together with a low-level light for localization and visualization of the affected site. The instrument contains a quartz optical fiber that directs the low-level light beam to the lesion site, followed by the coagulating argon beam; thus, the diagnosis and immediate treatment of the lesion are combined. The optical conduit, shown in Fig. 10-7, further provides a coaxial flow of carbon dioxide gas to remove excess blood from the lesion preparatory to the treatment. Both the argon laser radiation and the carbon dioxide gas flow are controlled by a foot pedal, leaving the hands free to localize the lesion site and direct the laser beam accurately to the lesion.

In this technique, a laser beam from a Nd-YAG generator can also be used; however, because the shorter blue-green wavelengths from the argon laser are highly

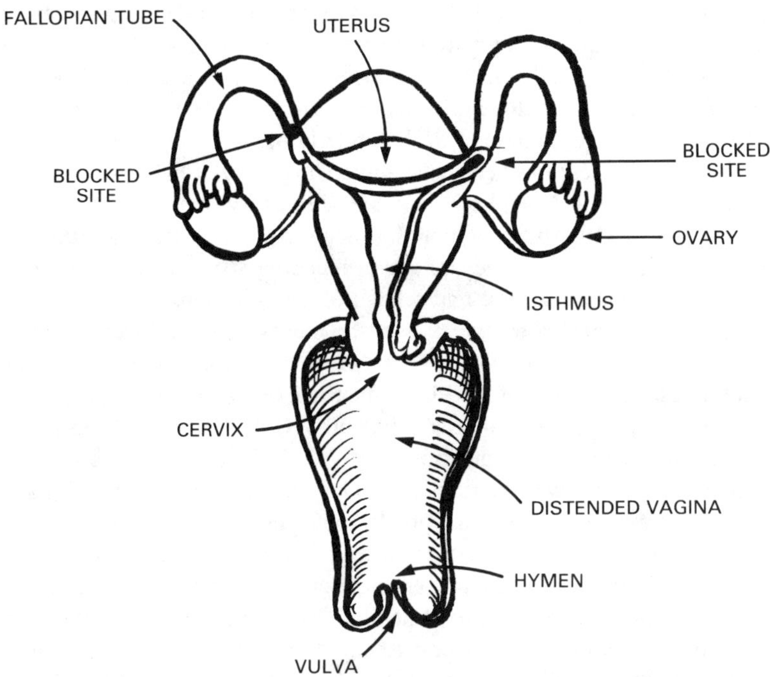

Figure 10-6 Section of female reproductive system.

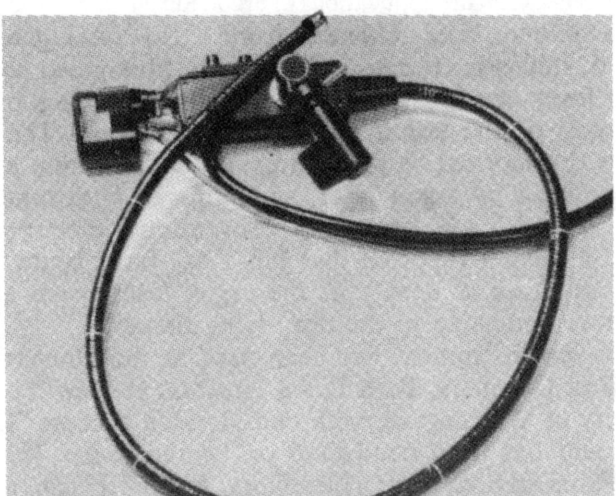

Figure 10-7 Endoscopic laser coagulator Model 770 used for the treatment of internal bleeding (Courtesy of Spectra-Physics, Inc.).

absorbed by the bleeding organ, coagulation of the lesion is effectively achieved. The manufacturer reports that the normal tissues contiguous to the lesion are not significantly affected by the argon laser beam, as compared with other types of laser beams, such as those from a Nd-YAG system, and conventional electrosurgery. It is further reported that the Nd-YAG laser beam penetrates the tissue four times as deeply as the argon laser beam and therefore is more damaging to the underlying tissues.

The argon ion laser instrument has been used at the University of Erlangen in West Germany, Beth Israel Hospital in New York, and the University of Washington with considerable success. The coagulative treatment has been applied to bleeding tumors, intestinal angiomatosis (bleeding), angiodysplasia (a bleeding structure), gastric erosion, gastric carcinoma, duodenal ulcers, and related bleeding lesions.

Apparently, there is a medical controversy over the merits of argon laser radiation versus Nd-YAG laser radiation for endoscopy, between the U.S. manufacturers of argon ion lasers and investigators at the Max Planck Institute of Biochemistry, Munich, as reported in the journal *Progress of Surgery*. The authors, Kiefhaber and Moritz of Max Planck, reported that when the output from a 50-W Nd-YAG laser was applied repetitively to the bleeding site for short periods in 110 patients, including hemophiliacs (bleeders), bleeding was controlled in 106 patients. In some patients, however, bleeding could not be restrained completely because of an obstruction in the canals traversed by the endoscope. The authors further reported that argon laser treatment was unsatisfactory because of the high susceptibility of the blood hemoglobin to absorption of the laser beam, which inhibited coagulation at the bleeding site. However, they considered that application of the argon ion laser would be useful at sites with less severe bleeding.

From discussions above and in the earlier sections of this book, it is apparent that no line of demarcation can be drawn between the applications of argon ion and Nd-YAG lasers in the control of bleeding of internal organs. The laser output power and the technique by which the radiation is administered to the particular lesion play important roles in the treatment. It is true that high power from an argon ion laser commensurate with that from a Nd-YAG laser cannot be achieved; however, the high-power radiation from a Nd-YAG laser may not be indicated in all cases. Also, to control the beam intensity at the focal point it is necessary at times to apply the radiation defocused in order to cover a significant area of the lesion. Thus, either an argon ion or a Nd-YAG laser beam can be administered to the lesion sites to control bleeding by employing properly devised laser treatment techniques.

Detection of Carcinomatous Tumor

While the argon ion laser is finding preferred use for cauterizing gastrointestinal bleeding, the combination of radiation from both an argon ion laser and a Nd-YAG laser when used selectively from the same laser conduit appears to be an attractive technique. The neodymium-YAG laser beam cuts the affected tissue, and the argon laser beam cauterizes the fissure. In accord with this procedure is a fiberoptic bronchoscope that transmits a beam from a krypton ion laser into an affected lung to detect a tumor. The tumor is visualized by use of a dye that has been injected via the bloodstream and fluoresces on irradiation by the krypton laser beam. Since the dye fluoresces in the visible spectrum, it permits visualization of the tumor through the fiberscope followed by the usual coagulation treatment. The scheme is also used for the detection of cervical cancer, which is then treated by use of present surgical techniques.

A photochemotherapy technique developed by Quentron Optics Pty., Ltd., South Australia,* employs a photosensitive chemical solution, known as hematoporphyrin derivative (HpD), which is injected into the patient's body and allowed to remain there for several days. When a tumor is present in any part of the body, the drug is absorbed by the tumor cells, which, when irradiated with a laser beam, interact with it and die. This procedure does not affect the normal, healthy cells adjoining the tumor tissue.

In operation, a gold vapor or copper vapor laser beam is conducted through a fiberoptic conduit to the site of the tumor to irradiate the tumor tissue, which becomes coagulated and then is removed mechanically from the tumor site. The laser beam is delivered at a peak power of 10 kW and a wavelength of 6260 to 6360 Å at a frequency of 10,000 Hz. The cytotoxic effect is achieved by a cumulative power of 300 mW/cm^2. A gold vapor laser is preferred to an excimer or CO_2 laser because either of the latter damages the terminal portion of the optical fiber. The only side effect of the gold vapor laser procedure is that the skin area treated has an increased sensitivity to sunburn for several months following the therapy.

Ophthalmic Surgery with Excimer Laser

Recent research in ophthalmic surgery has shown that refractive eye disorders, such as myopia (nearsightedness), hyperopia (farsightedness), and astigmatism can be corrected by use of a 1930-Å excimer laser beam. This surgery may in time replace the use of contact lenses or prescription spectacles. The conventional technique for correcting refractive eye disorders has been radial keratotomy, using a diamond-edged knife, in which a number of radial incisions are made peripherally to the cornea to ablate a certain amount of the tissue and change the radius of curvature to the cornea. However, it has been found that with this procedure the corneal tissue becomes susceptible to trauma. Also, the operated eye enters into regression after a year or so. In contrast, the excimer laser has eliminated the disadvantages inherent in refractive surgery with the diamond instrument.

The advantages of the excimer laser are that (1) it can remove corneal tissue to a precise tolerance, (2) basic research can be carried on in ophthalmology, and (3) the surgery will not impair the transparency of the corneal tissue, as has happened with other conventional methods. One other distinct advantage is that with the excimer laser beam only 10% of the corneal tissue is incised, in contrast to 90% with the conventional surgery.

Cooper Vision, Inc. of Santa Clara, California, is the first company that has developed an ArF excimer laser system operating at 1930 Å that will also be used in

* U.S. Patent No. 4,614,190, Photoradiation Therapy Technique, by Alexei Stanco and James A. Piper. Patent assigned to Quentron Optics Co.

basic research in ophthalmology. This instrument has an ablation accuracy of less than 1 μm, operating at repetition rates up to 20 Hz at an ablation site 1 to 6 mm in diameter. All beam delivery parameters are automatically calculated by a computer. The total corneal thickness removed is between 25 and 50 μm. Although excimer lasers of other wavelengths, such as 2480 Å, have also been employed in the surgery, the shorter wavelength of 1930 Å is found to offer quicker and cleaner healing.

Detection of Healing of Corneal Sutures

It has been reported that more than 6000 corneal transplants are made annually in the United States. Although this appears to be a routine procedure, a major problem resides in the timing of the removal of the sutures postoperatively. It has been found that when the sutures are removed prematurely severe astigmatism of the eye occurs, and when they are retained too long the eye becomes susceptible to infection, requiring additional treatment by the surgeon in attendance.

To overcome this handicap, a technique employing a holographic procedure has been developed by researchers at Johns Hopkins Hospital and used successfully by the attendant physicians there. The technique permits determination of the strength of the transplanted cornea prior to removal of the sutures. The procedure consists of using a beam of visible laser light, such as He-Ne type, to produce interference with a reference laser beam reflected from the transplanted cornea and capturing the two beams in interference on a photographic plate, as illustrated in Fig. 10-8. The resultant hologram, when examined with a laser beam, indicates greater deformation when the wound is incompletely healed.

In Fig. 10-8, a He-Ne laser beam is directed from the source L to a partially coated mirror M_1, which splits the beam into two rays and projects one ray to mirror M_2 and the other ray to a beam expander E_1, which expands the laser beam to a sufficient diameter to cover the corneal suture at the eye. The laser beam from mirror M_2 is projected to a similar beam expander E_2, which projects the expanded beam on mirror M_3, from which the beam is directed onto the surface of a photographic plate P; this is the reference beam. Laser beam reflections from the eye region are also directed to

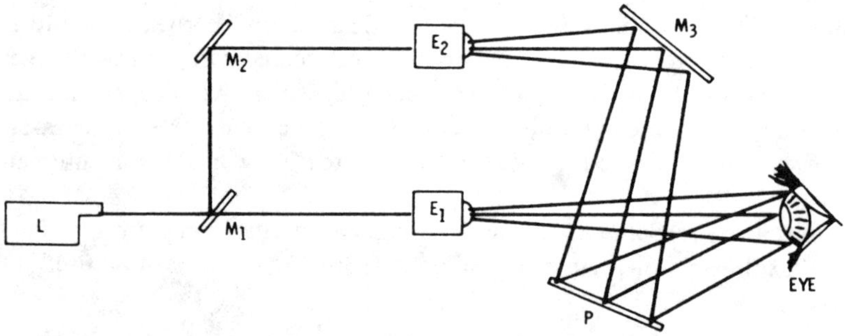

Figure 10-8 Holographic setup for making a corneal suture hologram.

the photographic plate P so that the two beams interfere with each other to form an interferometric pattern on the photographic plate. Any deformation of the corneal structure provided with a suture is recorded on the plate P, and after development with a laser beam directed from the opposite side of the plate, the three-dimensional configuration of the corneal suture will be delineated. Measurements are made of this configuration compared to the normally expected configuration of the cornea.

Laser Treatment of Genital Herpes

Genital herpes is an infectious disease, similar to syphilis, that occurs on the skin and moist tissues contiguous to the genital organs of both males and females. It is said to be painful and difficult or impossible to cure. In fact, there seems to be no conventional medical treatment for herpes. However, a 20-W carbon dioxide laser beam is now being tried by several gynecologists to destroy the diseased cells and the herpes virus by vaporizing them with the heat from the laser beam, which is directed to the lesion through a microscope known as a colposcope. The lesion is treated without bleeding or scarring, and the healing appears to be rapid and complete if the treatment is applied in the early stages of the disease. To date, several hundred patients with herpes infection have been treated successfully. Those with severe infections of several years duration may also be provided with some relief and apparent cure. However, it cannot be assumed that the condition has been cured completely, since the treatment is new and there has not been sufficient time to observe those treated with the laser beam for any recurrence of the disease.

10-4 PHOTORADIATION TREATMENT WITH LASER BEAMS (ACUPUNCTURE)

A medical controversy exists with regard to the usefulness of laser acupuncture (photoradiation) in the treatment of organic ailments. Nevertheless, in many countries the medical profession is beginning to employ the treatment with certain patients. Pain due to disorders of many internal organs of the body travels away from the source of disorder and is experienced at sites remote from the ailing organ. Since pain manifests itself due to irritation of a nerve trunk, its root, or its terminal ends, there are a number of areas in the human skin where pain from the organs appears and is known as referred or reflex pain. For example, the pain from a decaying tooth may manifest itself in the temple; inflammation of the gallbladder or liver produces pain in the right shoulder blade; and angina pectoris pain radiates to the left shoulder and down the arm, as illustrated in Fig. 10-9. Such pains have been shown to have definite relations to certain areas of the skin. Thus, advanced medicine in the Orient identifies up to 700 skin sites where acupuncture treatment is admitted to be a useful panacea. Acupuncture, in its traditional sense, is performed with sharp-pointed needles, which are inserted by medical experts into the skin to alleviate or overcome pain. The latest method is the use of a laser beam instead of needles.

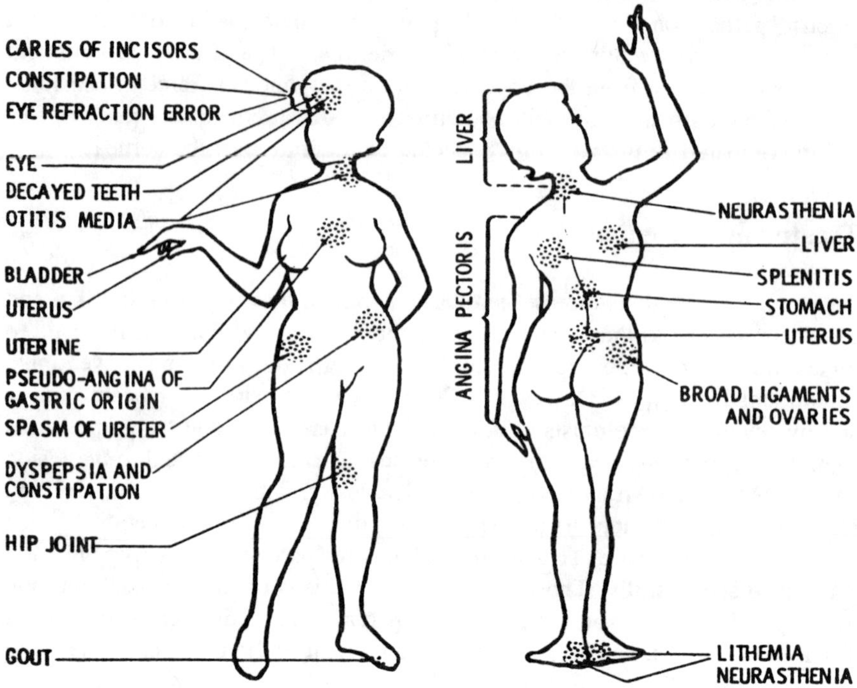

Figure 10-9 Cutaneous areas that are the seats of referred pain.

An acupuncture therapy unit using a laser beam for the treatment of various ailments of the body organs has been developed by Messerschmitt-Bolkow-Blohm of Germany. The system contains a Siemens 100-mW He-Ne laser adjustable in 1-mW increments. Since the skin is significantly transparent to red light, the He-Ne laser beam can penetrate it to 3 to 10 mm, depending on the texture and shade of the skin. Light skin offers the greatest transparency to red light.

In operation of the system, the laser beam is transmitted through a glass fiber conduit with a hand-held stylus located at the terminal portion of the conduit. The end of the stylus is provided with a 1-mm-diameter exit aperture. The point of the terminal end of the stylus is placed on a reflex area of the body, and the laser radiation is turned on for a period of several seconds to 60 seconds at a nominal laser power of 2 mW. The treatment is said to be painless as well as aseptic and can be employed to treat larger areas. Perceptible therapeutic effects have been obtained with patients having various organic ailments; the treatment has a further psychological effect on the patient because of the impressive laser instrumentation system. It is evident, however, that techniques may also be devised to use the treatment directly on the primary pain sites.

A number of physicians are using the laser acupuncture (photoradiation) technique for healing of wounds from burns, abrasions, and accidental punctures of the flesh. The beam is delivered to the area of disorder through a fiberoptic conduit having electrodes at its distal end peripheral to the laser exit port. The electrodes are used to measure the resistance of the patient's skin and thereby to locate the referred pain area. In some cases, the electrodes are used in combination with the laser beam to stimulate the affected area; the laser beam consists of He-Ne radiation from a 2- to 4-mW laser unit.

Various cosmeticians are reported to be employing a He-Ne laser beam to remove wrinkles from aged people's faces. The treatment consists of pulling the folds of the wrinkles apart by means of tweezers and applying the laser beam to the groove of the wrinkle for intervals of 30 to 50 seconds per treatment. The treatment is repeated for a period of several weeks, once a day, until a discernible effect on the "patient's" skin is noted.

10-5 ULTRAVIOLET LASER THERAPEUTIC ACTION

For many years, rhinologists have treated the nasal cavities of persons with rhinal disorders by use of an ultraviolet beam from a mercury discharge tube designed to be readily accommodated in the nasal passages of the patient. Because of the germicidal effect of ultraviolet radiation, sinusitis or gingivitis in some cases can be cured with the radiation given periodically in small doses. Since ultraviolet radiation of any specified wavelength can be obtained from a laser system, the procedure is apt to become routine in rhinology clinics and dental offices when laser equipment is designed that is suitable for the purpose. Because the present ultraviolet lasers can be focused or defocused to a specific area, unlike earlier ultraviolet beam instruments, the potential of the beam's application in both medical and dental fields appear to be great. Figure 10-10 depicts a cross-sectional view of the nasal cavities together with the oral structures where the ultraviolet beam could be utilized.

10-6 MEDICAL LASER SYSTEMS

At present, there is no standard medical or dental laser system for use by these professions, although some manufacturers are beginning to redesign their industrial laser systems to adapt them to the needs of the medical profession. Each laser system that is to be employed by an institution or a hospital is designed and developed in accordance with the requirements of the particular customer. Only one or two manufacturers in the United States and an equal number in Europe are producing laser equipment built expressly for the medical field. Accordingly, the medical laser systems that will be discussed below may be considered as typical in design and application.

Lack of Versatility of Laser Systems

One common problem that all researchers have encountered in the administration of laser beams in the medical and dental fields is the lack of versatility of the laser equipment used. This shortcoming has deterred many investigators from achieving the

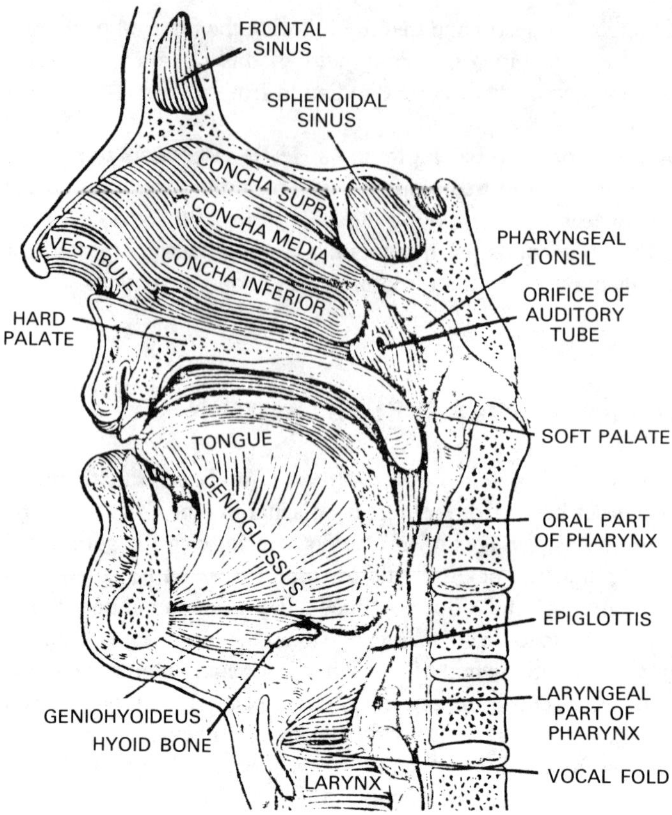

Figure 10-10 Saggital sectional view of the human mouth, nose, sinuses, and auditory tube.

maximum use and effectiveness of the laser beam. These investigators have been searching for a laser instrumentation scalpel that will have the flexibility of a conventional metal scalpel but greater effectiveness and utility. The general use of tubular conduits with reflectors, prisms, mirrors, or similar contrivances for delivering the laser beam from its source to the work area, as exemplified in Fig. 10-11, has been cumbersome and time-consuming. The enormous size of the apparatus, the irregularity of configuration, and the inflexibility that makes it difficult to reach corners or curved medical sites have limited the use of the instrument as standard equipment in medical work.

In a recent International Congress on Applications of Lasers and Electro-Optics, a number of medical surgeons emphasized the need for a flexible laser instrumentation system that is easily maneuverable at the surgical site and has an aiming light other than a He-Ne light, which has provided poor visibility of the laser focus when operating on a bleeding organ with a CO_2 or Nd-YAG laser beam. The physicians, in fact, have condemned the use of any articulated-arm device, although that type is the best available at present. They also wanted provision at the laser waveguide terminal for removing the smoke from burning tissue. Some of the physicians discussed the provision of several laser systems in a single medical unit, whereby they could easily change the beam—for instance, from a carbon dioxide laser to an argon ion laser—at the press of a pushbutton.

Foreseeing these problems, the author's facility has taken steps to develop and patent laser aids, auxiliary devices, and medical systems that should ultimately contribute to the versatility and effectiveness of the laser beam as a tool for medical and dental applications. The accompanying illustrations only restrictedly display the principle of the new developments for medical use of multiple-beam laser generators that will supply four laser beams with provision for frequency-doubled wavelengths. By pressing a button on the instrument panel, the surgeon will be able to select any of the four types of laser beams instantly.

Laserkinetics Medical Laser System

The medical instrumentation system shown in Fig. 10-12 is a Laserkinetics Model ML-102 series unit, which principally contains an optical waveguide that is capable of transmitting ultraviolet, visible, and infrared laser beams. The waveguide has an elongated stylus with buttons for adjusting the laser beam and switching it on and off. The waveguide also conducts a gaseous fluid that can be either a superficial anesthetic or a cooling fluid. Through the same conduit, suction of extraneous or smoky media from the tissue incision can be achieved entirely from the controls provided on the stylus. The system can be used in either medical or dental instrumentation. It has a laser power capability of several kilowatts, either mode-locked or Q-switched or both.

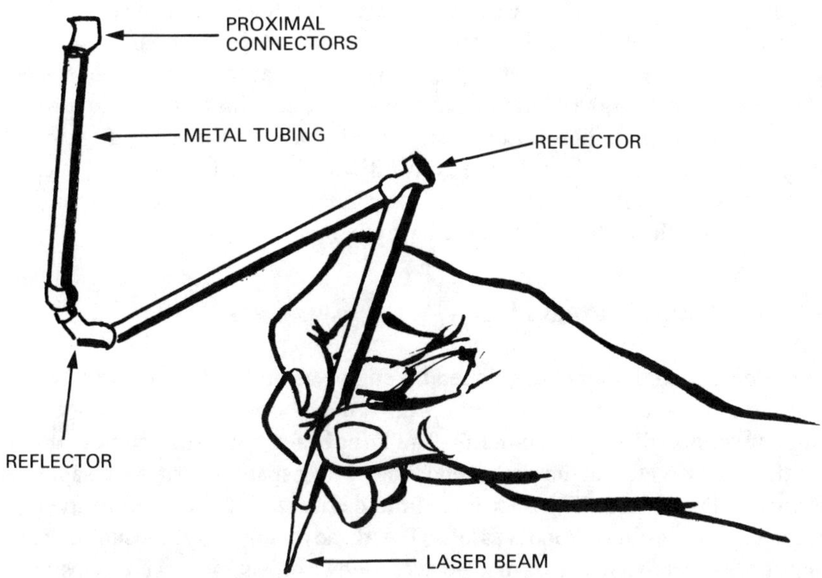

Figure 10-11 An experimental CO_2 articulated waveguide currently in use. (After Bell Laboratories.)

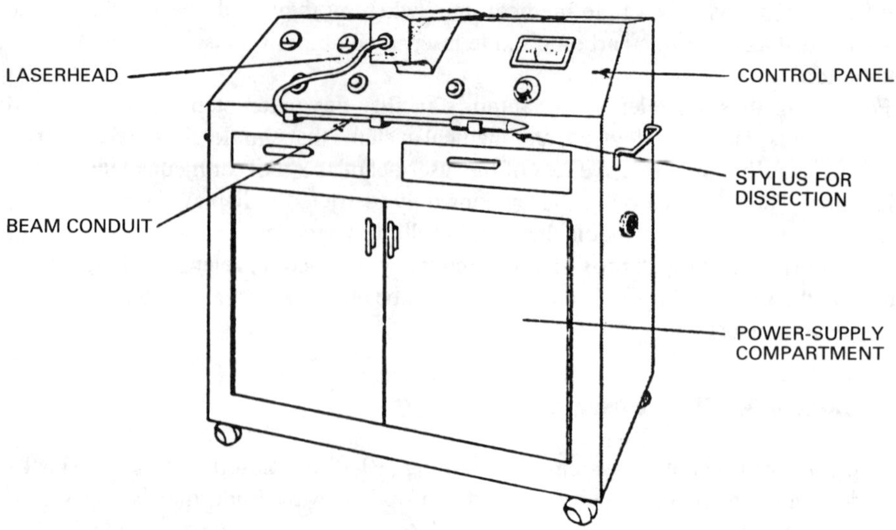

Figure 10-12 Laserkinetics Model ML-102 medical laser.

Meditec GmbH Nd-YAG Laser for Eye Surgery

A medical laser system manufactured by Meditec GmbH of Haroldsberg, West Germany, consists of a mode-locked and Q-switched Nd-YAG laser that operates with an energy of 4 mJ and provides a 30-nsec train of 30-psec pulses during the removal or vaporization of affected tissue in the eye. The radiation from the system is effective on both colored and transparent tissues. A safety feature is provided in the system so that if the laser beam is misfocused into the vitreous tissue behind the aqueous humor, the laser energy forms a tiny bubble that rapidly dissipates, according to the system operators. To date, nearly 15,000 patients have been treated with this technique. It is also reported that the manipulation for cataract surgery does not cause any fogging of the lens capsule, which is prevalent during conventional surgery for removal of cataracts. Such a laser system is shown diagrammatically in Fig. 10-1. Note that the surgical site is being visualized through a binocular microscope.

Laser Instrumentation Waveguides

An argon ion laser–assisted scalpel developed by engineers at the University of Washington is employed in the treatment of serious burns; the device reduces the blood loss that occurs in conventional surgery on burns. The laser beam is transmitted through an optical fiber that is provided at its distal end with a transparent quartz or sapphire blade, as shown in Fig. 10-13. The excision of the devitalized tissue is performed by simultaneously cauterizing the blood vessels. The disadvantage of the usual electric scalpel is that it prevents discharge of morbid secretions from suppurative tissues or of bacterial growth in dead tissues. Thus the condition may lead to sepsis or blood poisoning, necessitating multiple surgical treatments, especially for large-area burns, over a period of several days. The 25-W argon ion laser operates with a transmission efficiency of 80%. In this connection, other investigators have found that the He-Ne laser used with an argon-ion laser accelerates the healing of surgical incisions with the laser by irradiating the wound in 10-minute exposures per day for several weeks.

Laserkinetics Surgical Laser Waveguide with Stylus

A unique waveguide, developed by Laserkinetics Corporation for use in medical surgery and dental restorative work, is shown in Fig. 10-14. The versatile laser waveguide is provided with a novel conduit that conducts both a laser beam and a fluid for anesthetizing the area to be treated by the laser beam. Radiation emitted from a ruby, Nd-YAG, alexandrite, carbon dioxide, argon, or nitrogen laser can be transmitted through the waveguide because of the compatibility of the waveguide fiber with the wavelengths of these laser beams. With selective use of waveguide materials, no degradation of the waveguide filaments or solarization effects have been noted.

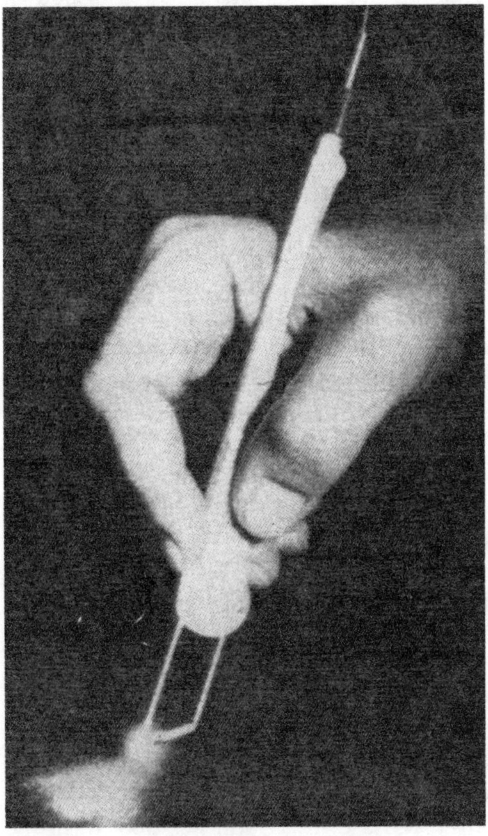

Figure 10-13 A laser scalpel with sapphire blade for excision of devitalized wounds. (Courtesy David C. Auth)

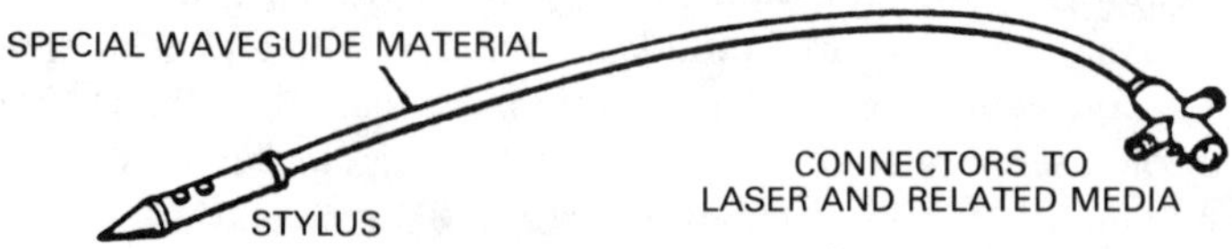

Figure 10-14 Laserkinetics surgical laser waveguide with stylus.

The device can be connected to any conventional laser source and can be used with the ease and flexibility of an ordinary mechanical scalpel. The surgery performed with it is almost bloodless, and the work area is sterilized and cauterized simultaneously. A single waveguide tube, containing an optically transparent material, guides the laser through the device with very low beam attenuation.

A special means of cooling, which is requisite for optimum operation of the surgical stylus, is provided so that energy levels up to several hundred watts or 80-J pulses without Q-switching are possible. Fingertip pushbuttons regulate the focal point of the laser beam incident on the work area and the flow of the fluid jet through the instrument into the affected site in the visceral, nervous, or blood-containing organs of the body. The focus of the laser plasma on the work area can attain a temperature up to 14,000°C. The fluid flowing peripherally to the fiberoptic material can also blow off debris from the wound. The conduit may also be connected to a suction vessel to remove the debris from the operated site if necessary.

Multiple-Laser Source System

An extension of the laser Model ML-102 is the Model MVL-4000, schematically illustrated in Fig. 10-15. The system consists of two beam-forming sections—a pulse-forming network and a continuous-wave beam generator. The control panel of the system console has two identified pushbuttons. One of the buttons leads the rectified high-voltage power to the pulse-forming network section and the other to the continuous-wave mode section, so that either of these modes can be selected as desired. In addition, the console contains three types of laser generators, which can be selected individually or in combinations of two lasers such as He-Ne and Nd-YAG.

All the physician has to do is to press the particular laser button to obtain the laser beam required for the surgery to be performed. The stylus of the waveguide is provided with two control buttons. One of the buttons controls the focus size of the beam and the other controls the flow of gas onto the lesion area. The system is also provided with a suction mechanism connected to the waveguide, which can either deliver a gas to the area of surgery or suck vapors evolved from the burning tissues, in addition to guiding a focused laser beam to the area to be operated.

Sterilizable Laser Beam Probe

A sterilizable laser beam probe has been developed at the University of Cincinnati Medical Center. The probe employs a Nd-YAG laser that furnishes 200 W in the 10,640-Å infrared range. The beam is used for the surgery of profusely bleeding organs,

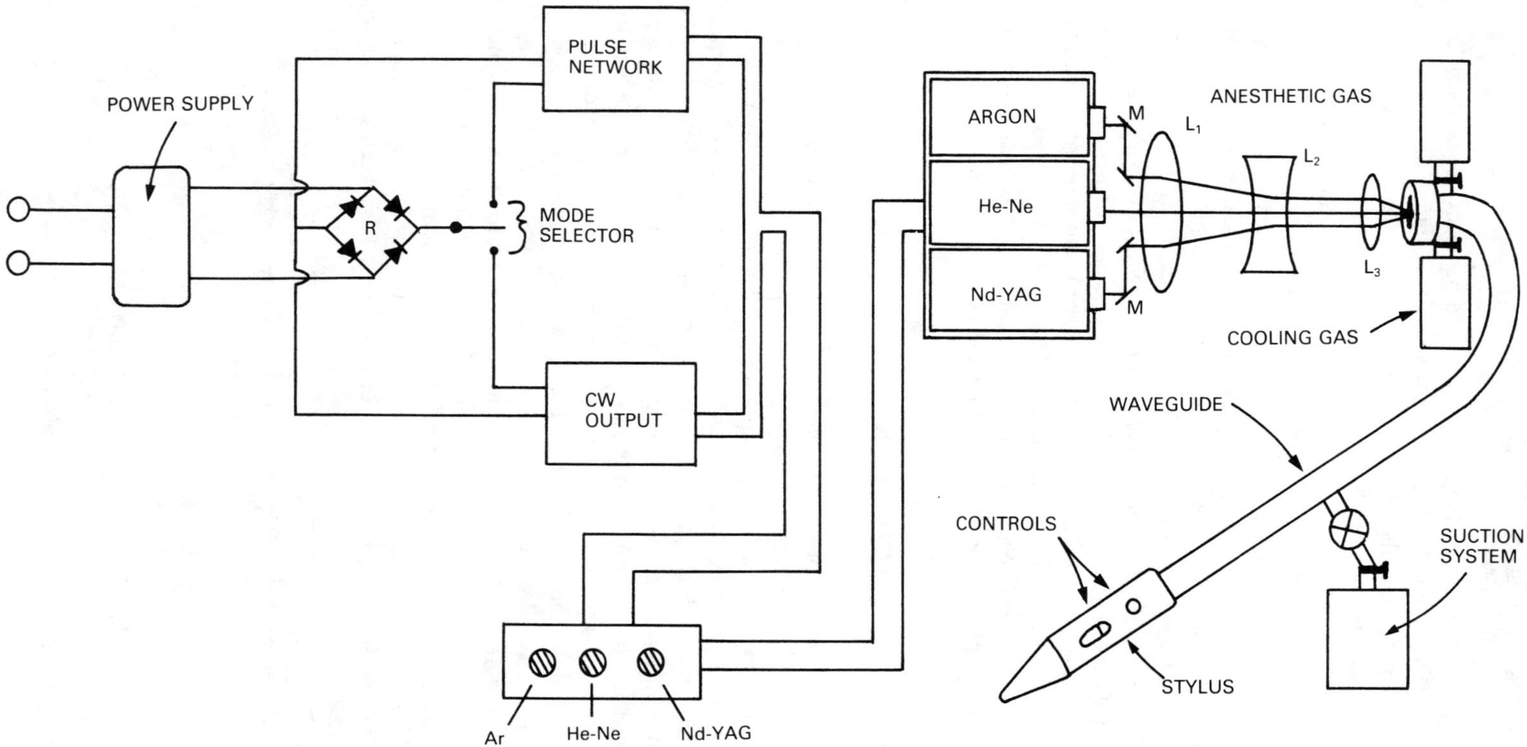

Figure 10-15 Multiple-laser-head medical instrumentation system Laserkinetics Model MVL-4000.

such as the liver, heart, lungs, stomach, and pancreas. Because of its high absorption coefficient for colored tissues, tattoos, and black tumors, the beam selectively destroys the affected tissue cells without severe damage to the surrounding healthy tissues.

The laser probe is a fiberoptic conduit with a transmission capability of 80% and a beam divergence of 10 mrad. Its developers claim that the laser probe can deliver to the work area energy that is several orders of magnitude greater than that available with ordinary fiberoptic materials. The focused power density of its radiation is about 700 W/mm^2 at the 0.4-mm focus of a lens of 50-mm focal length. For aiming the infrared laser to the work area, a He-Ne laser beam is used through the focusing probe to locate the point of incidence of the Nd-YAG laser beam.

The users of the probe also claim that the Nd-YAG laser is superior to a CO_2 laser in that its transmission through the fiberoptic conduit is more efficient. Also, by Q-switching the Nd-YAG and operating the laser beam at its longer wavelength of 13,000 Å, its cutting capability is improved. Thus, it is anticipated that the laser probe will someday be an indispensable tool in medical as well as dental surgery. An illustration of the fiberoptic probe using a He-Ne aimer is shown in Fig. 10-16.

10-7 GENERAL COMMENTS

This chapter is not intended to instruct the reader on how to administer laser treatment to the human body but rather to present an overview of the applications of laser beams in the treatment of various human ailments. The principal objectives of the text are to point out (1) the need to design and develop a laser system of maximum utility for the medical professional, (2) the need to incorporate ease and flexibility of use into equipment used for the delivery of the laser beam from its source to the work area, (3) the importance of including laser safety provisions in the equipment to eliminate the protective apparel now worn by the surgeon, and (4) the importance of the scientist and the physician working together as a team to formulate proper laser treatment techniques and dosages.

The author further calls the reader's attention to the progress made with lasers in the medical field and gives examples of the investigative and routine work now being conducted by the foremost medical and laser research workers, whose contributions to the field of laser surgery have formed the basis for the medical aspects of this text.

The various discussions of laser procedures have been intended only to provide

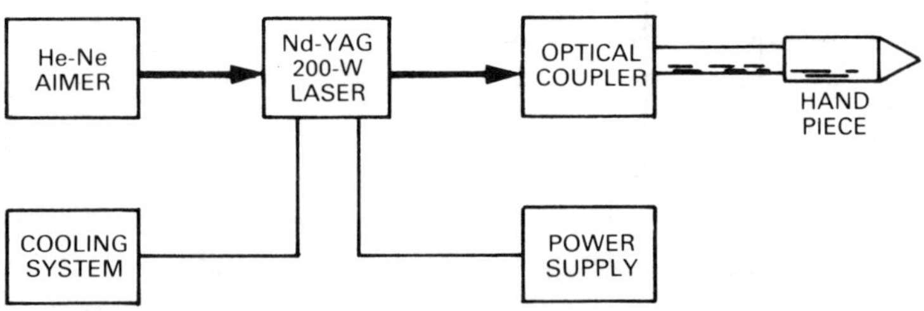

Figure 10-16 The laser probe with He-Ne aimer.

information and not recommendations by the author for their implementation. In this connection, the author and the publishers of this book take no responsibility for any damage incurred by any laser operator who is not expert in the pertinent medical field or is not aided by the medical professional in the application of laser beams to surgery or medical treatment.

QUESTIONS AND PROBLEMS

10-1 What laser materials were found to be effective in medical surgery? Do all laser wavelengths behave the same way on human tissue? What differences were found to exist among different laser wavelengths with respect to their action on human tissue? Which laser beams are absorbed better by the blood cells? What type of laser beam is found to make incisions without permitting much bleeding? The incision made by which type of laser beam heals faster than the other types? What is the difference in cutting power between a pulsed laser beam and one that is Q-switched and mode-locked?

10-2 What medical characteristics does a carbon dioxide laser beam show that other types of laser beams lack? Which beam, a Nd-YAG laser beam or a CO_2 laser beam, is more effective for surgery of the liver, stomach, pancreas, etc.? What lenses can be used with a CO_2 laser beam for effective transmission? In what branches of the medical profession can laser beams be used? Can laser beams treat diseased tissue as well as they can cut it? What type of laser beam is used for the attachment of a loosened retina? Can a laser beam be focused through the eye lens?

10-3 What types of laser beams are used at the University of Cincinnati for the removal of cancerous tissues, growths, nodules, etc.? Why do tissue areas treated with a carbon dioxide laser beam heal more slowly than those treated with other types of lasers? Is Nd-glass more effective than argon for a given tissue site? What did comparative experiments using conventional scalpel, electroexcision biopsy, and laser beam therapy indicate with respect to their effectiveness in therapeutic manipulations on the body? Was there any difference among the different laser beam therapies? What method was most successful in treating bleeding organs?

10-4 Which laser beam is mostly used in the removal of skin blemishes? Describe the surgical system employed at Beth Israel Hospital. How are polyps and growths on vocal cords removed? What type of laser beam is used in their removal? What problems did Swedish researchers encounter in their microsurgery work? How does the otological surgery instrument operate? What laser beam does it use? About what amount of laser power is used in tonsillectomy? What is the advantage of using a laser beam in tonsillectomy?

10-5 What type of laser beam is used in the treatment of sprained joints? How is the laser beam delivered to the joint of the patient? Is the joint opened for the treatment? Discuss. What type of laser beam is used and at what power is it used for coronary bypass treatment? Describe the method used in coronary bypass treatment. In brain surgery, must the skull be opened before the brain is treated with a laser beam? What laser beam is used for opening a blockage of the fallopian tube?

10-6 What type of laser beams can be used for gastrointestinal surgery? What laser beam is used for incision? What laser beam is used for cauterizing the treated area? What is the function of CO_2 gas in the manipulation of gastrointestinal surgery? How useful is a Nd-YAG laser beam in this type of surgery? Discuss. What is the controversy over the use of an argon laser versus a Nd-YAG laser in this type of surgery?

10-7 How are two laser beams employed in the detection and cutting of a cancerous tumor? How is a krypton ion laser beam used in the detection and removal of an ulcer from lung tissue? How can a laser beam detect the healing of a corneal suture? Describe the technique. What is genital herpes? Is it curable by conventional medical treatment? Is it curable with a laser beam? Discuss. How successful is the laser treatment of herpes?

10-8 Explain what is meant by acupuncture treatment with a laser beam. What laser radiation is used in this treatment? Is the treatment painful? What is referred pain? Does exposure of the skin at the referred-pain site cure the pertinent organ disorder in the body? What power of laser beam is used in this treatment? If there is pain from the shoulder down the left arm, what type of disorder does this indicate? How is the laser beam transmitted to the referred-pain site?

10-9 What is the problem with the present laser systems for medical instrumentation? How can this problem be corrected? Are there any laser radiation conduits that transmit a carbon dioxide laser beam? An ultraviolet beam? What are they? Explain. How does the Laserkinetics medical laser system differ from conventional laser equipment used for medical instrumentation? Discuss.

10-10 How does the University of Washington laser scalpel operate? How is it constructed? What is its main purpose? How does the Laserkinetics laser waveguide operate? What laser beams can the Laserkinetics waveguide transmit? How would it be possible to transmit these laser beams? Discuss. Describe the construction of the Laserkinetics waveguide.

10-11 What is a sterilizable laser beam probe? What type of laser beam does it use? Can it be used on profusely bleeding organs? How? What is its transmission capability? What focused-power density can it transmit? What focal length lens does it employ for focusing the laser beam? Can a CO_2 laser beam be transmitted through the laser probe? How is the beam through it focused if it is invisible, such as a carbon dioxide laser beam?

10-12 What is the technique for removing wrinkles from the face with a laser beam? What type of laser beam is used in this procedure? Is there any other agent that is used in combination with a laser beam for removing wrinkles from the face? What is the laser power used in this procedure? What types of professionals are now using this technique in the United States? Is the technique effective?

10-13 Show diagrammatically a laser system for delivering a laser beam to the site of a wrinkle. Include a fiberoptic waveguide with terminal optics for directing the laser beam. State what type of fiberoptic material is to be used in the waveguide. Also, state the beam exit aperture size at the distal end of the waveguide.

SELECTED SOURCES OF RELATED DATA

Alfano, R. R., ed. 1982. *Biological Events Probed by Ultrafast Laser Spectroscopy*. New York: Academic Press.

Carr, R. E., and I. M. Siegel. 1982. In *The Application of Laser Light Scattering to the Study of Biological Motion*. Adv. Sci. Inst. Ser., Vol. 59. New York: Plenum.

Davis, C. C. 1982. *Laser Science and Technology (Medical Section),* Vol. 2. Boca Raton, Fla.: CRC Press.

DeShazer, L. 1987. Advances in tunable solid-state lasers (their respective medical applications). *Laser Focus* February, pp. 54–58.

Earnshaw, J. C., and M. W. Steer. 1983. In *The Application of Laser Light Scattering to the Study of Biological Motion*. Adv. Sci. Inst. Ser., Vol. 59. New York: Plenum.

Epstein, M. 1985. Endoholography can give new view of surgical site. *Bio-Laser News* January.

First laser-stimulated laser firing reported. 1983. *Lasers and Applications* August, pp. 26–28.

Goldman, L. 1967. *Biomedical Applications of Lasers*. New York: Springer-Verlag.

Goldman, L. 1976. Medical instrumentation. *Med. Instrum. J.* 10(2):125.

Goldman, L., et al. 1987. New developments with the heavy metal vapor lasers for the dermatologist. *J. Dermatol. Surg. Oncol.* February.

Higgins, T. 1987. Computer control helps Nd-YAG laser fuse tissue. *Lasers and Applications* January, p. 22.

Hochstrasser, R. M., and C. K. Johnson. 1986. Laser spectroscopy in medical research. *Laser Focus* May, pp. 100–116.

Hofmann, R. 1986. Fragmentation of calculi with lasers. *Arch. Exp. Urol.* September 29.

Holmes, L. 1985. Ophthalmic treatments exploit novel laser/tissue interactions. *Laser Focus* March, pp. 20, 26–28.

Hora, H., and G. H. Miley. 1984. *Laser Interaction and Related Plasma Phenomena*. New York: Plenum.

Horn, G. 1984. Erbium laser: New tool for ophthalmology. *Bio-Laser News* February.

Iscoff, R. 1987. Excimer lasers in ophthalmology. *Lasers and Optronics* November, pp. 65–67.

Iscoff, R. 1988. Trends in medical imaging. *Lasers and Optronics* January.

Iscoff, R. 1988. Blood vessels welded by low-power lasers. *Lasers and Optronics* March, p. 28.

Kirchner, R. A. 1988. Lasers in medicine and surgery. *Photonics Spectra* January.

Koebner, H. K., ed., 1980. *Lasers in Medicine*. New York: Wiley.
Kreidl, J., and T. Higgins. 1986. CO_2 lasers remain in angiosurgery picture. *Lasers and Applications* December, pp. 40–44.
L&A Staff. 1983. Controversy surrounds laser facelifts. *Lasers and Applications* January, pp. 24–30.
L&A Staff. 1983. Carbon dioxide laser fiber delivers 2000-watt pulses for surgery. *Lasers and Applications* February, pp. 40–44.
L&A Staff. 1983. Laser treatment in pediatry. *Lasers and Applications* May, pp. 44–46.
L&A Staff. 1984. Laser fusion spins off synthetic vein material. *Lasers and Applications* April, pp. 36–38.
L&A Staff 1984. CO_2 laser superior for pacemaker surgery. *Lasers and Applications* December, pp. 28–30.
Lamping, S. 1986. Laser-blood interaction causes deadly clots. *Lasers and Applications* December, pp. 44–46.
Laudenslager, J. B. 1988. Excimer lasers adapt to angioplasty. *Laser Focus/Electro-Optics* May, pp. 57–68.
Martelluci, S., et al., eds. 1985. *Laser Photobiology and Photomedicine*. New York: Plenum.
McCaughan, J. S., Jr. 1983. Photoradiation treatment for cancer using HpD. *Laser Focus* May, pp. 48–56.
McCord, R. C. 1986. Laser-tissue interaction. *Lasers and Applications* November, pp. 61–65.
Moretti, M. 1988. Eye surgeons pursue corneal sculpting by laser. *Laser Focus* January, pp. 28–32.
Muncheryan, H. M. 1974. Medical Instrumentation Scalpel. U.S. Patent No. 3,821,510.
Parish, J. 1983. Surgical emphasis, new development. *Laser Focus* March, pp. 56–60.
Renault, G. 1987. Clinical applications of laser fluorometer. *Lasers and Optronics* December.
Saksena, S., et al. 1987. Argon laser ablation of malignant ventricular tachicardia. *Am. J. Cardiol.* January.
Shamberg, A. M., et al. 1987. Use of Nd-YAG laser in treatment of bladder cancer. *Urol. J.* January.
Shamberg, A. M. et al. 1987. Treatment of interstitial cystitis with Nd-YAG laser. *Urology* April 29.
Shirahama, T., et al. 1987. Application of Nd-YAG laser hyperthermia in urogenital tumors. *Gan To Kagaku Ryoho* March 14.
Spectra-Physics. 1979, 1984. Endoscopy with Argon Laser, technical brochure. Mountain View, Calif.
Srinivasan, R., et al. 1987. Dynamics of the ultraviolet laser ablation of corneal tissue. *Am. J. Ophthalmol.* March.
Von Bally, G. 1979. *Holography in Medicine and Biology*. New York: Springer-Verlag.
Wright, V. C., and M. A. Riopelle. 1982. *Gynecological Laser Surgery: A Practical Handbook*. Houston: Biomedical Communication.
Young, M. 1984. *Optics and Lasers,* 2d ed. New York: Springer-Verlag.

CHAPTER

ELEVEN

LASER INSTRUMENTATION IN DENTISTRY

11-1 INTRODUCTION

Since 1964 there has been considerable investigation of dental specimens, both in vivo and in vitro, using laser beams from a Q-switched ruby laser emitting at 6943 Å and a Q-switched carbon dioxide laser emitting at 106,000 Å. The investigators include laser scientists and dental professionals working in research institutions and dental laboratories both in the United States and abroad. The investigative work has covered the restorative anatomy of dental structure and dental diseases, employing pulsed laser beams. This endeavor has led to the laser-induced inhibition of caries (tooth decay) and is centered on the effect of the laser beam on dental soft tissues, dental bone structures, and buccal-cavity mucosa; the degree of laser safety necessary in the manipulations; and the potential applications of laser radiation in dentistry in general.

The principal drawback in current use of the laser beam in dental offices has been that of building suitable laser systems that offer the greatest versatility of instrumentation, compactness, efficiency, and effectiveness of the laser beam as a dental tool. This chapter describes the investigative work accomplished, difficulties encountered, and how these difficulties can be overcome to make the laser beam useful in clinical and prosthetic practices.

The discussions in this chapter further evidence the importance of the use of the laser beam for procedures in both restorative and therapeutic dental work and the benefits that can be expected. It is believed that certain types of laser instrumentation in the dental office could become standard practice as the recently improvised dental systems are further improved to make them practical. To this end, much work has been done by the author to advance the state of the art and the design of system integration.

11-2 THE STRUCTURE OF THE TOOTH

The alveolar processes of the maxilla and the mandible contain sockets covered by a dense, insensitive, fibrous membrane (the gums). The sockets or alveoli are lined by the periosteum, which, together with the gums, serves to attach the teeth to their sockets and to provide a means for nourishment for the teeth. Each tooth has three main sections: the root, consisting of one or more fangs; the crown, which is beyond the level of the gums; and the neck section, which occupies the region between the root and the crown, as illustrated in Fig. 11-1. These three structures are harder than the bones in the body, the enamel being the hardest.

As shown in Fig. 11-1, each tooth consists of a central structure—the dentin—which encloses an internal pulp cavity and is capped by a dense layer of enamel. The pulp cavity is continuous with a canal containing blood vessels and nerves. The enamel is composed of 98% inorganic and 2% organic matter. The dentin contains approximately 30% organic and 70% inorganic matter; it is attached to the enamel by means of the cementum, which interfaces the dentin, the enamel, and the gums. The enamel is the exposed part of the tooth and is used in mastication. The carious process usually starts in the enamel and in an exposed area of the cementum, in the case where the gums are receding.

A disease of the gums, such as pyorrhea, may affect the health of the tooth by permitting the passage of bacteria from the affected gums to the root canal, infecting the tooth. When this occurs, some of the bacteria may pass into the bloodstream and settle in the visceral organs, also infecting these organs. Caries may also start when the tooth enamel is scratched by rough dental abrasives or by other hard objects. Various methods are used at present to alleviate or cure caries. One of the most common techniques involves removal of the affected area and its restoration by a chemical, plastic, or metallic material. The permanence of such restorative work is dependent on the experience and competence of the dental professional.

When the disease of the teeth is left unattended and the bacteria from the mouth invade any of the vital organs, such as the liver, gallbladder, pancreas, kidneys, and heart, serious disorders of these organs can develop that require immediate medical care and treatment (see Fig. 11-2). Although the stomach can destroy many types of bacteria by means of its hydrochloric acid secretion, it may at times be infected by certain bacteria that can cause serious trouble. Many kidney and heart diseases may

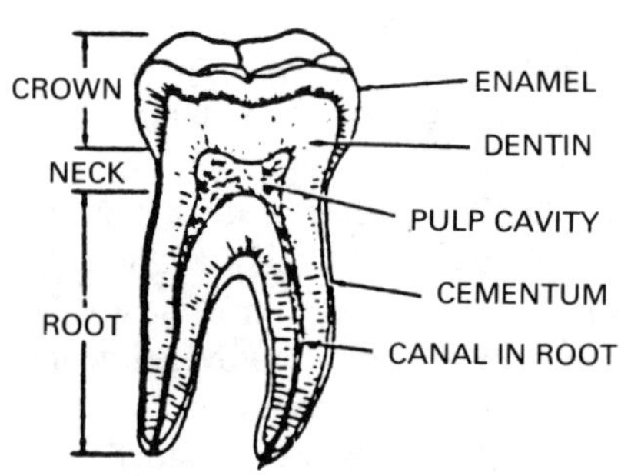

Figure 11-1 Longitudinal section of a human tooth.

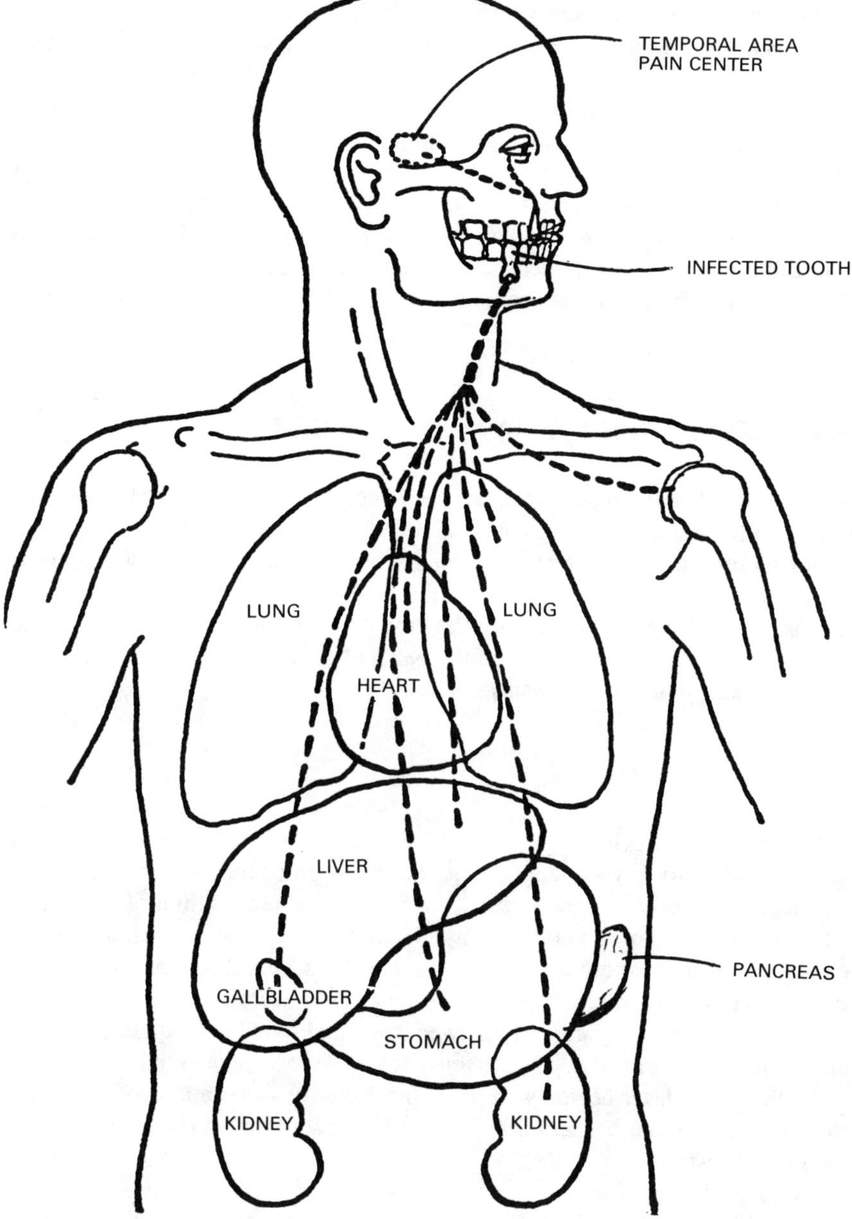

Figure 11-2 Vital organs of the body affected by dental disease.

have their origin from bacteria that have invaded these organs from infected teeth, or so-called tooth abscess. Consequently, it is readily seen that proper maintenance of dental hygiene is a very important responsibility of the person who intends to maintain a healthy body.

It has been reported that dental disease attacks about 98% of the population at some time during their lives. An average preschool child is said to develop at least two decayed teeth by the time he reaches school age. Approximately 68% of children between the ages of 12 and 17 have been found to have some form of gum disease. An average 20-year-old person is said to have 14 decayed teeth, filled or missing. An estimated 45 million adults are infected with some form of gum disease, and this does not include immigrants from various part of the world, 75% of whom have been found to have dental diseases.

11-3 EFFECT OF LASER RADIATION ON THE ENAMEL

During the past several years, much work has been under way on the use of laser beams in dental therapeutic and certain restorative applications. The beam can drill out the carious tooth structure, cauterize or sterilize the cavity, or fuse the enamel where there is slight porosity or a minute crack. The results of the experiments conducted have shown that laser "brushing" of an initially healthy enamel can prevent caries in the normal teeth, because the thermal effect of the beam alters the surface structure of the enamel to resist decay-causing elements.

Application of Ruby Laser

The earliest application of a laser beam on the enamel to prevent caries was from a ruby rod emitting at 6940-Å wavelength. The disadvantage of the ruby laser was that it created a diathermal action on the enamel, causing partial coagulation of the pulpal substance by its absorption of the laser energy. Thus, the investigation was diverted to other types of lasers in an effort to discover a laser beam that would be more compatible with the enamel without significantly affecting the pulpal substance within the dentin cavity. Both Nd-YAG and CO_2 laser beams were tried and found to be quite effective, and the absorption by the pulpal substance appeared to be inconsequential. Additional advantages of these two laser beams were their insensitivity to variations of color and shade from one tooth to another and the fact that the absorption of the beams by the enamel remained constant.

Extensive as well as intensive investigations were conducted on teeth both in vivo and in vitro, using ruby, Nd-YAG, and carbon dioxide lasers. The carbon dioxide laser beam was found to be superior to the ruby laser beam because of its longer wavelength. By using a laser beam intensity that is more than three times that of ruby, the invisible laser beam from carbon dioxide was found to be readily absorbed by the tooth tissue. An added advantage of the CO_2 laser beam was the negligible rise in the tooth temperature contiguous to the area being worked on, which made it more attractive for application in dental instrumentation.

In 1971, experiments conducted on the enamel by J. Vahl of Leipzig, Germany, indicated that a pulsed laser beam at 25 to 50 J/cm^2 melts the enamel to a depth of 0.5 μm (5000 Å). The findings further indicated that cariogenic substances preferentially attack the prism borders of the enamel. Thus, it was concluded that naturally occurring micropores in the enamel can be sealed, preventing invasion of cariogenic substances into the pores.

Researchers at New York University and at Eastman Kodak Park have studied and performed experiments on fusing tooth-colored hard materials directly to the enamel and have achieved high success. It should be possible to develop a tooth-colored cement that could be placed in a tooth cavity and allowed to set chemically. The cement could then be cured and glazed by laser radiation so that the cement-enamel interface is intimate. Work in this direction is in progress in several research institutions, where an inorganic solution is being used to seal the cavity and subsequently dry and glaze it, using either a carbon dioxide or a Nd-YAG laser beam.

Bioactive Tooth-Implant Material

Battelle Institute of Frankfurt, Germany, has developed a sintered calcium phosphate that is compatible with living tissue and forms an intimate bond with the jaw bone or the alveolar cavity of the tooth. The material can be made to dissolve in the bone tissue, and as the bone tissue grows it replaces the implanted chemical material. Bone defects can also be temporarily bridged by the material because the controlled porosity of the substance promotes tissue growth. The implant material has been used on laboratory test animals in plugging implantable tooth roots and bone cavities, repairing defects in the jaw and shank bones, and plugging cranial apertures that are surgically opened.

The most significant application of this sintered material, which is a ceramic composite, is in tooth implantation. The porous composite permits the bone tissue to grow into the pores of the material to form a tooth root firmly bound to the jaw bone. The researchers intend to investigate the suitability of the ceramic material for coatings on conventional prostheses combined with antibiotics in osteomyelitis therapy. The curing and promotion of growth of the bone tissue into the porous ceramic would be enhanced by periodically irradiating the material with a short-wavelength argon laser beam. The ultraviolet rays from the argon laser are not only germicidal but also effective in solidification of the combined bone-ceramic composite.

Lee Pharmaceuticals, El Monte, California, have pioneered a composite, called Enamelite, that is applied to a broken or defectively grown tooth (Fig. 11-3), shaped into a normal tooth configuration (Fig. 11-4), and then cured to simulate the natural tooth. Such a composite can be successfully glazed and hardened by means of a laser beam to a lasting, durable dental structure. The appearance of the structure is superior to that achieved with earlier methods of repairing the tooth with a noble metal, and the procedure is faster and less costly to the patient.

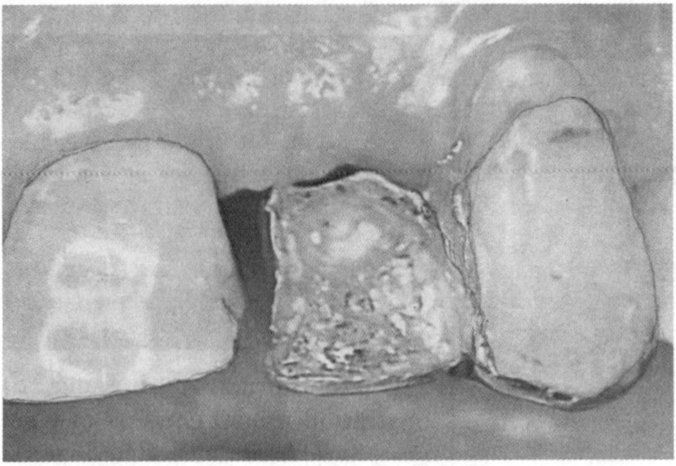

Figure 11-3 A vestigial, incomplete tooth structure.

Dental Drilling and Cauterization

Work is also in progress on drilling teeth in which the carious process has begun to destroy the impaired enamel. Although laser drilling of the enamel and part of the dentin (depending on the depth of the carious disorder) can be performed successfully, the ultimate result of the laser procedure depends on the skill of the operator of the laser system. In this regard, the procedure does not differ from the present technique using mechanical drills; that is, drilling too far into the pulp cavity might result in infection or destruction of the nerve endings, thus ending the life of that tooth. The laser beam must be controlled in the same way as the dentist's drill, so that the drilling

Figure 11-4 The vestigial tooth after having been repaired with Enamelite.

procedure can be arrested at any depth from the surface of the enamel. This requires knowledge of the spectral characteristics of the laser beam utilized as well as anatomical knowledge of the various elements of the particular tooth composition. The results obtained on the mechanism of tissue-laser interaction offer great promise in the use of the laser beam as a radiant tool in preventive dentistry.

During drilling with a laser beam, the unintentional effects of the laser radiation on the soft tissue surrounding the teeth, such as the mucous membranes, contiguous gingiva, and pulpal tissue cells, have also been considered. The findings indicate that, with or without preventive means of keeping the radiation away from the soft tissue, the effect of the laser beam on the soft tissue is inconsequential. With properly designed equipment and accurate administration of laser radiation to the work area, extreme precautionary measures would not be necessary any more than they are for mechanical drilling procedures.

The area in contact with the laser beam is immediately sterilized, owing to the rapid and extreme thermal conditions prevalent at the focal area. However, because of the instantaneousness of the laser thermal energy, the surrounding tissues are not affected. When there is a tendency to bleeding during the dental work, the characteristic cauterizing property of the laser beam offers a sophisticated and very effective means of cauterizing and treating the affected site.

Exposure of Oral Mucosa to Laser Radiation

The oral mucosa covers the gingiva, cheek, tongue, and labial membrane. Tests have been conducted with an argon laser to determine what effect a laser beam will have on mucous membranes unintentionally exposed to the radiation. Rubber dams have been employed while working on the enamel, and at times no precautionary measures have been taken. However, there is still the possibility of injuring the dental pulp by a diathermal effect through the enamel. Low-power infrared rays are less effective than short-wavelength radiation, for instance, from an argon laser beam. On the gingiva exposed to both types of laser beams, no significant change has been observed. However, moderate exposure of Caucasian skin to a ruby laser beam has resulted in exudation of the underlying cells and agglutination of the erythrocytes. This has led to the concept that coagulation of the blood cells by exposure to the laser beam could "weld" certain tissues together; an example of the latter action is the attachment of retinal tissue in cases of advanced diabetes.

Therapeutic Action of Ultraviolet Lasers

The gingiva of the teeth is susceptible to inflammation and infection unless the teeth are adequately cleansed and the plaque, which generally contains *Streptococcus mutans,* is removed frequently. Pyorrhea, which is inflammation of the gums, may also result from improper care of the teeth. At times it is necessary to remove the plaque by means of strong antiseptic preparations. However, in practice, the deep-rooted microorganisms are difficult to extract from the gums, especially when the inflammation has advanced to a state where suppuration from the dental root areas is predominant. In the usual method, the dentist applies a solution of silver nitrate to destroy the germs as well as to coagulate the tissue cells for replacement by new tissue growth. In the past, ultraviolet light from a mercury vapor discharge tube was also used with some success to alleviate the condition. However, this method did not provide a cure because (1) the radiation intensity and the wavelength could not be varied for different kinds of tissue under treatment, and (2) the instrument did not permit flexibility of usage because of its bulkiness and rigidity.

It is well known that the exposure to ultraviolet rays will destroy microorganisms such as staphylococci and streptococci in alveolar abscesses and sites harboring pus-forming bacteria in the gums. The rays will also destroy fungi, parasites, and many pathogenic and zymogenic bacilli. The body needs ultraviolet radiation to convert foodstuffs to vitamin D, a deficiency of which causes rickets, depression of the breast bone, and enlargement of the bones of the forehead. Heliotherapy in Great Britain and Switzerland has been practiced with marked effects on persons deficient of vitamin D. Radiation from a mercury discharge tube (ultraviolet rays) also has been used with success in treating dermal diseases such as scurvy, lupus, and eczema with carefully graduated radiation doses.

With the advent of laser radiation, copious emission of ultraviolet from various laser devices is available. This radiation can be applied with precisely controlled intensity, quality, wavelength, and directionality on any part of the body with the use of flexible quartz or sapphire waveguides. The source of ultraviolet laser radiation, as stated earlier, is argon, krypton, nitrogen, or cadmium vapor ions. These laser devices can produce emission of far-ultraviolet as well as near-ultraviolet wavelengths, depending on the wavelength necessary for a particular treatment. In the case of dental work, argon ion ultraviolet light of 4- to 10-W power is used in the continuous-wave mode for "brushing" gingiva that have been inflamed due to some form of infection. The dentist periodically treats the gums by passing a defocused beam of the radiation over the affected area. The advantage of this treatment over the silver nitrate method is that the patient does not feel pain and his daily food intake is not interrupted. The treatment will be most beneficial at the early stages of the gingivitis or pyorrhea. However, the author must caution that the treatment would be unsafe except when given under medical supervision.

Overdose of radiation therapy may lead to serious disorders. The dentist administering ultraviolet laser radiation as a therapeutic treatment must use considerable care to avoid injury to the tissues by overexposure. Accordingly, the operator of the therapy must have proper training and experience in administering proper dosages to specific body areas to be treated. At times, this procedure should be carried out in a hospital with the assistance of nurses specialized in this type of therapy.

Therapeutic Action of Carbon Dioxide Laser

Recent investigative work with laser beams both in the United States and abroad has shown that the carbon dioxide laser beam is in extensive use for the treatment of pathological tissues in the mouth cavity. Both microscopic morphology and macroscopic structures, such as hydroxyapatite, have been treated with significant effects by transformation of the composite by hydroxyl ions in the saliva. Sterilization of nonmineralized tissues with a carbon dioxide laser has also been achieved. Since the CO_2 laser can coagulate blood vessels with diameters as small as 0.5 mm and surface surgery can be performed bloodlessly, hypertrophies, tumors, leukoplakia, and similar structures are easily removed by means of the laser beam. The surgery with the beam is rapid and sterile, and healing takes place without the formation of fibrosis or sclerosis, which would be expected with conventional methods when performed without special care.

It must be kept in mind that the teeth are surrounded by saliva, which contains more than 10^{10} microbes per cubic millimeter when a tooth is infected. This infectious material can affect other tissues susceptible to the invasion of bacteria because of their weakened condition. In such an event, the calcified structures are softened, making them amenable to microbial invasion. The laser beam can sterilize the mineralized tissues, preventing further infection.

It is a proven fact that the laser beam not only can treat dental decay but also can produce healing in the affected tissues. When dentin is irradiated with a laser beam, its structure becomes hardened and its vulnerability to the acid environment of the mouth is substantially decreased by the formation of a microstructural film without thermally affecting the pulpar substance. In this connection, radiation from a Nd-glass laser has also been employed with distinct success. These tests and treatments in vitro indicate that with improved instrumentation systems, laser beam treatments of buccal cavity and related infections could be routinely applied in the near future.

Further Studies of Enamel Immunization

Current experiments on the immunization of the enamel against caries have shown that laser exposure of the enamel produces recrystallization and growth of the hydroxyapatite of the enamel, permitting diffusion of fluoride ions from the treatment solution into the enamel. Such a treatment may also preclude the use of acid etching of the drilled surfaces for ease of bonding of any resin filling. In another experiment, Nd-YAG emitting at 10,640 Å was used on the enamel because it does not produce diathermal effects on the pulpal tissues; however, the radiation from Nd-YAG is not readily absorbed by the enamel. For the latter reason, a radiation-absorbing substance, such as apatite or a carbonaceous chemical, is applied on the enamel surface before subjecting it to the laser irradiation. In still another study, fissures in the enamel were sealed with a phosphate mixture using a carbon dioxide laser beam. The phosphate mixture is sintered in the fissure by using laser energy densities up to 50 J/cm^2. Since the time of exposure to the radiation in sealing the fissure is extremely short (milliseconds) with 1.5 J of laser energy, the pulpal substance is not affected by the heat involved.

To facilitate enamel immunization and gingival treatment, a pair of cheek retractors have been developed by the L. D. Caulk Company, a division of Dentsply International, Inc., to provide a clear and unobstructed field in which the dentist can manipulate instruments for any dental laser treatment of the enamel, as illustrated in Fig. 11-5. The device further permits cavity preparation, surgical instrumentation, prophylaxis, restorative bonding, and occlusion examination in general.

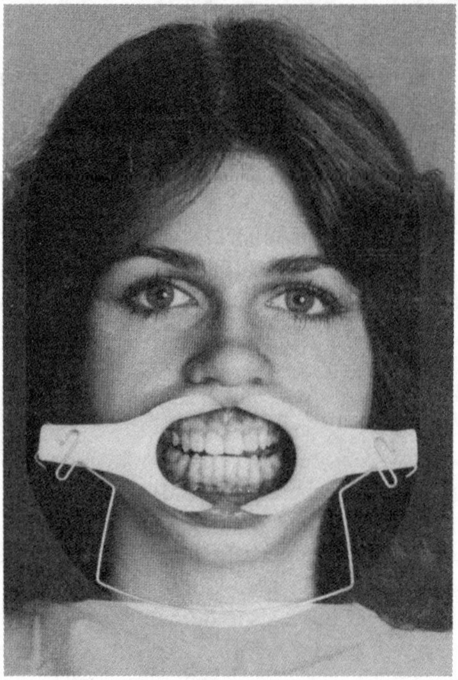

Figure 11-5 A typical technique for laser-brushing the enamel by retracting the cheeks apart.

11-4 ROOT CANAL TREATMENT WITH LASERS

Dentists at times encounter a patient with an innocent-looking tooth that has been filled with gold or silver and externally appears to be a healthy tooth, yet the patient may be suffering with ache and is unable to chew on that tooth. An x-ray reveals that the root of the tooth where the root canal joins the nutrition-feeding tissue has been infected. This infection may have entered the root through the unbonded interface between the tooth and the filling or through the inflamed gingival area. The dentist usually removes the old filling, cleans and sterilizes the root canal, and fills the cavity with a ceramic or similar material; subsequently, he crowns the tooth. Occasionally, the procedure requires the performance of minor surgery by an endodontic surgeon. In any case, the procedure is intricate and costly and sometimes may be beyond the financial ability of an average wage earner.

A method using a laser beam has been advanced by the author but has not gone beyond the experimental stage by dental professionals. The procedure consists of directing a fine laser beam focus from an argon ion laser or low-power CO_2 laser to the gingival area external to the root where the root canal infection has developed; a fine hole is drilled into the gingival wall to reach the infected cavity. The hole periphery becomes cauterized during drilling with the laser beam, and the accumulation of infected fluid can drain out through the hole into the mouth cavity, from which it is removed by suction with an aspirator used by the dentist. The hole is allowed to remain open for a few days while the patient is given antibiotics to help the residual infection to heal, whereupon the aperture closes naturally.

Recent experimental work conducted by Prof. J. Melcer and his colleagues and sponsored by the Atomic Energy Commission in Paris has confirmed the fact that the results obtained with a carbon dioxide laser are immediate and excellent. The drawback appears to be reinfection of the root canal by germs emanating from the surrounding chronically infected dentin. However, when the condition is arrested in an incipient stage by application of the radiation, significant sterilization of the infected areas is noted. This action seals the dentin canaliculi by creating a germproof thin growth over them. These investigators have shown that carbon dioxide–treated buccal tissue structures heal faster with denser new growth structures. Work on in-patient cases is said to be in progress.

Although the procedure appears to be relatively simple and inexpensive, no dental researcher has applied it to a living human patient routinely in a dentist's office. The technique should be tried and established with the assistance of a laser physicist specialized in laser beam compatibility with different human tissues and adapted to fine control of the laser beam.

In connection with the procedures just discussed, it would be proper to state at this time that the preceding laser treatment may not become a reality if the medicinal method investigated by Forsyth Dental Center, Boston, is put into operation prior to the routine application of laser beams in dental work. The center has been investigating the potency of a vaccine that will reduce cavities in the teeth of young and old persons. This work has been based on the theory that dental cavities (caries) are caused by bacteria known as *Streptococcus mutans*. The bacteria produces an enzyme, glucosyltransferase, that extracts the sugar from the plaque of the teeth and forms glucan. This substance assists the bacteria in colonizing the tooth surfaces; they then gradually work their way into the enamel to produce the carious disease.

In the dental restorative procedure, a vaccine produced by the bacterial enzyme is injected into the salivary glands of the test subjects (animals) to increase the antibodies normally present in the glands. The antibodies fight the formation of the carious germs. It has not yet been determined whether the vaccine will be administered by injection, by swallowing the material in a capsule form, or by applying it to the patient's dental or oral surfaces. Whichever method is selected, the procedure will immunize the patient for a period of time, after which the treatment is to be repeated. It is readily apparent that in the future we will see new approaches to caries treatment, whether by "laser brushing" the enamel surfaces or the diseased gingiva or by immunization of the patient with the projected vaccine procedure.

11-5 POTENTIAL PROBLEMS INVOLVED IN LASER INSTRUMENTATION

At present, there are many problems involved in carrying out dental work with laser beams. For instance, in the conventional method, the dentist presses a drill against the tooth and drills in any direction he deems necessary, being careful not to drill into the pulp cavity inadvertently. When a laser beam from conventional laser equipment is used, the laser head must be positioned by means of special fixtures to point the beam at the work area. To focus the invisible beam, a visible light beam must first be focused through a microscope which the beam traverses. The light beam must be turned off and the laser beam must replace the light beam path. Alternatively, a reflective conduit scheme as described earlier must be used; this is cumbersome and inflexible in operative procedures. The advantages gained by the laser beam method are defeated by the many existing disadvantages of the equipment.

It is thus evident that in order to utilize the useful characteristics of a laser beam technique, new and unique laser equipment should be designed and constructed that offers versatility, reduction of cost, permanence, and, above all, high-quality work that is superior to that achievable with any method presently known to the dental profession. Anticipating these problems, the author has designed and patented a dental laser instrumentation system that uses a laser beam manipulator of high versatility and flexibility to meet the demands of both the medical and dental professions. Several other companies are also currently following this concept.

11-6 DENTAL INSTRUMENTATION SYSTEM

Two types of dental laser instrumentation systems have been developed by the author. One type is designed for use on human teeth in vivo for drilling and sterilizing the cavity; it will also be used to restore the tooth cavity with a chemical substance and to glaze the tooth enamel, as well as being used for therapeutic purposes. The second type will be employed for repairing fractures in bridgework.

Both of these systems are relatively small and are provided with a laser beam conduit that can transmit the laser beam to any section of the tooth under manipulation. These systems permit the laser beam conduit to be employed in the same way as a dentist's drill or a surgeon's scalpel. The laser beam is automatically focused to the spot or area to be instrumented, thus offering the ultimate in flexibility and ease of operation. The work should be fast, clean, and highly effective. The dental instrumentation system is shown diagrammatically in Fig. 11-6.

The diagram in Fig. 11-6 is only a typical representation of the laser system without detailed design and construction features. Principally, the diagram illustrates the manner in which the laser beam is applied to the area of interest and the provision for preventing infection due to the patient and the dentist breathing each other's exhalations. Furthermore, a binocular microscope is provided for magnification of the area treated. The system consists of a flexible laser beam conduit (A) projecting from the laser generator (B) furnished with a power supply contained in the console (C). Also shown are the microscope (M) and the sanitary shield (S) connected to the intake of the aspirator conduit (D), from which the operator's exhaled breath is dissipated into a vessel contained within the console. An auxiliary stabilizing device may be attached to part of the chair (F) to immobilize the patient's head when necessary.

It should be noted that the equipment is self-contained and does not require an additional dental chair for conducting routine dental work. The flexible laser conduit (A) can be directed to any part of the tooth for drilling, sterilizing the drilled cavity, or cauterizing the gingival tissue if bleeding occurs. Furthermore, certain types of

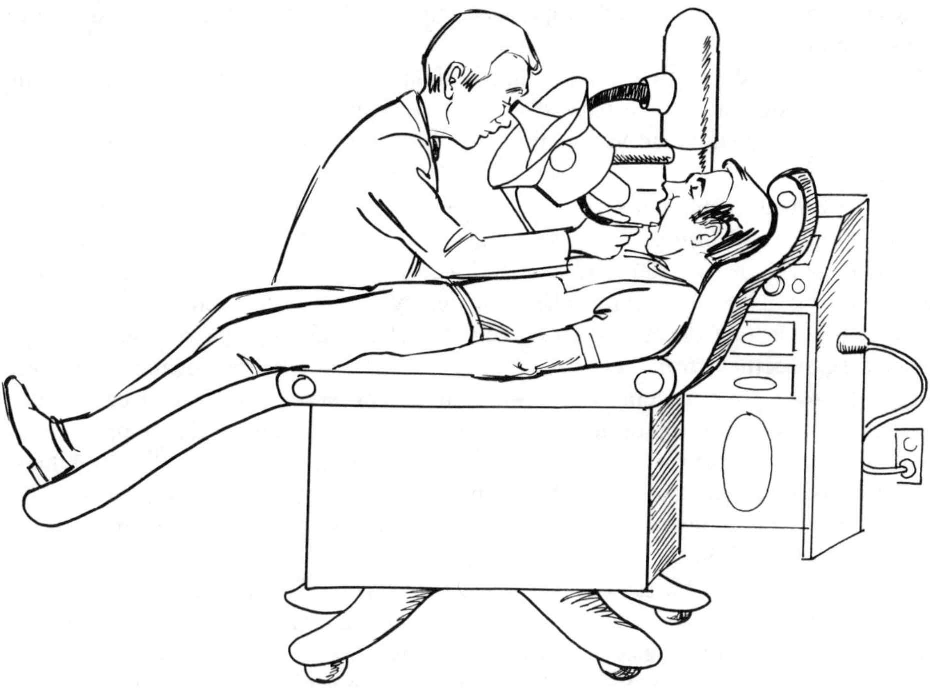

Figure 11-6 Dental instrumentation system developed by Laserkinetics Corporation.

dental fillings and bridgework repairs can be made in the mouth without inconvenience to the patient. The console is also provided with the conventional dental instruments (not shown in the figure) for other mechanical instrumentations now current in a dentist's office.

11-7 LASER WELDER FOR DENTAL BRIDGEWORK

For many years dental bridgework elements have been joined together by the use of gold or an alloy heated with a blowtorch. This process generally wastes material and is time-consuming, and the results obtained are said to be inconsistent from one bridgework to another. The process is costly, and the cost is invariably passed on to the patient. Investigations of laser beam welding of gold and other dental alloys have shown the following distinct advantages of laser welding over the torch welding method:

1. The laser technique yields superior joints and strength.
2. It reduces the time of manipulation 10- to 15-fold.
3. Distortion of the finished product due to heat from the gas torch is eliminated by the use of a laser beam.

4. The accuracy of the work approaches the theoretical possibilities of a joining method.
5. The work is completed without an additional appointment for the dental patient.
6. No additional alloys are needed with the laser method.
7. Laser beam manipulation simplifies the operational procedures.

A laser instrumentation system developed by Laserkinetics Corporation for dental bridgework, shown in Fig. 11-7, is a compact and portable unit designed for welding of dental specimens. It offers the same ultimate versatility as the Model DL-3000 unit shown in Fig. 11-6. The principal use of this system is in the dental laboratory for welding and drilling dental workpieces. The system can also repair numerous other articles and microelectronic devices used in aircraft and space applications. Accordingly, the use of the system is not limited to dental laboratory work; it can also be employed for the processing of industrial materials. For instance, the equipment can weld, drill, and cut microelectronic devices and materials; drill holes in thin- and thick-film circuits; and scribe dielectric and ceramic materials, including certain plastics. The system can also solder or desolder microcircuit leads, such as transistors, diodes, capacitors, and hybrids.

The laser system Model DL-202 consists of a bench instrumentation unit with a switch in the foot treadle, which triggers the laser beam after the laser head contacts the workpiece. The driving of the laser head to the workpiece is also performed by the motorized treadle. If the workpiece is too large to be accommodated under the laser head, the laser beam can be directed to the workpiece through a laser waveguide designed for power loads up to 30 J or 8 kW/cm^2. To use the laser conduit, the distal portion of the laser head is removed from the console and the proximal end of the laser

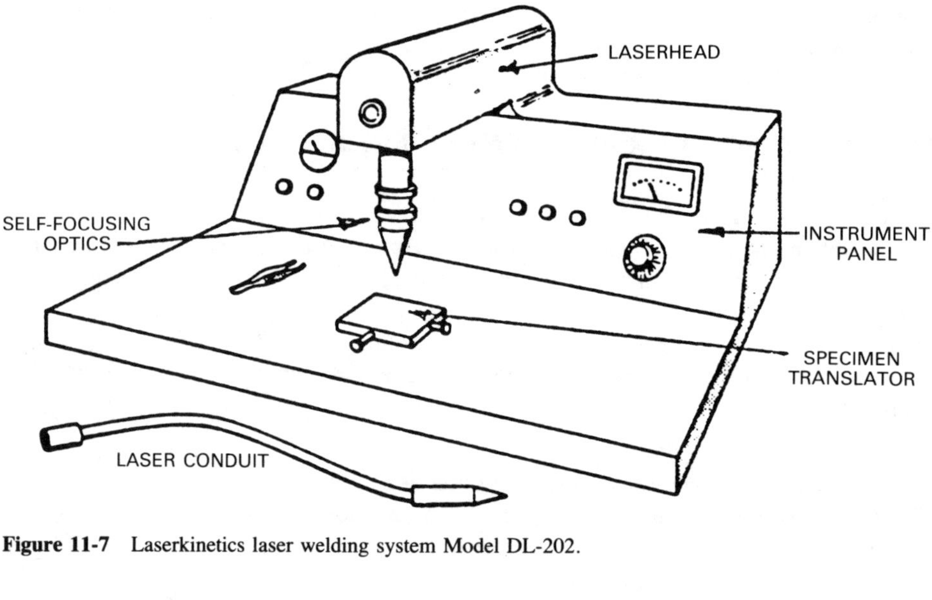

Figure 11-7 Laserkinetics laser welding system Model DL-202.

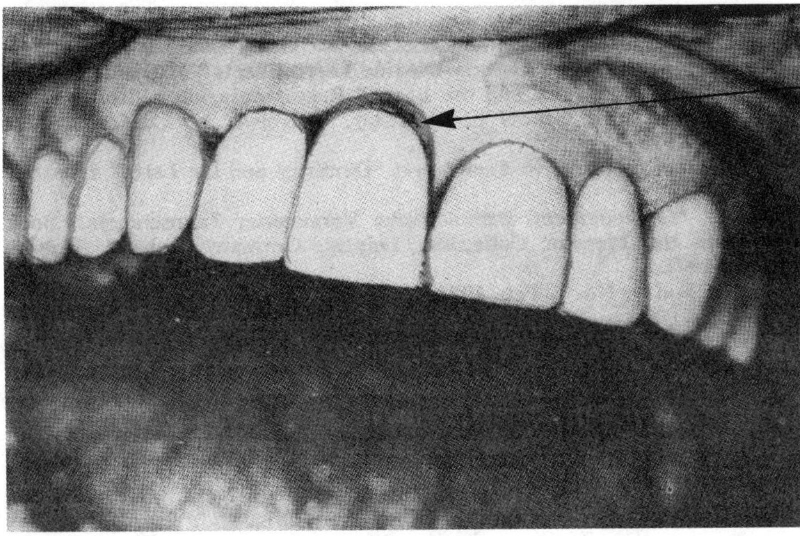

Figure 11-8 A laser-welded dental bridgework.

conduit, provided with an adaptor, is screwed on the laser head; this operation can be performed within 30 seconds. The laser conduit then can be operated as easily and conveniently as a ballpoint pen. It is anticipated that in the future, when application of laser beams becomes prevalent in dental offices, the Model DL-202 will be used with its waveguide directly on the tooth to weld broken parts of the fillings or denture inlays in the patient's mouth without removal of the parts.

A laser-welded dental bridge is shown in Fig. 11-8, which shows a photograph of the upper front incisor caps. The metal was cast and finished and was then butt-welded in an area 0.7 inch long and 0.04 inch thick. A series of 8-msec, 16-J pulses was used, and the pulse penetration was 4 mils. The work was performed by Thomas E. Gordon, Jr., D.D.S., head of the Laser Research Department, Orange Memorial Hospital, Orlando, Florida.

When laser implementation in a dental office becomes a reality, the laser waveguide will be manipulated as illustrated in Fig. 11-9, in which a laser waveguide is employed to convey the laser beam directly to the patient's mouth. The work is expected to be fast and painless, since it will not require injection of anesthetic into the gingiva prior to the instrumentation. In the figure, the technique of laser brushing is also illustrated.

In conclusion, it may be stated that the treatment of periapical infection with a carbon dioxide laser has produced excellent results when undertaken clinically. Preventive therapy is still at the stage of laboratory research. It was shown that the emission from a carbon dioxide laser has favorable effects on the dental tissues, although Nd-glass will also be a useful dental tool when proper equipment is available to the practicing dentist. It is evident that with improved and less costly dental equipment, the laser beam will prove to be an indispensable instrumentation device for the dental profession.

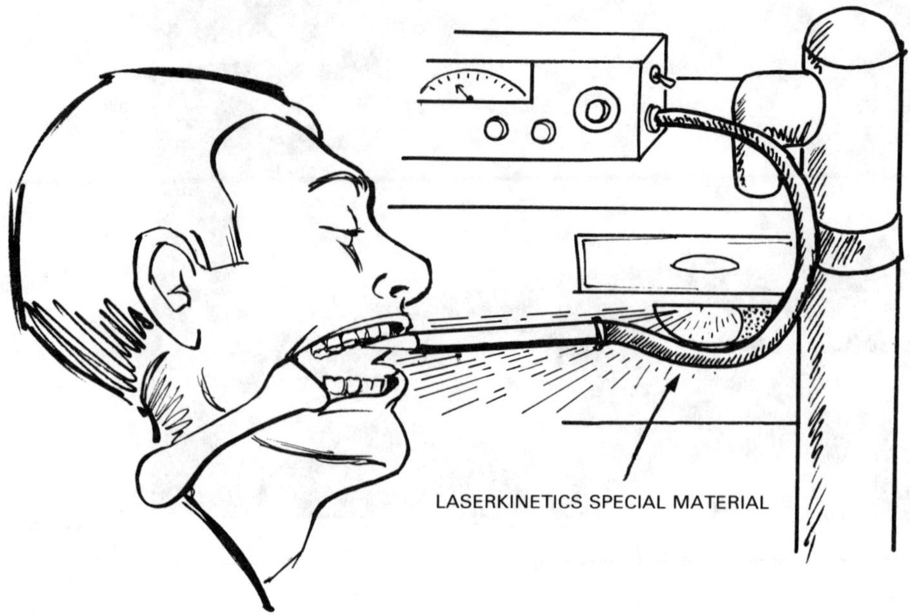

Figure 11-9 Scheme illustrating the technique of transmitting a laser beam to the teeth in vivo.

11-8 COMMENTS

The text on dental experiments and investigations contained in this chapter is based principally on the results obtained by a large number of workers in the dental profession, particularly that accomplished by the late Dr. Ralph H. Stern, School of Dentistry, University of California at Los Angeles, and his co-workers, Prof. Reider F. Sognnaes, Prof. Johanna Vahl, Dr. Harold F. Eastgate, Dr. Donald Rounds, and many others both in the United States and abroad. Dr. Stern and his colleagues have spent many years in the study of teeth both in vivo and in vitro, using laser beams from solid-state and gaseous laser sources. Their work has led to the discovery of laser-induced inhibition of caries, which means that for the first time the human tooth can be laser-beam brushed to prevent caries in the enamel.

Although the author does not believe that laser radiation is a panacea for all dental disorders and will supersede all mechanical dental work now being performed in dental clinics, he strongly contends that with properly designed laser equipment and instrumentation technique, the laser will eventually become an indispensable dental tool for both restorative and therapeutic dentistry. Any procedures described in the text were presented only for information and should not be construed as recommendations by the author. This study is principally intended to motivate scientists and engineers to innovate, design, and develop the dental laser instrumentation systems and devices so direly needed in this profession, which has been ignored by the laser industry in general in the past.

QUESTIONS AND PROBLEMS

11-1 What has been the principal drawback to date regarding designing and building suitable laser systems for dental instrumentation? What can be done to improve this situation? Are there any laser types that are suitable for dental work? What are they? What was the first discovery with regard to inhibition of dental caries by the use of laser beams?

11-2 Describe the structure of a tooth, naming all its component parts and their function. What percent inorganic matter is in the enamel of the tooth? What percent inorganic matter is found in dentin? On what part does the carious disorder start on the tooth? Where does pyorrhea occur? What is the effect of pyorrhea on one's health? Describe what organs are affected and how.

11-3 What is the effect of a laser beam on the enamel of the tooth? Can caries be inhibited by means of a laser beam? If so, how? What types of laser beams are effective in inhibiting caries in enamel? Does incident heat from the laser beam affect the pulpal substance? Which laser beam does not have a significant effect on the pulpal substance when the enamel is irradiated with it? Which of the two laser beams, Nd-YAG and CO_2, does not produce a diathermal effect on the pulpal cavity? Discuss.

11-4 Which laser beam is readily absorbed by the tooth tissue? Should the laser beam be used pulsed or continuous? How can a tooth-colored cement be cured and glazed? What type of laser is used for this procedure? What is a bioactive tooth implant? What is the mechanism of dental bone growth in the presence of a tooth implant? Why should one know about the spectral characteristics of laser beams? What is the effect of the laser beam on the soft tissue during drilling of the enamel? What advantage has the cauterizing property of a laser beam with respect to various dental instrumentations?

11-5 What effect do laser beams have on gingiva? When working on the enamel with a laser beam, should any precautionary measures be taken? What property of the laser beam gave rise to the concept that it can join tissues together? What is the effect of an ultraviolet beam on diseased gingiva? What types of bacteria are killed by exposure to ultraviolet rays? What are zymogenic bacilli? What laser beam is a source of copious ultraviolet radiation? Can an overdose of ultraviolet exposure cause a serious disorder in the tissues? What is hydroxyapatite? Can fissures in the enamel be sealed with a laser beam? Discuss how.

11-6 What is root canal disorder? Can this disorder be treated with a laser beam? Discuss the procedure whereby a laser beam can be used to treat root canal disorder? Is it being used now? If a hole is made in a human tissue and the beam cauterizes the periphery of the hole, how can this hole be closed if needed? What is *Streptococcus mutans?* How does it attack the enamel?

11-7 What are the potential problems to be overcome before laser beam instrumentation becomes routine in a dental office? What are the advantages of the Laserkinetics dental instrumentation system? Can this system be employed in vivo and in vitro? Can it be used for prosthetic work? Discuss how. How does this system sterilize the drilled cavities? What are the advantages of the dental bridgework welder designed by Laserkinetics?

11-8 What intensity of laser power is required for dental instrumentation in general? What laser power would be required for welding metal bridgework parts? How is a Laserkinetics beam conduit constructed? If this conduit (waveguide) can transmit both infrared and visible radiation, can it also transmit ultraviolet radiation? Discuss how this can be accomplished. What is laser brushing of a tooth? Is the beam focused or defocused during laser brushing? Discuss this procedure.

11-9 If you were asked to design a dental instrumentation system using a laser source, what type of laser would you incorporate in the system? Draw a diagram and discuss the operation of your system.

SELECTED SOURCES OF RELATED DATA

Holmes, L. M. 1986. Carbon dioxide laser applications. *Laser Focus* March, pp. 72–84.

Laser Focus Staff. 1984. Recent developments in laser dentistry. *Laser Focus/Electro-Optics* May, pp. 22–32.

Lasers used for dentistry in France. 1985. *Laser Focus* June, pp. 26–30.

Melcers, F., J. Melcers, and J. Goutier. 1981. Proceedings of the 4th International Congress for Laser Surgery, Issue 12-1.

Mirzabekov, O. 1987. Lasers and Soviet dentistry. *Soviet Weekly* January 24.

Muncheryan, H. M. 1974. Laser Instrumentation Waveguide with Surgical Scalpel. U.S. patent No. 3,821,510.

Muncheryan, H. M. 1983. Laser applications in dentistry. In *Laser Technology,* chap. 14. Blue Ridge Summit, Pa.: TAB Books.

Pick, R. M. 1987. Lasers in dental surgery. *CDA J.* February, pp. 10–14.

Stern, R. H. 1973. Dentistry and the Laser, final report. School of Dentistry, University of California at Los Angeles.

Stern, R. H., and R. F. Sognnaes. 1972. Dental investigations with lasers. *J. Am. Dental Assoc.* 85:1087–1090.

Stern, R. H., J. Vahl, and R. F. Sognnaes. 1972. Laser enamel: Ultrastructural observations of pulsed carbon dioxide laser effects. *J. Dental Res.* 86:867–901.

Vahl, J. 1971. Geesunder und Pathologische Vernaderter Zahnschmelz. Booklet 43, Meusser Collection, Leipzig.

Willenborg, G. C. 1986. The evolution of lasers in dentistry. *Laser Focus* October, pp. 82–90.

CHAPTER

TWELVE

LASER FUSION TECHNOLOGY

12-1 INTRODUCTION

Laser fusion programs now under way at several institutions involve extraction of the intrinsic energy of the atom by transmuting it under controlled conditions. A similar effort was started several decades ago on a laboratory scale by the use of cyclotrons and particle accelerators. In a cyclotron, the atomic components—electrons, protons, or neutrons—are projected at high speeds through magnetic and electric fields and are made to collide at the end of their transit with a target consisting of transmutable low-atomic-weight atoms. The target then is either disintegrated or integrated into a larger atomic structure with the colliding particles. During this transmutation, a large amount of energy is released, which, when acquired under controlled conditions, could be converted into electric power. To date, nothing of commercial significance has been produced by the cyclotron, although the energy technology has advanced.

With the discovery of laser radiation, a new era in atomic transmutation was opened; energies in terawatts could be obtained from the laser radiation for use in the fusion of atoms to release energy. The effort is being centered on transmuting matter into energy by means of the enormously high temperatures that can be attained with the use of properly designed laser systems. In addition to this advantage of laser radiation, the energy produced by the transmutation of matter by this method would be waste-free, almost inexhaustible, and as clean as electricity. The fuel used to generate the energy is readily available as deuterium and tritium, which are heavy isotopes of hydrogen.

Fusion of matter with the use of laser thermal energy involves high-power laser radiation in the pulsing mode, together with application of compression on the target material to many times atmospheric pressure. The pressure heats the target, which is a tiny pellet of a mixture of deuterium and tritium, to millions of degrees to cause their nuclear combination (fusion) with formation of the heavier element helium (which has four protons and four electrons). In other words, the transmutation forms an alpha particle with two additional electrons. During the synthesis of helium from the two elements, a large amount of energy is released, nominally 6.5×10^{18} ergs, or 6.5×10^{11} joules, per atom of helium. It is this energy in which the investigators are interested. The amount of energy released depends on the type of target material, as target materials other than those mentioned can be used, as well as on the design of a laser system capable of furnishing the energy necessary for the atomic transmutation.

The present status of the atomic fusion program is not without discouraging problems. These include the difficulty of acquisition of uniform-intensity beam synchronization, consideration of the use of multilayer mantle fabrication, decrease in Coulomb collisions in operative status, tendency of the laser beam to be absorbed by the outer shell, occasional encounter by nonlinear force of plasma interaction, and many more. Most of these problems have been overcome with the establishment of incidence of a monochromatic laser beam on the target, resulting in the projected rise of current densities as given by

$$F_d = \frac{1}{c} jH + 4\frac{1}{\pi} EVE - 4\frac{1}{\pi} V(1-n)EE \qquad (12\text{-}1)$$

where F_d is the nonlinear force density, c is the speed of light, j is the current density, E is the electric field (electron volts/cm), V is the voltage, H is the magnetic field (gauss/cm), and n is the refractive index with change of plasma density.

It was found in the investigations that the electrostatic field could limit the gain to low values. In addition, human errors made in measurements compared with theoretical yields have further retarded progress. Another classical error has been the consideration for some time that magnetic confinement was the only method for achieving fusion. Investigations of astrophysical phenomena indicate that the Gaussian density and velocity profiles can be retained when adiabatic compressional states are conserved on conversion of the kinetic energy of the plasma into elevated-temperature states, with restitution of the kinetic energy by expansion of the imploding media.

The ultimate target design adopted is the spherical configuration, as illustrated in Fig. 12-1. The 0.95-μm-diameter deuterium-tritium (D-T) pellet is surrounded in a 300-μm-diameter mantle; the sphere is the target constituting the inner D-T fuel core. The mantle contains two diametrically opposed apertures drilled into it at grazing angles. In such an aperture configuration, production of undesirable x-rays is greatly reduced. Two laser beams are injected through these apertures into the space (known as the *hohlraum*) within the mantle and peripherally to the spherical D-T fuel. As the beams strike the fuel ball from opposite directions, they cause explosion of the fuel (target). A radiation pressure is developed in the space within the mantle. This pressure compresses the fuel pellet in a way dependent on the compression velocity of the radiation. If the latter velocity is larger than the plasma thermal velocity, the fuel pellet will be compressed up to 10^{25} ions/cm^3, which will require more than 1 million joules of laser energy. The shorter the pulse duration of the laser energy, the higher the radiation power, which runs up to 10^{16} W at an approximate gain of 50. For the

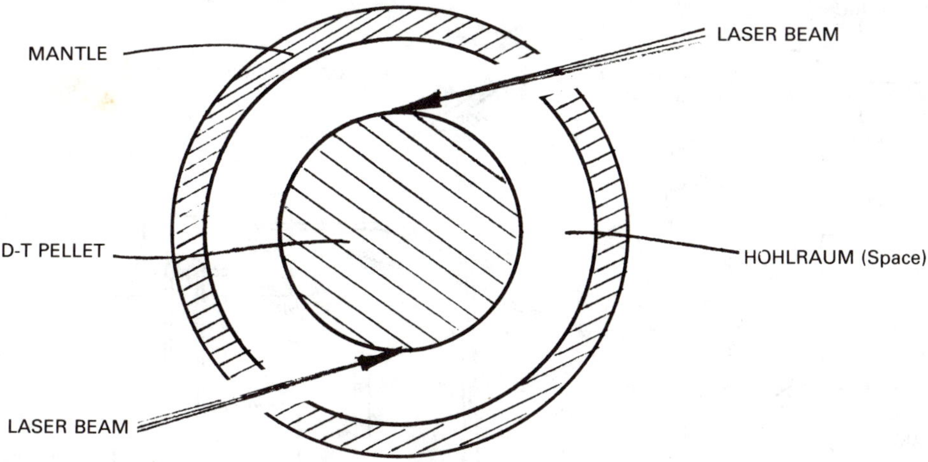

Figure 12-1 The D-T pellet with its outer mantle provided with oblique apertures.

final fusion system integration, the energy generation scheme is expected to take the configuration shown in Fig. 12-2.

12-2 NUCLEAR TRANSMUTATION

In our study of the generation of laser radiation from atoms, we were concerned only with the extranuclear phenomena in which electrons of an excited atom made transitions from a higher to a lower quantum energy level, giving rise to emission of radiation. The higher the energy involved in the transition, the higher the frequency and shorter the wavelength of the radiation thus produced. Thus, radiant transitions that occurred between two widely separated quantum states in an atom gave rise to ultraviolet radiation, whereas transitions between quantum energy levels that were closer together resulted in the longer-wavelength and lower-frequency infrared radiation at the other extreme of the spectrum. Transitions involving quantum energies intermediate between these two extremes produced visible radiation.

In the case of nuclear energy transitions, the transmutation of two or more atoms occurs; either disintegration (fission) of the larger atoms into smaller atoms occurs with emission of energy, or nuclear integration (fusion) of two or more atoms occurs to form a larger atomic structure with emission of energy. Our concern in this text is with the nuclear fusion of atoms to produce larger amounts of energy than are applied to the atoms in the reaction. In the transmutation of the nuclear structure of an atom into a different nuclear structure by fusion, we shall study several different methods of bringing about this interaction. The most prominent methods now being investigated are by the use of short-wavelength laser radiation, x-ray preionization of gas discharges, excimer radiation, inertial confinement, and electron beam plasmas.

Using any one of the above-stated schemes, the ultimate result is the emission of

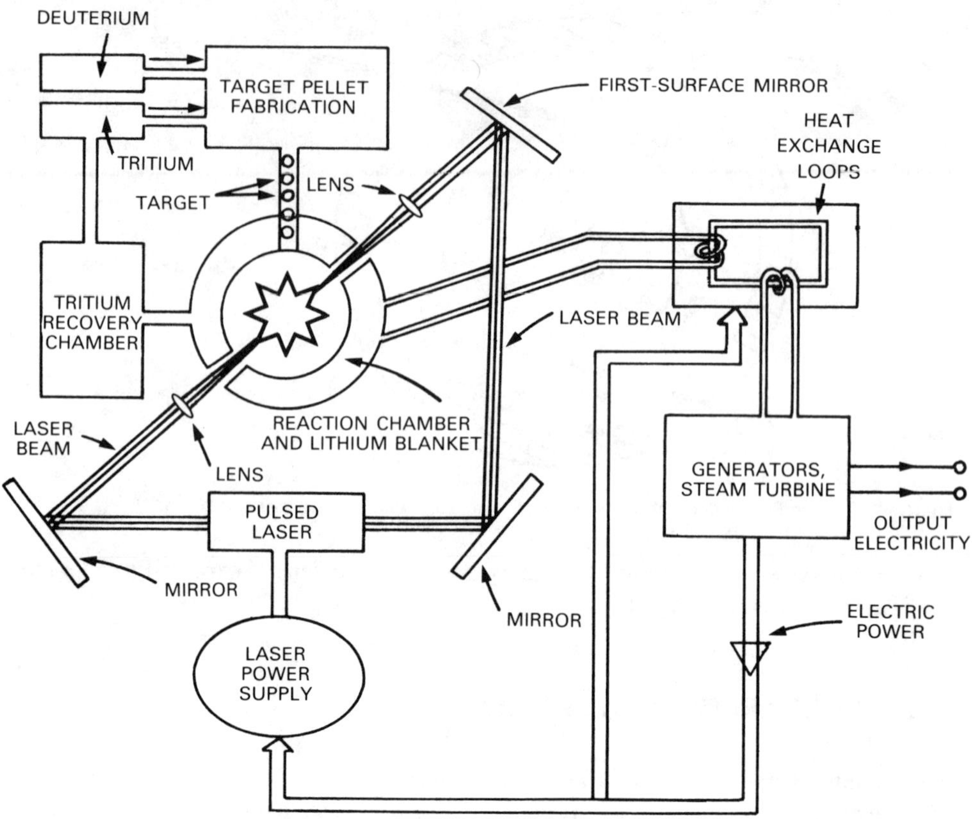

Figure 12-2 Schematic diagram of target implosion process, showing the manner of feeding of the pellets into the reaction chamber and the generation of electrical power.

energy *E,* which varies with the type of procedure employed; various research institutions are using different transmutation procedures in their facilities. To explain the nuclear reaction in the simplest manner that can be understood by the uninitiated, two examples are given here—one illustrating the disintegration (fission) of the atoms in interaction and the other the fusion of the atoms. Here again, our main concern will be with the fusion of the two atomic nuclei with emission of energy. As examples, we will illustrate the disintegration of a lithium atom bombarded with deuterium to split the lithium atom into two helium atoms [Eq. (12-2)] and the fusion of a lithium atom with deuterium to form a beryllium atom of atomic weight 8 [Eq. (12-3)]:

$$_3\mathrm{Li}^7 + {}_1\mathrm{H}^2 \rightarrow {}_2\mathrm{He}^4 + {}_2\mathrm{He}^4 + {}_0\mathrm{n}' + E \qquad (12\text{-}2)$$

$$_1\mathrm{Li}^7 + {}_1\mathrm{H}^2 \rightarrow {}_4\mathrm{Be}^8 + {}_0\mathrm{n}' + E \qquad (12\text{-}3)$$

Here the superscripts correspond to the atomic weights of the atoms and the subscripts correspond to their atomic numbers. The transmutation in Eq. (12-2) is a fission reaction, and Eq. (12-3) is a fusion reaction; the latter is considered to be a more efficient reaction with laser irradiation and to emit larger amounts of energy than do nuclear fission reactions.

In fusion of deuterium with tritium, the energy released is expected to be many times the applied laser energy. Actually, what occurs here is the conversion of matter into energy during the fusion reaction. The energy thus released is equivalent to the energy that it would take to build the new atomic structure from the component elements.

Formation of Nuclei

From mass spectrographic data, the atomic masses of different elements are determined on a scale in which oxygen is taken to have a mass of 16.0000. This unit is called the *atomic mass unit*. The data furnished by chemical methods on the mass of oxygen are based on the average mass of the mixture of oxygen isotopes of atomic masses 16, 17, and 18, the average chemical mass of oxygen being taken as 16.002, which is given in chemical mass units. Therefore, the conversion factor from the chemical mass unit to the atomic mass unit is 0.000125.

In an earlier section, it was stated that the nuclei of atoms contain a mixture of protons, neutrons, and alpha particles, the presence of which is dependent on the size of the atom. For example, the nucleus of lithium contains one alpha particle, three protons, and two electrons. From Aston's work, the mass of the hydrogen atom is 1.00778 atomic mass units and that of the electron is 0.00055 atomic mass unit. Hence the mass of a proton is 1.00778 − 0.00055, or 1.00723 atomic mass units.

Now, in the structuring of a helium nucleus (the alpha particle) from four protons and two electrons, there is a loss of 0.0288 atomic mass unit (a.m.u.), as illustrated in the following equations:

$$\text{4 protons} + \text{2 electrons} \rightarrow \text{1 alpha particle}$$

$$4 \times 1.0072 + 2 \times 0.00055 - 4.0011 = \Delta m_a$$

$$\Delta m_a = 4.0299 - 4.0011$$

$$= 0.0288 \text{ a.m.u.}$$

where the numerical quantity 4.0011 is the observed mass of the alpha particle and Δm_a is the loss of mass in the reaction. For a Li^7 atom, the mass loss in the nuclear formation may be given as

$$\text{1 alpha particle} + \text{3 protons} + \text{2 electrons} \rightarrow Li^7 + \Delta m$$

$$4.0011 + 3 \times 1.0072 + 2 \times 0.00055 + 3 \times 0.00055 \rightarrow 7.012 + \Delta m$$

in which 7.012 is the observed mass of the Li^7 atom, hence the addition of three electrons to the left-hand side of the equation.

Computing the products of the terms and rearranging the equation, we have

$$4.0011 + 3.0216 + 0.0011 \rightarrow 7.01035 + \Delta m$$

$$\Delta m = 0.01345 \text{ a.m.u.}$$

which quantity is the loss of mass during the formation of the Li^7 nucleus from electrons, protons, and one alpha particle. The total loss of mass in the formation of a Li^7 atom in the reaction of seven protons and seven electrons will therefore be

$$\Delta M = 7(1.0072) + 7(0.0005) + 7.012$$
$$= 0.04255 \text{ a.m.u.}$$

which value may also be found by adding the separately computed losses of mass in the Li^7 formation as follows:

$$\Delta M = \Delta m_a + \Delta m$$
$$= 0.0288 - 0.1345$$
$$= 0.04255 \text{ a.m.u.}$$

The loss of mass is known as the "mass defect" of the atom or, more generally, as the "packing effect." The packing effect for the atom, then, may be expressed in the following form:

$$\Delta M = P_n H_m - M_o \tag{12-4}$$

in which M_o is the observed mass of the atom, P_n is the number of protons in its nucleus, and H_m is the mass of the hydrogen atom, given as 1.00778 a.m.u.

Applying relativity theory to the phenomenon of the mass defect and computing the energy liberation ΔE due to the mass defect ΔM, we may write the mass-energy equivalence by the following general equation:

$$\Delta E = \Delta Mc^2 \tag{12-5}$$

where c is the speed of light and the energy ΔE is numerically equal to 9×10^{20} ergs when ΔM is 1 gram. In practice, the energy ΔE is expressed in electron volts. Then we have

$$\Delta E = \Delta Mc^2 = Ve \tag{12-6}$$

where V is given in electrostatic volts and e is the charge on the electron, equal to 4.8×10^{-10} electrostatic units. Converting V into practical volts, the expression becomes

$$1.649 \times 10^{-24}\,\Delta Mc^2 = \frac{V \times 4.8 \times 10^{-10}}{300} \tag{12-7}$$

where ΔM is expressed in atomic mass units when multiplied by 1.649×10^{-24}, the mass of one atom of hydrogen, given in grams.

Solving for V, the energy released in loss per atomic mass unit will be

$$V = \frac{1.649 \times 10^{-24} \times 300 \times (3 \times 10^{10})^2 \times \Delta M}{4.8 \times 10^{-10}}$$
$$= 9.33 \times 10^8 M \text{ volts}$$

and

$$\Delta E = 9.33 \times 10^8 M \text{ electron volts}$$

We thus note that in the formation of one lithium nucleus from electrons and protons the energy liberated for a loss of 0.04225 a.m.u. is

$$\Delta E = 9.33 \times 10^{8} \times 0.0225$$
$$= 39.42 \times 10^{6} \text{ electron volts}$$

The last quantity is the binding energy of the lithium nucleus. The binding energy of an atom is the amount of energy that is set free during the formation of the atomic nucleus and is equivalent to the mass defect of the atom. The greater the energy released during nuclear formation, the greater is the stability of the atom. For example, a helium atom is more stable than a carbon atom because there is a greater loss of mass in the formation of the helium nucleus. Hence, we may conclude that the mass defect, packing fraction per atom, and binding energy are closely related to each other in their equivalence of magnitude in a given atom.

Principal Investigators in Fusion Programs

The general investigative work in fusion programs to date is confined to four or five key centers, including Lawrence Livermore National Laboratory, Los Alamos National Laboratory, the University of Rochester Research Center, and KMS Fusion Facilities. The first two laboratories are engaged in inertial confinement fusion programs, Lawrence Livermore having the facilities Novette and Nova, which has apparently replaced Novette. Los Alamos is engaged at the Antares facility, and Rochester Research Center is operating the 24-beam, 12-terawatt Omega laser facility. KMS, which was one of the earliest advocates of fusion work, is principally using a frequency-doubled Nd-glass laser facility.

According to the Los Alamos Scientific Laboratory (LASL), a laser system to be useful in atomic transmutation must generate in the neighborhood of or in excess of 1 million joules per pulse of at least 1-nsec duration. Any spatially coherent radiation from the ultraviolet through the visible to the infrared spectrum having a focused beam not larger than 1 mm in diameter would be necessary for the fusion process, which must operate with an efficiency of at least 5%. On a commercially practical scale, an average of 100 terawatts (TW) must be generated from the laser source. To date, this energy of laser radiation has not been achieved from present laser systems; however, experiments are in progress.

The LASL Shiva facility, which superseded Argus and incorporated the technology learned from Argus, was a 20-beam laser system with 20-cm-diameter amplifiers (Fig. 12-3) that achieved 10 kJ with a pulse duration of 1 nsec. It accommodated hohlraum targets (shown in Fig. 12-1); however, excessive soft x-rays produced within the mantle interfered with the expected implosion process. This facility finally gave rise to the Nova facility at Lawrence Livermore National Laboratory (LLNL). The Nova laser facility is expected to produce 100 kJ at 3-nsec pulses. The objective of this facility is to attain a fuel (D-T) compression of 1000 times its liquid density, with a laser-plasma interaction at 1-cm diameter. This magnitude of energy will heat the target to emission of neutrons to enable measurement of the compression and efficiency of the system

Figure 12-3 An early fusion facility at LASL, operating under the name Shiva, was a 10-kW facility producing more than 20 TW in 1-nsec pulses.

performance. A forerunner of Nova, the Novette, is pictured in Fig. 12-4, which shows the parallel laser amplifier arms; the target fusion chamber is located in the far background.

Atomic fusion experiments conducted toward the end of 1983 included a 10-TW pulse from a carbon dioxide laser projected at a gold-plated copper sphere enclosed in a spherical plastic mantle about the nominal size of a pea. A small star was produced at the point of focus, where the target vaporized. Gold plating was used because of its high atomic number, so that x-rays could be generated from the beam-plasma collision. Experiments conducted toward the close of 1984 used the 24-beam laser of Antares, the world's largest CO_2 laser system, to supply 40 TW of laser power to targets filled with deuterium-tritium fuel and demonstrated the success of the inertial confinement technique. Further tests and studies are still under way.

Lately, laser radiation employed in atomic fusion consists of 10,640-Å-wavelength energy from a Nd-glass laser. KMS Fusion has been using this radiation frequency-doubled to obtain 5320 Å, which is in the green spectrum. By bombarding a pellet of D-T with a 150-psec pulse of 0.5 TW, KMS obtained 5×10^4 neutrons during the early stages of their investigation. They have also irradiated the target mass, compressed to 5 g/cm^3, with a 10,640-Å beam from a Nd-YAG laser and obtained 7×10^7 neutrons. Researchers at KMS have also considered freezing the target medium to eliminate the back pressure produced by the thermal energy from the laser; the thermal energy prevents the occurrence of target implosion.

The experiments described are designed to heat the atomic target with laser radiation to boil away (ablate) the surface of the spherical pellet of deuterium-tritium target; to compress the sphere inward to a density that is 1000 times its liquid density; and to cause implosion of the sphere, which then releases thermal energy to heat a circulating fluid in a thermal generator. The heated fluid is then transformed into steam to drive electrical generators, as illustrated in Fig. 12-2. Slight radioactivity will result from the transmutation energy. However, it is said that the relatively harmless radiation waste will be easier to dispose of than the radioactive waste from present nuclear generators of equivalent energy potential.

In similar projects conducted in France at the École Polytechnique Palaiseau, using inertial confinement procedures, the researchers are also attempting to create densities from 100 to 1000 times the solid-state density of the fuel at pressures of up to 1000 megabars. They have found that short-wavelength laser power is less effective in producing an interaction between the target and the laser beam, with resulting generation of superthermal material particles; the condition has limited the preheating of the plasma preparatory to compression. The researchers initially used a Nd-YAG laser at its fundamental frequency, producing 30-J, 200-psec pulses at 150 gigawatts (GW). This was followed by the use of frequency-doubled Nd-YAG, which has been superseded by a Nd-YLF oscillator that currently produces laser pulsewidths of 20 to 350 psec at 10,530 Å using phosphate glass rods. Phosphate glass is found to have a greater cross section for stimulated emission than the silicate glass rod originally used. This material emits radiation at a wavelength of 10,500 Å in pulses of 115 J and 600 psec, with a power output of 200 GW. Using a two-beam laser, with beams of the same power, the resulting emission power has been doubled.

The fusion experimenters have concluded that short-wavelength laser radiation has absorption efficiencies up to 95% at 2500 Å with a density of 10^{15} W/cm^2. Furthermore, as the wavelength is decreased the ablation power of the beam increases, as confirmed by preliminary tests with microballoons. To further improve the performance of their technique, the researchers have used six Nd-glass amplifiers with spatial filters. The

Figure 12-4 A view of LLNL's Novette facility for atomic fusion.

new system is said to operate with 200-psec pulses at 1.5 TW peak power. Further tests are to be conducted using the second-harmonic wavelengths of this laser emitting 600-psec, 1-TW peak pulses. The third and fourth harmonics of this laser are also projected for tests with spherical targets.

12-3 FURTHER DEVELOPMENTS IN ATOMIC FUSION PROGRAMS

With the preceding laser facility successes, the compression of a D-T target to 100 times its liquid density has been demonstrated. From the results of the effort, the concept has evolved that shorter-wavelength laser light could be produced by frequency multiplication of Nd-glass lasers. The necessity of employing shorter wavelengths (harmonics) has been predicated on the concept that plasma interaction with the laser beam creates higher densities in the target. Furthermore, laser absorption is found to be more efficient (up to 90% at 3500 Å) when inverse bremsstrahlung shorter wavelengths are impressed. The Nova facility, with a 74-cm aperture and nine KDP segments with the direction of phase matching precisely aligned, is completed. This system is expected to yield 100 to 200 kJ in 3-nsec pulses; in shorter pulses in a 100-psec format, it will produce a power density of 100 TW/cm^2.

Nova has a 10-arm laser in which each arm is similar to that of Novette. Each arm of Novette had a 20.8-cm-diameter amplifier producing the terawatt laser pulses. Additional amplifiers added to each arm in Nova are to produce 10 TW of infrared laser power. This power will be transmitted through KDP crystals for frequency multiplication at a conversion factor of about 70% in both the ultraviolet and visible regions.

KMS Fusion is currently advocating the use of a free-electron beam laser because of its tunability to wavelengths that are optimum for interaction with the fusion target, with an efficiency in excess of 20%. The laser pulse is emitted by the interaction of the free electrons with microwave beams transmitted in the opposite direction to the electron beam within the resonant cavity and is expected to yield the maximum radiation necessary for target fusion. This study is being carried out by several other institutions both in the United States and abroad.

At Los Alamos National Laboratory, the 12 CO_2 laser beam fusion system for Antares is also operational. This facility, shown in Fig. 12-5, is capable of delivering 40 kJ of laser energy at a power density of 60 kW/cm^2 in 1-nsec pulses to be projected to a target the size of a tiny dot. Two diagonally opposite duplicate systems bombard the target simultaneously. This facility produces copious amounts of hot electrons of several hundred kiloelectron volts, which tend to decrease the efficiency of absorption of the laser radiation by the target material. The projected goal of this facility is nuclear weapons simulation; electric power generation is a secondary objective. In line with this laboratory's effort, Sandia National Laboratories are actively investigating the capabilities of a light-ion pulse-power driver to achieve 3.5×10^6-J, 100-TW laser output power.

Experiments at the Lawrence Livermore and Rochester laser facilities have shown

Figure 12-5 The Los Alamos KrF laser facility for fusion research delivers 20 kJ.

that converting the infrared radiation at 10,500 Å to the 3500-Å ultraviolet wavelength produces target absorption of the energetic radiation of up to 90% at power densities of 200 to 300 TW/cm^2. The x-ray conversion efficiencies of the shorter wavelengths also increase up to 50%; this action considerably reduces the hot-electron production experienced in previous investigations. The harmonic wavelengths from either CeF or KrF in the ultraviolet region can also be used as optical drivers.

At the Greco Fusion Research Facility of École Polytechnique in France, the researchers also confirmed the use of high-power pulses of Nd-glass harmonic wavelengths at 5300 and 2600 Å, using silicate rods and later phosphate glass rods (Hoya ZHG8); the Nd-YAG laser oscillator was also replaced by a Nd-YLF oscillator, which generates pulses 20 to 350 psec wide at a wavelength of 10,530 Å. At the latter wavelength, the system can produce 60-J, 200-psec pulses at a peak power of 300 GW. A frequency-conversion efficiency of 56% has been achieved with 600-psec pulses. A number of basic physical questions raised in these procedures are currently being studied, and some of them have already been answered. A view of the Greco Laser Fusion Facility is shown in Fig. 12-6.

These advances in the fusion effort have been complemented by the latest improvements in damage-resistant optical coatings and in LiF spatial filters, the achievement of accuracies to within 1 μm in optical surface fabrication, and the achievement of alignment precision of the optical elements down to 20 mrad. Furthermore, an increase in the damage threshold by use of Al_2O_3-SiO_2 and SC_2O_3-MgF_2 for shorter wavelengths and efficient phase conjugation in the ultraviolet range has prolonged the lifetime of the optical elements. Transcending these achievements, efficient radiation absorption, pellet compression, and control of the laser plasma are simultaneously being improved.

Other approaches to the achievement of the target fusion involve the use of krypton fluoride and xenon chloride, beryllium fluoride, and other excimer sources. Because of their short wavelengths, generally in the ultraviolet spectral range, their coupling efficiency with the target pellet is very favorable. One other approach is the application of heavy-ion accelerators in conjunction with the conversion of their radiation into long-wavelength x-rays prior to driving the fusion material. Using a carbon dioxide

Figure 12-6 The Greco Fusion Facility at École Polytechnique in France.

laser as the pulse driver, energies up to 1 megajoule are estimated to be available to perform the expected fusion operation.

The frequency-doubled and -tripled Nd-glass of Nova is the preferred laser source to date for weapons physics research. The frequency-multiplied radiation has a high affinity for coupling with the fusion fuel. In addition, the spatial and temporal characteristics of these harmonic wavelengths can be converted into long-wavelength x-ray radiation to irradiate the target and cause implosion.

QUESTIONS AND PROBLEMS

12-1 What is meant by laser fusion of atomic nuclei? What is the purpose of the laser fusion program now being conducted? What is meant by transmutation of atoms? What laser energies can be produced at the present time for fusion work? What is the function of laser radiation on atomic nuclei? What fuel is used to generate atomic energy? In what mode is the laser beam utilized in the fusion of atomic nuclei? What is the function of the pressure applied on the atomic fuel? On what factor does the amount of nuclear energy released depend? How much energy is released in the synthesis of a helium atom?

12-2 How does the electron transition energy affect the emitted radiation frequency and the wavelength? Discuss. What is the disintegration of laser atomic nuclei into smaller nuclei called? What is nuclear integration into larger nuclei called? Discuss the present method of producing nuclear fusion.

12-3 Write an equation for producing an atom of beryllium of atomic weight 8. What type of transmutation is this production method? In the fusion of deuterium and tritium, where does the released energy come from? Explain. What is an atomic mass unit? How does it differ from a chemical mass unit? In what atomic transmutation is 0.0288 atomic mass unit lost? What is this loss equivalent to in terms of electron volts?

12-4 What is the mass defect? What is the packing effect? Are these two terms equal in energy? What is meant by the binding energy of the nucleus? How is nuclear binding energy related to the stability of an atom? Discuss. Are the energies involved in the mass defect, packing fraction, and binding energy equal in a given atom? Discuss.

12-5 What is inertial confinement fusion? How many joules should a laser system generate to be useful in atomic transmutation? For what duration should this energy be expended? At what efficiency should this energy be generated? What solid-state laser element is used in nuclear fusion? Is this radiation used frequency-doubled or in its fundamental mode? What should the deuterium-tritium density be so that it can be used as a fusion fuel? Explain.

12-6 What institution uses Nd-glass for fusion work? How many neutrons have they produced in fusion and at what laser power? How much energy can Argus produce? How much energy will Shiva produce when completed and in operation? What is the fusion mechanism in the release of energy in the deuterium-tritium reaction? What must the density of the atomic fuel be for fusion? How is that density obtained? Is the atomic reaction an explosion or implosion?

12-7 Why is some radioactivity emanated in the fusion of atoms? Is this emanation harmful? Explain. In the latest fusion effort, why is it necessary to employ short-wavelength laser beams? From what source can short-wavelength laser beams be obtained? At what wavelength is absorption efficiency the greatest? What is that efficiency as a percentage? How is the Nova facility expected to operate? At what energy level? At what power density?

12-8 Describe the construction of the Nova facility. How will the energy be multiplied in the Nova facility? At what percentage of conversion factor will Nova operate? Why is a free-electron beam laser advocated by KMS? What purpose is served by the tunability of a laser beam in fusion work? What amount of energy is projected for Antares? What detrimental effect do hot electrons have on fusion?

12-9 How are ultraviolet wavelengths produced from infrared radiation? What is the effect of ultraviolet on target absorption? What purpose is served in producing high x-ray conversion efficiencies? What is the purpose of phase conjugation in the ultraviolet range? Which excimers show good coupling efficiencies with the target? How is CO_2 used as a pulse driver? How are spatial and temporal characteristics of the harmonic wavelengths controlled?

12-10 Write a chemical equation for the conversion of carbon-12 into nitrogen-13 if a neutron and energy are released. Now calculate the energy in electron volts, using the procedure given in the text.

12-11 How does Novette differ from Nova? Compare the structure of Antares with that of Nova. Which one produces more energy? Give the reason for your answer. Why is a lithium blanket used in the atomic implosion process? How does the heat from the reaction chamber get to the heat exchangers? If tritium is recovered from the reaction chamber, where does the emanated energy come from? Discuss.

12-12 Write a chemical equation for the conversion of a deuterium-tritium pellet into energy by means of a laser beam. What is the product left behind after the implosion of the pellet? The harmonic wavelengths from CeF can be used for what purpose? In what part of the spectrum are these harmonic wavelengths? Discuss.

SELECTED SOURCES OF RELATED DATA

Ahlstrom, H. G. 1982. *Physics of Laser Fusion,* Vol. II. Springfield, Va.: National Technical Information Service.

Aladel, S. L., et al. 1979. Laser fusion reactors. *Nucl. Technol.* 43(1).

Bromberg, J. L. 1982. Thirty Years of Fusion: A New Energy Source. Cambridge, Mass.: MIT Press.

Channell, P. J., ed. 1982. *Laser Acceleration of Particles*. New York: American Institute of Physics.

Chen, F. F. 1984. *Introduction to Plasma Physics and Controlled Fusion*. New York: Plenum.

Dance, B. 1987. German researchers develop high-energy iodine laser for the study of laser fusion. *Laser Focus/Electro-Optics* December, pp. 78–83.

Dolan, J. 1982. *Fusion Research*. New York: Pergamon.

Electro-Optics Staff. 1983. KrF will trigger fusion. *Electro-Optics Magazine* July, pp. 9–10.

Forrest, G. T. 1988. Laser fusion—inertial confinement fusion emphasis shifts to driver development. *Laser Focus/Electro-Optics* May, pp. 42–46.

Godwin, R. O., and W. W. Simmons. 1985. How industry help build NOVA, the world's largest high-precision optical project. *Laser Focus* April, pp. 78–88.

Hecht, J. 1986. Laser-aided fusion to generate energy. *Lasers and Applications* December, pp. 20–24.

Holzrichter, J. F. 1983. Laser fusion update (1961–83). *Lasers and Applications* March, pp. 65–70.

Hora, H. 1984. New avenues to success in laser fusion. *Laser Focus/Electro-Optics* February, pp. 59–71.

Hora, H., and G. M. Miley. 1986. Success of new avenues in laser fusion. *Laser Focus/Electro-Optics Technology* June, pp. 94–110.

KMS Fusion, Inc. 1986. Method of Controlled Release of Thermonuclear Energy. U.S. Patent No. 4,608,222.

Laser Focus Staff. 1983. Selected laser research and applications (fusion programs). *Laser Focus* January, pp. 33–36.

Laser Focus Staff. 1985. OMEGA sets neutron record in laser fusion. *Laser Focus* June, p. 10.

Mittleman, M. H. 1982. *Introduction to the Theory of Laser-Atom Interactions*. New York: Plenum.

Occidental Research Corp. 1980. Fusion driven by neutral particles. *Laser Focus/Electro-Optics* February issue.

Schwarz, H. J. 1981. *Laser Interaction and Related Plasma Phenomena*. New York: Plenum.

Stark, H. 1982. *Applications of Optical Fourier Transforms*. New York: Academic Press.

Tarasov, L. V. 1983. *Laser Physics*. Chicago: Imported Publications, Inc. (MIR Publishers).

U.S. and USSR revive laser fusion meetings. 1988. *Laser Focus* February, pp. 26–30.

Yariv, A., and P. Yeh. 1984. *Optical Waves in Crystals*. New York: Wiley-Interscience.

CHAPTER

THIRTEEN

MODE LOCKING AND CAVITY DUMPING

13-1 INTRODUCTION

Mode locking is amplitude modulation of a laser beam in the optical cavity at the Fabry-Perot frequency, which is equal to the laser resonator mode spacing. In this state, each mode contains symmetrical sidebands that interact with contiguous modes. Because of the constant relationship between a mode and its individual sidebands, this effect produces phase and amplitude locking of the mode, resulting in a series of short pulses at the chosen frequency of modulation. In an intracavity system of this type, the mode locker and the wavelength-selection prism are usually combined to reduce the intracavity losses to a minimum. The index of refraction of the prism is modulated by means of an ultrasonic transducer bonded to the prism and operated in the radio-frequency range.

Within a given transition, laser radiation is capable of oscillating at several frequencies; mode locking is utilized to produce extremely short pulses, although it lacks the energy per pulse generated by Q-switching. Thus, because a multitude of photons can be condensed to a minimum of space, cavity dumping will lend itself to the expulsion of enormous laser power from the cavity with minimum loss, with the timely arrival of the photons at the output mirror. By proper choice of the repetition rate of this reaction and the relaxation time period, the radiation is cavity-modulated. For this reason, mode-locked laser radiation is driven most efficiently when triggered externally.

For the purpose of discussion of this important phase of laser generation and dispensation, we use as an example a sophisticated system that combines both mode-locking and cavity-dumping techniques, the Spectra-Physics Model 166/366 Combo Mode-Locked and Cavity-Dumped Ion Laser. Because of its versatility and the fact that it is representative of standard mode-locked and cavity-dumped laser systems, the following discussions and circuit treatments will be based on this model. The theory discussed hereafter is directly related to all types of mode-locked and cavity-dumped laser systems as applicable. Accordingly, the intent of this chapter is to present the fundamentals of the pertinent modes.

Mode Locker

A mode locker provided in a commercial laser system is basically an intracavity amplitude modulator operating at the Fabry-Perot resonant frequency of $C/2L$, in which C is the radiation speed and L is the distance between the feedback mirrors. The quantity $C/2L$, known as the axial intermode spacing, represents the phase-locked axial modes. The output energy in the time domain then consists of short pulses with $2L/C$-second separations. Such a device functions as a pulsed regenerative amplifier, sweeping across the Doppler width. In such a state, each mode has two symmetrical sidebands separated by $C/2L$, as shown in Fig. 13-1. Each sideband will therefore interact with an adjacent mode.

Since the independent oscillators in a nonlinear medium tend to synchronize each other, each mode has a fixed phase relationship with its sideband, whereby general locking of all the modes in a definite phase and amplitude relationship will occur. In the time domain, the Fourier transform of this frequency spectrum is a series of short pulses, shown in Fig. 13-2, propagating at a repetition rate of $2L/C$.

Within a given quantum transition, certain lasers oscillate at more than one frequency because of their inhomogeneous broadening. Consequently, use of mode locking will produce extremely short pulses, although it will not generate the high-energy pulses developed with Q-switching. To determine the mode-locked pulse relation in a Gaussian envelope, the following approximation may be utilized:

$$I_g = I_c \exp\left\{-4 \ln\left[2\frac{(nB_c)^2}{B}\right]\right\} \qquad (13\text{-}1)$$

where I_g is the mode intensity at the signal gain coefficient, I_c is the mode intensity at line center, n is the mode number, B_c is the center frequency at the linewidth, and B is the linewidth covering the modes with greater than one-half of the center-mode ampli-

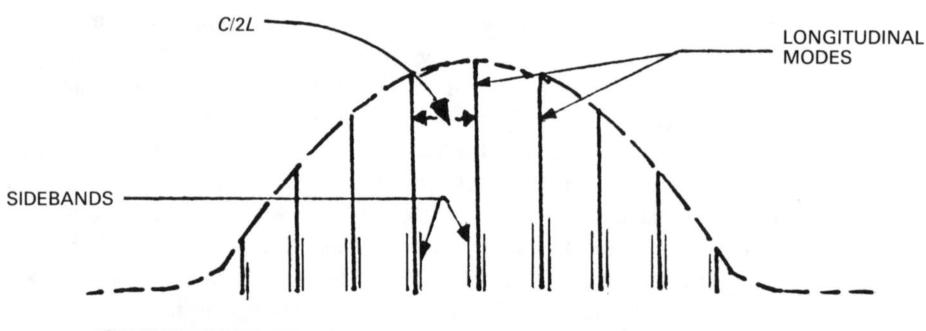

Figure 13-1 Frequency and symmetrical sidebands.

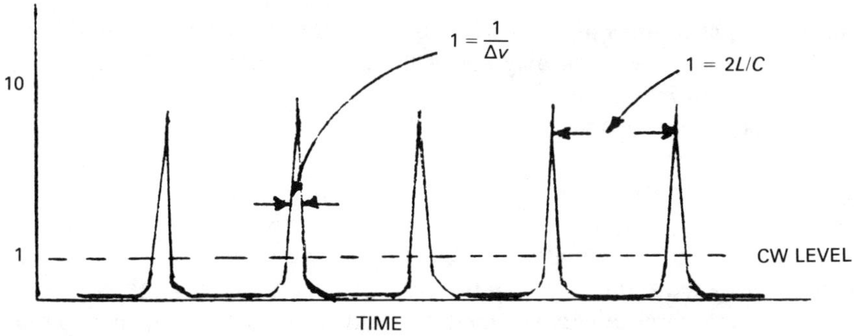

Figure 13-2 Time domain waveforms.

tude. The radiation dominant beyond the threshold recurring modes at the center would be larger than the leading and trailing modes of the envelope.

Cavity Dumper

The cavity dumper is an acousto-optic deflector that operates in the Bragg mode, as illustrated in Fig. 13-3. When a light beam is incident on a quartz crystal at the Bragg angle, the diffracted beam is in the first order and at the Bragg angle. The Bragg angle is defined as

$$\sin\theta = \frac{\lambda_i}{2\lambda_a} \tag{13-3}$$

where λ_i is the wavelength of the incident light beam, λ_a is the wavelength of the acoustic wave front, and θ is the Bragg angle in degrees.

The reflected beam is Doppler-shifted by the traveling acoustic wave, and its frequency is given by

$$W_d = W_i + W_a \tag{13-3}$$

where W_d is the Doppler-shifted deflection wave frequency, W_i is the incident wave frequency, and W_a is the acoustic wave frequency.

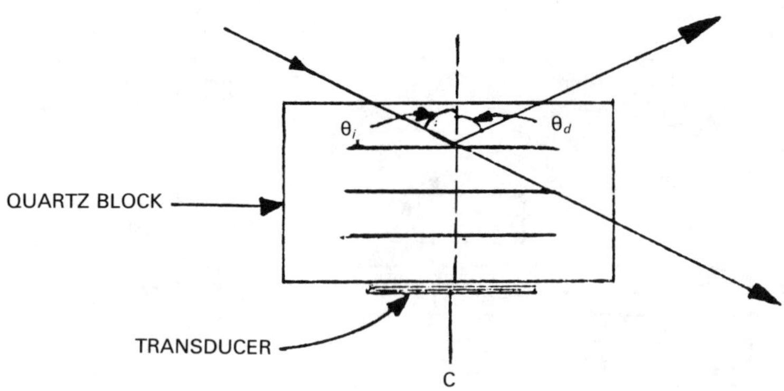

Figure 13-3 Acousto-optic Bragg deflector.

If the incident light beam is in the reverse direction, the frequency is downshifted by the same amount (W_a). A block diagram of the mode-locked cavity-dumped ion laser system is shown in Fig. 13-4.

In a double-pass optical geometry, the modulator is positioned at the focus of the optical cavity; as the laser beam passes through the modulator, part of the beam is diffracted and shifted in frequency by an amount equal to the frequency of the acoustic exciting field, and the remaining part of the beam is unaffected. The two beams then are allowed to pass through the modulator a second time, which causes the undiffracted beam to become diffracted and mixed with the diffracted and frequency-shifted beam. This state produces interference between the two beams and results in modulation of the output amplitude. (See Section 8-8.) The amplitude-modulated output may be represented by

$$A_o = \sqrt{1-\eta}\,\sqrt{\eta}\,E_c \cos(\omega - \Omega)t + \sqrt{1-\eta}\,\sqrt{\eta}\,E_c \cos(\omega + \Omega)t \qquad (13\text{-}4)$$

where A_o is the amplitude-modulated output, and the remaining quantities represent the standard physical terms as generally defined.

Folded Cavity

The cavity-dumper modulator is positioned at the center of a folded cavity, shown in Fig. 13-5, which is configured in such a manner that the centers of curvature of the mirrors M_1 and M_2 coincide. This implies that the modulator is centered at the confocal point of the two centers of curvature of the mirrors, whereby the optical power is directly concentrated on the modulator.

As optical power traverses in both directions inside the optical cavity, the extracted deflected beam becomes composed of two collinear beams, one of which is upshifted

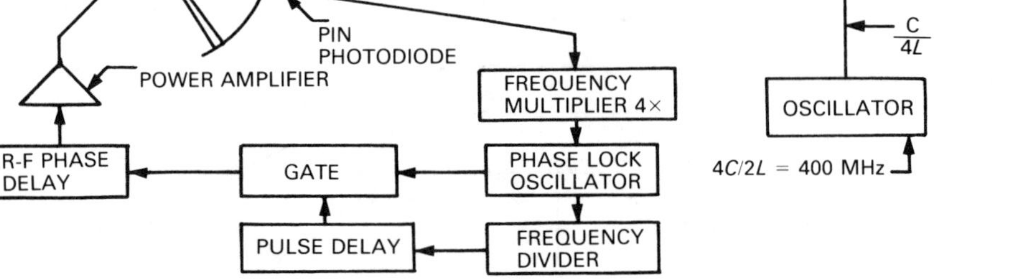

Figure 13-4 Model 166/366 mode-locked cavity-dumped ion laser combination.

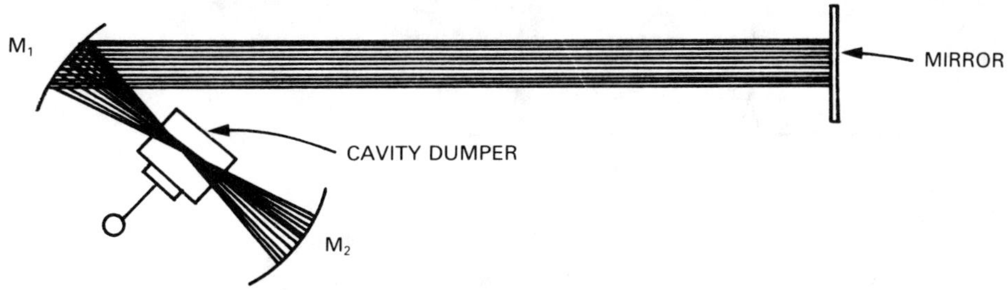

Figure 13-5 The cavity dumper in a folded cavity.

in frequency by the quantity W_a and the other of which is downshifted in frequency by the same amount. A fast photodetector thus will detect $2W_a$ beats between the two superimposed beams.

The average efficiency of the diffracted beam in the folded cavity is given by

$$E_{ff} = 2K(1 - K) \tag{13-5}$$

where E_{ff} is the average efficiency of the diffracted laser beam in the folded cavity and K is a function of the acoustic power.

When K is equal to 0.5, for example, E_{ff} will be equal to 50%, one-half of the intracavity power in the cavity dumper. Since the loss is much more than the cavity can sustain, the deflected power behaves as an exponentially decaying pulse. The next pulse of equal amplitude can be extracted after the oscillations have built up to a maximum in the cavity. The typical buildup time for stimulated emission in ion lasers is of the order of 0.5 μsec. Accordingly, the optimum dumping frequency becomes 2 MHz. Figure 13-6 illustrates the energy flow in the cavity dumper.

13-2 MODE LOCKER AND CAVITY DUMPER COMBINED

For maximum diffraction efficiency, a mode-locked pulse must be in phase with the acoustic wave traveling in the cavity-dumper crystal. This implies that the acoustic

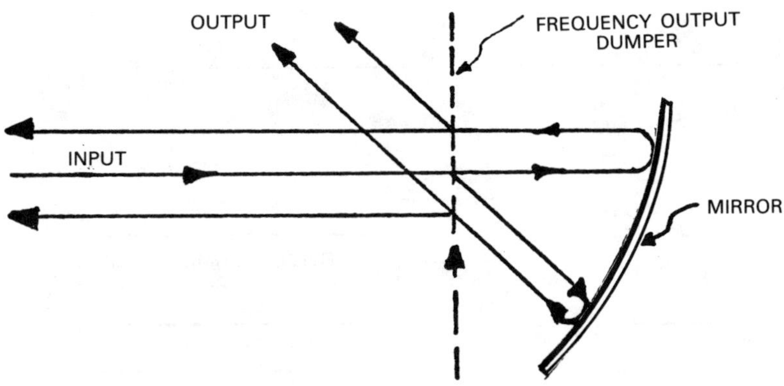

Figure 13-6 Energy flow in cavity dumper of the laser system.

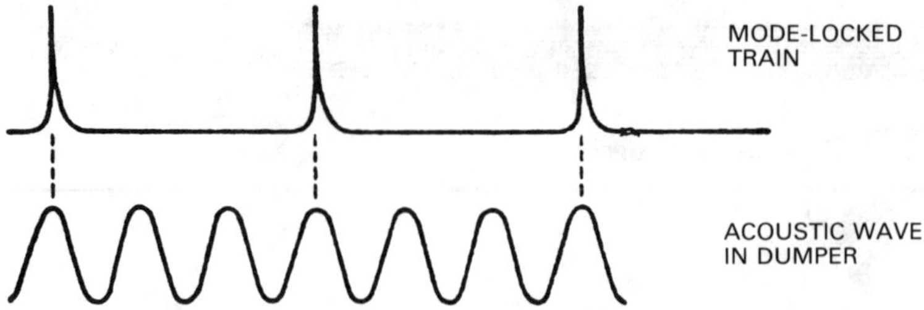

Figure 13-7 Relationship of mode-locked train of pulses to acoustic waves in dumper.

dumper frequency must be an integral multiple of the mode-locked frequency, as illustrated in Fig. 13-7. This is achieved by detecting the frequency of the mode-locked pulse with a fast photodetector. The output of the photodetector is then fed to a frequency multiplier to synchronize the phase-locked oscillator at 400 MHz, which becomes the acoustic frequency, as illustrated in Fig. 13-8. The phase-locked oscillator output power is amplified, by means of a power amplifier, to drive the acoustic transducer on the dumper modulator.

As stated earlier, the phase-locked oscillator beam consists of two beams superimposed on each other. One of the beams is upshifted by an amount equal to the Bragg frequency W_a and the other is downshifted by the same amount W_a, making the output 100% modulated at $2W_a$, or twice the Bragg frequency. For maximum output, the mode-locked pulse at the center of the dumper crystal must be in phase with $2W_a$ modulation, as illustrated by the waveforms shown in Fig. 13-9. Under such conditions more than 50% of the mode-locked pulse can be extracted, depending on the width of the pulse relative to the 800-MHz modulation period. A variable radio-frequency phase shifter is inserted between the phase-locked oscillator and the power amplifier (Fig. 13-8) to ensure that the mode-locked pulse is in phase with the second harmonic of the acoustic wave.

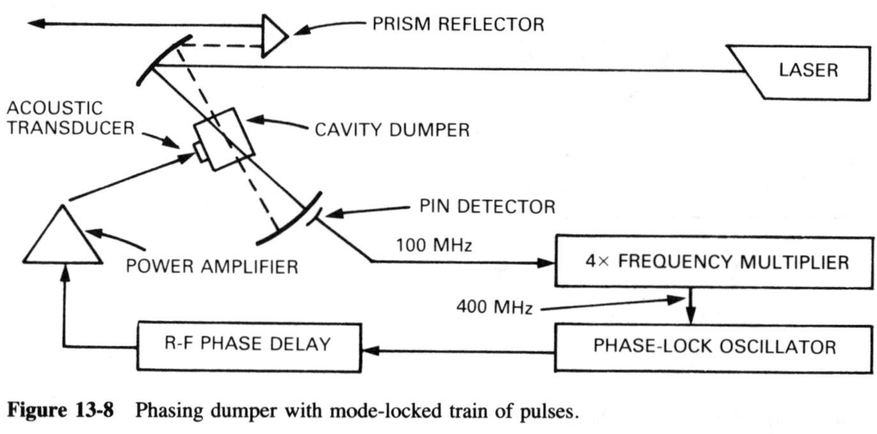

Figure 13-8 Phasing dumper with mode-locked train of pulses.

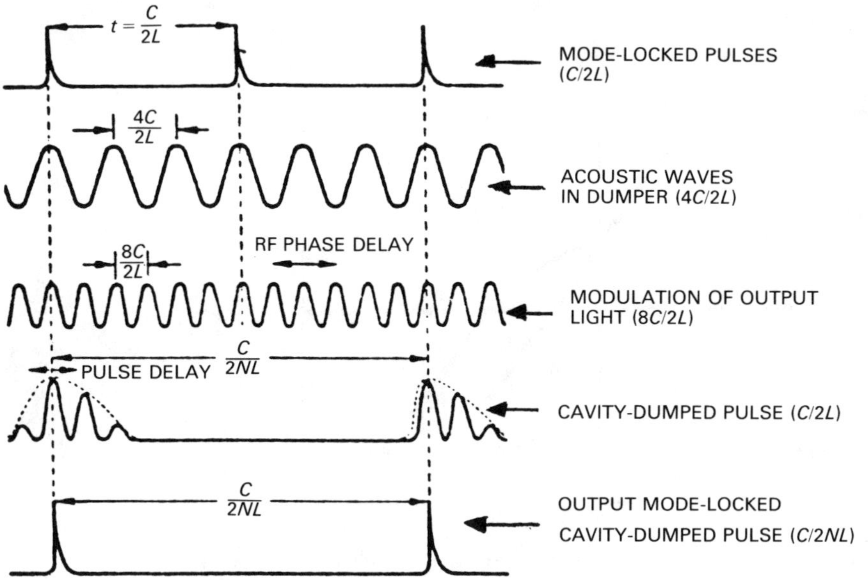

Figure 13-9 Relationships of various waveforms in the mode-locked cavity.

13-3 EXTRACTING A MODE-LOCKED PULSE AT A DESIRED RATE

To select a mode-locked pulse at a desired rate, the loop shown in Fig. 13-10 has to be pulsed at submultiple frequencies of the mode-locked frequency. The minimum pulsewidth is limited to 10 nsec, which is approximately the rise and fall time of the system, as illustrated by the waveforms shown in Fig. 13-9.

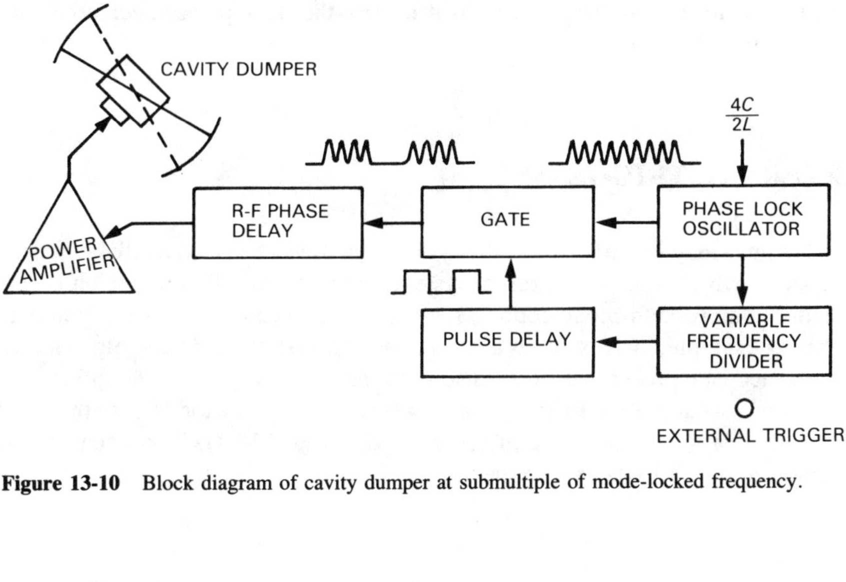

Figure 13-10 Block diagram of cavity dumper at submultiple of mode-locked frequency.

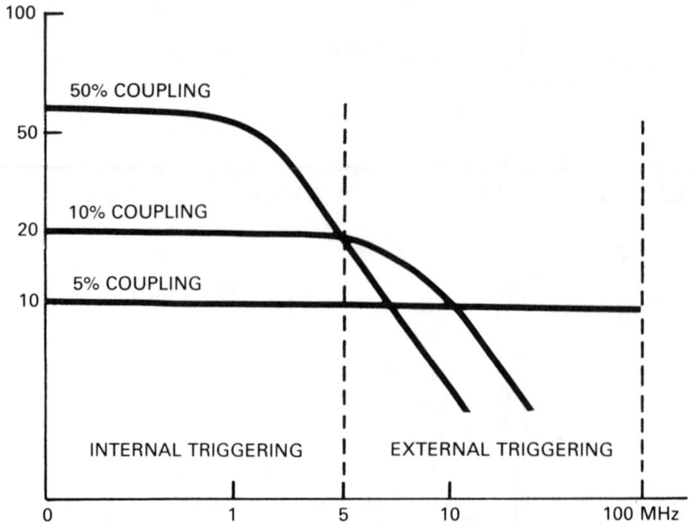

Figure 13-11 Peak power as a function of acoustic coupling in the cavity dumper.

To achieve pulsing at the desired rate, a gate is inserted between the phase-locked oscillator and the power amplifier (Fig. 13-10). A variable frequency divider network and a pulse delay line are disposed between the phase-locked oscillator and the gate. The time delay is adjusted so that the peak of the cavity-dumped envelope coincides with the mode-locked pulse, as shown in the figure.

The variable frequency divider provided in the system permits a choice of multiples of mode-locking frequency ($C/2L$ = 100 MHz) from a single shot up to several megahertz, the upper limit for the combined mode-locked cavity-dumped operation being 5 MHz. Above the latter frequency, the cavity dumper extracts power much faster than the cavity oscillations can build it up and the peak power decreases very rapidly, as illustrated in Fig. 13-11.

13-4 EXTERNAL TRIGGERING

An external trigger may be provided in the system to extract the "next available pulse" from the system, which makes the present system very versatile. This provision is also ideal for pulse compression-mode coding, range gating, distance measuring, and data storage. As an example, if it is desired to operate the system at a very high rate of repetition, the acoustic power is decreased so as to sustain only 5 to 10% coupling; this type of coupling is equivalent to that of a standard output reflector. Thus, the laser oscillation will always be maintained inside the cavity (Fig. 13-11). It will be possible then to dump up to 100 MHz in a sequence.

13-5 CAVITY-DUMPED OPERATION

When the mode locker is switched off, it is possible to obtain practically any pulsewidth desired. If internal triggering is used, the cavity dumper will deflect 12-nsec high-peak-power pulses. Peak power depends on the gain and saturation characteristics of each individual line and varies from 20 to 50 times the average continuous-wave power. On the other hand, if the external triggering mode is employed, any pulsewidth can be developed from 12 nsec per pulse to continuous-wave format. The pulse repetition rate can be varied from a single pulse to more than 10 MHz. The acoustic power drive can be varied to optimize the peak power of the pulses as a function of repetition rate.

13-6 CONTINUOUS MODE-LOCKED OPERATION

The combination mode-locked and cavity-dumped system can be operated in continuous mode-locked format. This is accomplished by switching the cavity dumper to the

Figure 13-12 Synchronously pumped, mode-locked, cavity-dumped dye laser by Spectra-Physics.

continuous-wave mode. Then the cavity dumper will operate continuously and will dump a train of pulses at 100 MHz. The peak power and the average power will vary from one wavelength to another. The peak power is then 5 to 15 times the continuous-wave power, while the average power is 0.1 to 0.3 times the continuous-wave power.

13-7 SYNCHRONOUSLY PUMPED, MODE-LOCKED, CAVITY-DUMPED DYE LASER

The process of mode locking and cavity dumping may also be employed with dye lasers. A synchronously pumped, mode-locked, and cavity-dumped dye laser system is shown in Fig. 13-12. The system offers the ultimate in versatility and performance. It operates at an average power of 25 mW and a peak power of 1.5 kW, at a pulse repetition rate varying from 400 Hz to 4 MHz. The energy per pulse varies up to 30 nJ, with a pulsewidth up to 20 psec. The system contains a high-efficiency acousto-optic modulator, which is positioned at the focal point of a double-pass optical resonator that is astigmatically compensated. The diffraction efficiency approaches 80%, with an extinction ratio greater than 500:1 for suppression of pulse trains.

The losses within the cavity are minimized by reduced numbers of reflective surfaces; this feature maximizes the conversion efficiency of the dye laser. Any synchronously pumped mode-locked dye laser can be converted to cavity-dumped operation by incorporating the cavity-dumper head shown in Fig. 13-13 into the system. A typical

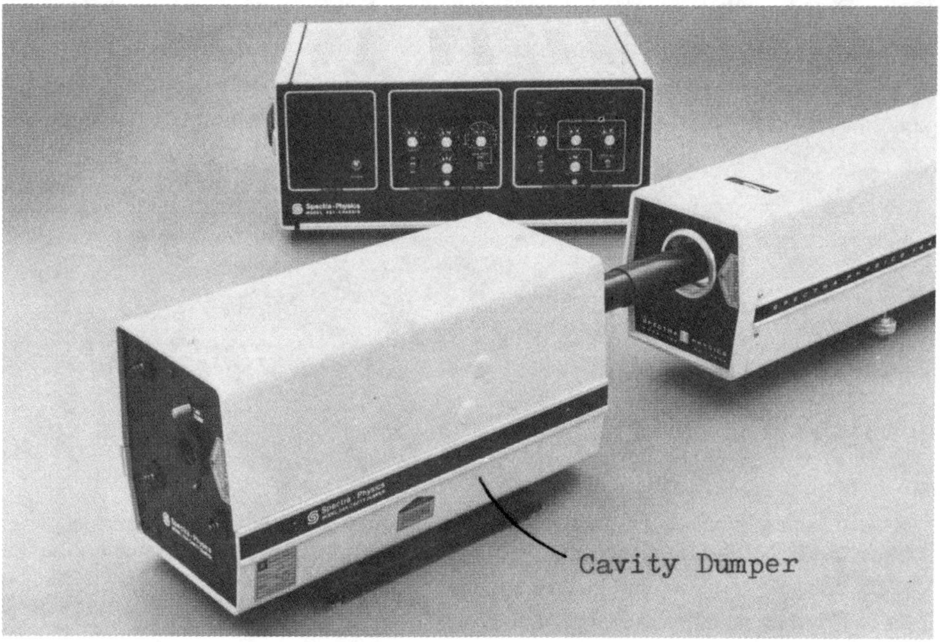

Figure 13-13 Spectra-Physics cavity dumper.

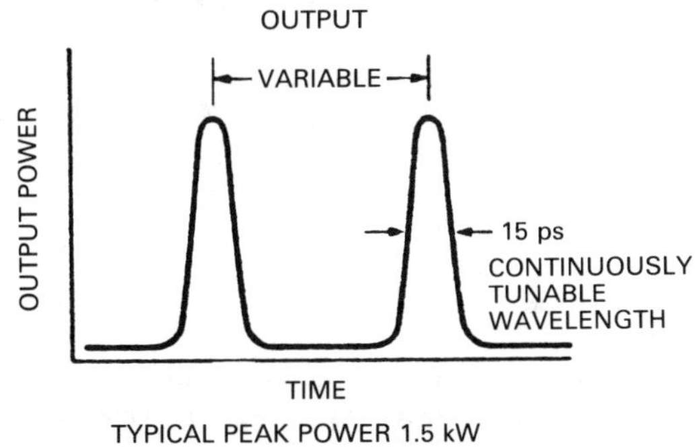

Figure 13-14 The waveform of peak power.

output-power waveform is shown in Fig. 13-14; the power shown by the waveform is obtained by the use of rhodamine 6G. Other peak powers can also be achieved by using other dyes as desired.

13-8 MODE-LOCKED LASER FOR OPHTHALMOLOGY

Comparative investigations of mode-locked laser beams versus Q-switched laser beams have shown that a mode-locked Nd-YAG beam used for ophthalmological surgery is superior to a Q-switched beam. This conclusion is predicated on the reduction of risk of injury to the underlying tissue structures during surgery. Furthermore, a mode-locked laser beam can be employed selectively at high peak powers. Q-switched laser beams have formed plasmas and caused optical breakdown in the ophthalmic site. It is claimed that the first pulse arriving at the tissue results in the interaction, with subsequent pulses blocking the energy transmission. The mode-locked pulses do not seem to have any tendency to block energy, especially at lower pulse energies. Damage thresholds are also reduced for mode-locked laser beams.

The reader should now review the discussion of Q-switching in Section 2-6 to fully understand the difference between a Q-switched laser beam and a mode-locked, cavity-dumped laser beam. Differences in the time domain of pulses, pulse energies, and photonic emission characteristics exist, and these should be clearly understood and considered in designing a laser system. Also, the type of lasing element in the system should be determined prior to selection of the performance modes.

QUESTIONS AND PROBLEMS

13-1 What is mode-locking of a laser beam? What significance do a mode and its individual sidebands bear with respect to mode locking? How are intracavity losses reduced to a minimum? In what respect does mode locking of a laser beam differ from Q-switching? Why is mode-locked laser radiation driven most effectively when triggered externally?

13-2 What relation does the Fabry-Perot resonant frequency of *C*/2*L* have to the intracavity amplitude modulator? What quantity represents the phase-locked axial modes? How can such a device function as a pulsed regenerative amplifier? What is the Fourier transform of the frequency spectrum of all the phase-locked modes? What is a sideband relative to a pulse?

13-3 How is the intracavity wavelength selection accomplished? How is a standing wave within the quartz crystal produced? What is a cavity dumper? How is it brought about? What is the relationship of Bragg angle to cavity dumping? What is the function of the traveling acoustic wave in Doppler-shifted deflection?

13-4 If a laser beam of 4200 Å is incident on a Bragg deflector at an angle of 8°, what will be the wavelength of the acoustic wave front? What will be the Doppler-shifted deflection wave frequency?

13-5 What is meant by folded cavity? In what optical respect is a cavity dumper disposed in a folded cavity? What does a photodetector detect in the folded cavity? What is the typical buildup time for stimulated emission in an ion laser cavity?

13-6 How can one achieve maximum diffraction efficiency in the cavity dumper? What is the relationship between the acoustic dumping frequency and the mode-locked frequency? How can this relationship be achieved in the mode-locked cavity?

13-7 How many beams does a phase-locked oscillator consist of? How can 100% modulation be achieved in the cavity? How can one generate an acoustic frequency in the phase-locked oscillator? How is the acoustic transducer driven? How is a maximum output from the dumper crystal obtained?

13-8 What is the purpose of a variable radio-frequency phase shifter? How does one select a mode-locked pulse at a desired rate? What is the purpose of a gate in a cavity dumper? With what does the cavity-dumped envelope coincide?

13-9 What is the function of a variable-frequency divider in the present system? What does the cavity dumper do above the 5-MHz mode-locked operation? What is external triggering? What is internal triggering? How does one accomplish each of these triggering procedures?

13-10 How does one accomplish cavity-dumped operation? How does one obtain any desired magnitude of pulsewidth? How does one obtain high-peak-power pulses from the cavity dumper? How does one obtain 12-nsec pulsewidth up to a continuous-wave power? How does one optimize the peak power of the pulses in the dumper?

13-11 Discuss what cavity-dumped operation is. Discuss continuous-wave mode-locked operation. What is the peak power range? What is the average power range in the dumper? How does one obtain each type of operation? Discuss.

13-12 What are synchronously pumped, mode-locked, cavity-dumped operations? Discuss. What is the function of the acousto-optic modulator in this system? What significance is borne by the diffraction efficiency in such a system? Draw a block diagram of a complete mode-locked and cavity-dumped ion laser system.

SELECTED SOURCES OF RELATED DATA

Allocca, J., and A. Stuart. 1984. *Transducers: Theory and Applications*. Englewood Cliffs, N.J.: Prentice-Hall.

Campbell, J. C., et al. 1982. Optical AND gate. *IEEE J. Quantum Electron.* 18(6):992–994.

Chin, S. L., ed. 1984. *Multiphoton Ionization of Atoms*. New York: Academic Press.

Javan, A., Laser Sciences, Inc. 1983. Transient Regenerative Amplification (Laser). U.S. Patent No. 4,410,992.

L&A Staff. 1982. Mode-locked Nd-YAG laser for eye surgery. *Lasers and Applications* October, pp. 24–26.

Laser Focus Staff. 1984. Laser/E-O research at NRL. *Laser Focus/Electro-Optics* May, pp. 74–84.

Mittleman, M. H. 1982. *Introduction to the Theory of Laser-Atom Interactions*. New York: Plenum.

Stark, H., ed. 1982. *Applications of Optical Fourier Transforms*. New York: Academic Press.

Weber, M. J., ed. 1982. *Handbook of Laser Science and Technology*, Vols. 1 and 2. Boca Raton, Fla.: CRC Press.

Yao, S.-K., Rockwell International Corp., 1982. Optical Cosine-Transform System. U.S. Patent No. 4,354,247.

CHAPTER

FOURTEEN

UNIQUE LASER SYSTEMS

14-1 VARIOUS APPLICATIONS OF LASER SYSTEMS

Two types of laser systems are presented in this chapter—systems that have already been developed and are in commercial use and those that remain to be developed. Of those that are already developed, only systems that are commercially successful and unique in their applications are discussed in this section.

The innovative ideas have been selected for those who have gained a basic knowledge of laser principles and are capable of applying their knowledge in practice. In suggesting the ideas, the author has also given a brief plan of approach to the problem and in some cases has outlined any background history necessary for understanding the problem for adequate solution.

The applications of various laser types and their systems given in this section are based on the principles that have been learned in the preceding chapters; accordingly, some of the details of operative concepts may not have been given fully, and the reader should use his wealth of knowledge in determining the fundamental knowledge employed. In this respect, these system applications should be construed as review study dependent on the information acquired in the preceding chapters.

Laser System for Personnel Verification

A laser system that provides instant confirmation of fingerprints on identification cards has been developed by KMS Industries, Inc. The system scrambles the print to make substitution of unauthorized fingerprints almost impossible. To use the system, which is shown in Fig. 14-1, an ID card is prepared by using a laser beam to make a hologram of the identifying finger on a microdot in the usual ID card. The microdot is encapsulated together with the ID card in a plastic or metal substrate. The system is provided with two slots. The ID card is inserted into one slot, and the subject's matching finger is inserted into the other. The pattern on each image is scanned by a laser beam, and the two images are compared to verify or deny the identification automatically.

The elapsed time for the verification is 3 to 5 seconds. The system performs with an accuracy of within 0.01%. The unit is used in top-security clearance in industry and in government installations. A potential application would be verification of credit cards to prevent the use and falsification of stolen cards.

Fingerprint Identifier

A laser-aided fingerprint identifier, capable of identifying a fingerprint within 1 second, has been developed for the Federal Bureau of Investigation. The system employs a subsystem that preprocesses the fingerprints by optically scanning them and using image enhancement techniques to identify the fingerprint patterns that are characteristic of each person. The processed information is relayed to a DEC PDP-15 section, which compares and counterchecks the fingerprint data already on file. The system is expected

Figure 14-1 A fingerprint identifying system shown in operation (Courtesy of KMS Industries, Inc.).

to reduce the number of people now employed in manually checking the 25,000 or more fingerprint inquiries received daily and to verify the fingerprints against the 20 million prints on record.

A more recent laser technique using the visible light from an argon laser has been developed in Canada by Xerox Research Center and Ontario Provincial Police to detect and identify fingerprints by the fluorescence produced by the fingerprint ingredients. The system can recover the latent image of a print that would be undetectable by conventional methods, which use either a sprayed powder or a chemical reagent to interact with the fingerprint constituents to develop the image. The new method is reported to indicate the ridge detail immediately with high definition when the fingerprint mark is illuminated. The fluorescence of the mark is of sufficient intensity for photographing the print on a color film. The photograph can then be transmitted to any distance over digitally operated transmission lines for display or for storage in the memory bank of a central station computer.

In operation, the argon laser system emits continuous radiation of 5145 Å in the green spectrum, which is partially filtered and expanded to cover an area of 8 × 10 inches. The expanded argon laser beam excites the freshly marked fingerprint to a yellow-green fluorescence. Fingerprints that have been left on any substance, such as glass, paper, walls, metals, and wood, for a considerable time fluoresce with an orange color. It is claimed by the developers that the age of the print degrades its luminescent power but sufficient definition is still left for identification purposes. The print retains its configuration even after exposure to relatively higher temperatures than the room ambient for several years. Even moisture does not affect it significantly during that period. Because of the degradation of the fingerprint with age, it is believed that this characteristic may be utilized some day, by further research and experimentation, to determine the date of the print.

Tire-Mold Roundness Checkout System

An innovation, first reported from Germany, is a laser system that checks and measures the roundness of tire molds. Developed by Siemens Aktiengesellschaft, the system uses a helium-neon laser with a 1-mW output at 1.6 kV and checks by the triangulation method to reduce the length measurement to time measurement. The time gap between a reference plane and a working plane is translated to the error in roundness and is immediately displayed, giving the location and degree of the error. With this machine, the operator can measure parts of all kinds on lathes, milling machines, and presses. The complete equipment includes a laser probe, a photodiode that detects error, and a digital and analog display with a control unit. The system operates from a 115-V, 48- to 60-Hz current.

Laser Microphone

A laser system has been developed in England that can "listen" to conversations carried out in a room. The system operates by projecting a laser beam onto a window, analyzing the beam vibration patterns on the window, and then decoding the patterns into speech.

The window glass acts as a microphone and vibrates in accordance with the acoustic vibrations of the spoken words within the room. However, use of the system is restricted to rooms having a glass window without a shade or a curtain, which prevents the sound waves being transmitted from the window glass by absorption or interference.

Gyroscope Balancing

Rotating devices, such as gyroscopes or jet-engine turbine blades, must retain their balance almost perfectly at all times in order to accomplish their jobs effectively. In the past, gyroscopes have been balanced mechanically under very cumbersome and difficult conditions, and this has been time-consuming as well as costly. A laser pulse emanating from a laser welding or drilling machine can be used to balance a gyroscope in one-tenth of the time required by mechanical methods. Another advantage is that the gyroscope can be balanced while it is rotating, which is the correct way to balance a gyroscope.

In operation, while the gyroscope is rotating, the laser beam removes a small increment of metal at the rate of a few micrograms to one milligram per pulse. By pulsing 20 to 30 times per minute, the beam can remove up to 30 mg of the metal per minute.

Different types of laser radiation may be used in the removal of metal in balancing the gyroscope. The commonly used lasers are CO_2 TEA, Nd-YAG, Nd-glass, and ruby lasers; they are used with or without Q-switching. However, since Q-switched lasers pulse in nanosecond durations, they do have a drawback. The total laser energy is limited by the formation of high-density plasma, which dissipates the energy by absorption when a certain energy level is reached. The amount of metal that is removed by each pulse is limited because of low overall efficiency. For this reason, a TEA-type laser beam with a repetition rate of 20 to 30 pulses per minute has been preferred. All removed metal is separated from the gyroscope while it is spinning, avoiding any possible damage to the gyroscope.

In the procedure, the gyroscope is supported on a spinning shaft and is rotated at speeds of 10,000 to 40,000 rpm. The gyroscope is adjusted in front of the laser head, and its extent of unbalance is sensed by two detectors placed at right angles to each other on vertical and horizontal planes. When the sensors detect an imbalance, a laser beam of 40- to 100-μsec pulsewidth is instantly directed at the unbalance point, removing an increment of the metal from the surface. When a balanced condition is attained, the gyroscope balancer automatically cuts off the laser beam from the workpiece.

Algae Mapping with Laser System

A new algae-mapping laser instrument, known as a laser fluorosensor, has been developed by the National Aeronautics and Space Administration (NASA). It consists of a low-repetition-rate, pulsed, liquid rhodamine 6G dye laser. Mounted on the underside of a helicopter, it projects a 5900-Å laser beam onto the surface of water located about

30 m below the craft. This causes the algae in the surface to fluoresce and reflect a red light at 6850 Å. The intensity of the light beam is recorded on a linear analog-to-digital converter, which compares the natural intensity with the retroreflected intensity and indicates the amount of algae present in a given area.

Cattle and Salmon Branding

Cattle merchants claim that when cattle and livestock are branded with a heated iron, each animal loses from 20 to 30 pounds of its weight because of the suffering it undergoes and the aftereffect of the branding. Since branding with a laser beam is faster and the animal does not suffer afterwards, a carbon dioxide laser system operating at 100 W and up has been tried with favorable results. The beam scans a path of about 18 inches within 1 second and completes the work before the animal takes a second breath. The laser beam follows a continuous path designed in the branding template.

Hatchery biologists find that laser beam tagging of salmon and other edible fish is harmless to the fish. The radiation is directed to a template or stencil and a mark is instantly made on the side of the fish for future identification. By controlling the frequency of the laser beam, only certain pigments are removed from the fish without burning the fish.

Laser-Equipped Cane for the Blind

A laser-equipped guidance cane for the blind is being manufactured by Bionic Instruments, Inc., under the sponsorship of the Veterans Administration. The cane contains a transmitter near its crook end, with three GaAs laser diodes connected in series. The diodes emit 0.1-μsec pulses at a 9040-Å wavelength at a rate of 40 pulses per second. The diodes form three light probes projecting forward, downward, and upward to warn the blind user of obstacles 3 to 12 feet ahead, above, and below, respectively. A receiver containing three photosensors is located about 9 inches below the transmitter. The retroreflected signals are caught by the respective photosensors located behind objective lenses. The photosensors produce signal outputs that create a poking sensation in the blind person's fingers to warn him of obstacles lying ahead. The sensors also provide auditory sounds for objects located above or below the plane of the projected laser beam. The cane is about 50 inches long, and the guidance system is battery operated and has a laser output of 3 to 10 mW.

Stenographic Laser Erasers

As explained earlier in the chapters on industrial applications of laser beams, a focused laser beam can vaporize metals, ceramics, plastics, and other materials. The extent of vaporization depends on the energy level of the beam and on the density and optical characteristics of the workpiece. The same principle is employed in erasing errors from typewritten matter and from drawings without damaging the substrate on which the matter is printed. This is possible due to the absorption properties of a controlled-energy laser beam. The beam is absorbed by dark, black, and colored surfaces and is reflected by white and shiny surfaces.

The principal advantages of a laser eraser are that it is easy to use, does not damage the paper on which the character is written, and instantaneously erases the typewritten error. Another characteristic of the present eraser is that it employs a defocused divergent laser beam to afford coverage of a larger area than when the focus of the beam is used. Since the energy density of the defocused laser beam is much smaller per square area than that of a focused beam, it is easier to control. For instance, a laser beam of 1-J energy, when focused to a point of 5 mils on a written character, will drill a 5-mil hole in the character. However, when the beam is defocused and expanded to a 1-mm square area and then directed to a 1-mm square character, the character will disappear instantly due to oxidation of the ink of the character.

One of the laser erasers developed and patented by Laserkinetics Corporation consists of a hand-held erasing stylus suitable for use on a typewriter. The stylus has a coniform erasing head with an aperture at its tip for ejecting a laser beam from the stylus onto the character to be erased. The laser generator consists of an array of semiconductor laser diodes energized from a pulse-forming network in the power supply unit. A beam-integrating element, positioned adjacent to the diodes, integrates the laser radiation from the array, with the radiation reflected from a parabolic reflector enclosing the diode array. The integrated radiation is directed onto a converging lens. The laser beam emerging from the lens is focused within the coniform section of the stylus and then exits from the aperture of the stylus tip as a divergent beam. When a thumb switch on the stylus is depressed, the laser beam is triggered in one pulse.

The divergent laser beam becomes attenuated in intensity as it radiates from the tip of the stylus. Since it will be difficult for the user to determine what part of the beam should be used for erasing, the lens is adjusted within the coniform section so that the energy of the emergent laser beam at the exit port of the tip is just sufficient to vaporize the character to be erased when the tip is placed in contact with the character. Thus, for most erasures, the diameter of the emergent beam at the exit port is adjusted to 2 mm using 1 J of laser energy. For erasing larger characters, the laser energy is increased from the power supply and the diameter of the diverging beam at the exit port is also increased by adjusting the lens distance from the tip, so as to cover the entire area of the character. Since the laser beam energizing current is pulsing, a single pulse is all that is necessary to erase the character. The diameter (d) of the emergent beam cone at the exit port can be determined by the relation

$$d = \frac{Ds}{S} \tag{14-1}$$

where D is the diameter of the lens optical plane, s is the distance between the laser beam focus in the coniform section and the tip of the stylus, and S is the distance between the optical plane of the lens and the laser beam focus.

The diameter of the exit beam can be adjusted by internally adjusting the lens distance from the exit aperture. Markings are printed for various diameters of focus size at the exit port. A photograph of the laser eraser and the power-supply unit is shown in Fig. 14-2.

Provision has also been made in the power supply unit to convert a 115-V AC

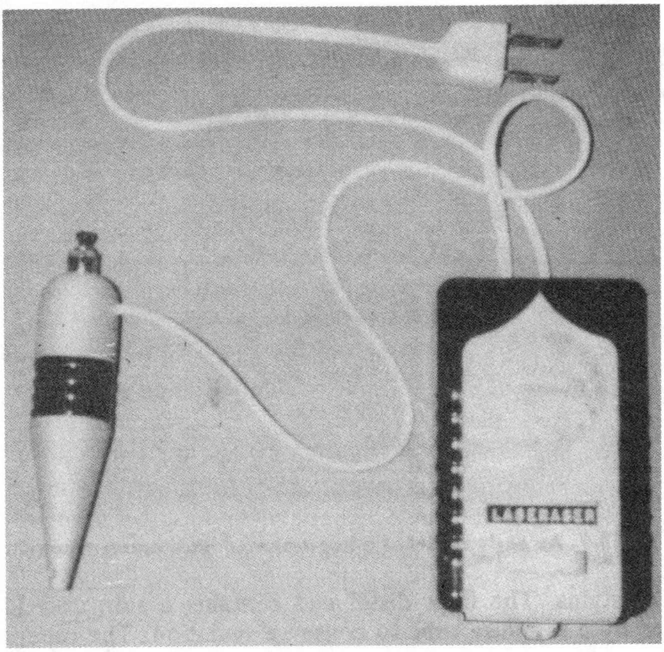

Figure 14-2 Laser eraser and power-supply unit.

current (any household current) to a low-level operating voltage as desired. A pulse-forming circuit raises the frequency to 1000 Hz or more, as necessary, for the operation of the laser generator. In another model of the device, provision is made so that the device can operate with direct current, using rechargeable batteries.

Safety of the user has been the prime requirement of this device so that neither an electronic shock nor any laser radiation can harm the user or any person in the immediate vicinity. In this connection, a modified version of the device uses no external switch. The operation of the eraser is made automatic; when the tip of the stylus is placed on the error character and pressed on it, the laser pulse is instantly triggered, making the generation of a laser pulse impossible until the tip of the stylus is pressed against the character.

In another version of the laser eraser, the laser head is miniaturized so that it can be mounted permanently on a typewriter key. A switch is incorporated under the key so that when the key is depressed the typewriter carriage moves back one space and simultaneously the laser eraser moves to the printing area. This action causes the switch to close, and a pulse of laser beam impinges on the error character and instantly vaporizes it.

Laser Eraser and Microwelder

This device, developed by the author, is a laser eraser as well as a microwelder. The system, illustrated in Fig. 14-3, consists of a power supply unit connected to a stylus through a fiberoptic conduit, which receives laser radiation from the generator located

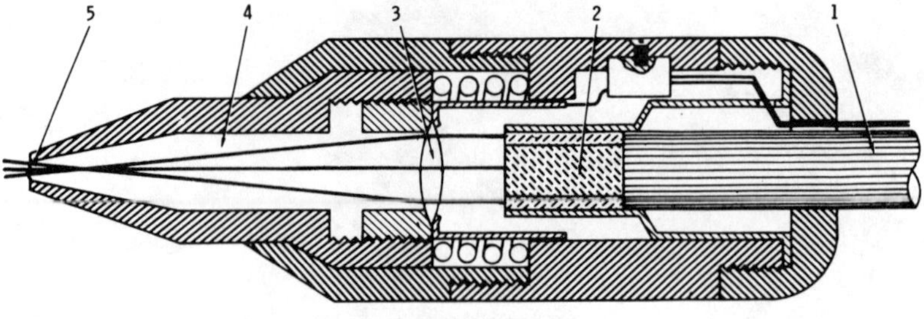

Figure 14-3 Cross-sectional view of the laser stylus.

in the power supply unit and transmits it to the stylus. The distal end of the fiberoptic conduit (1) contains a ruby rod (2) surrounded by a sapphire tube to conserve radiation. The outer face of the ruby rod is made partially reflective by depositing a multilayer dielectric material to form a resonant cavity between the ruby and the laser source, such as Nd-YAG or Nd-glass. Either continuous-wave or pulsed laser radiation can be obtained from the power supply circuit. The pulse network circuit furnishes a 1-msec pulse during each trigger; this pulse can be used for erasing or for a single spot weld. In the continuous mode, the beam can weld fine wires, thin films, small microelectronic components, watch parts, and jewelry and can erase long lines in drawn diagrams. The device can also drill holes of diameters from 1 mil to several mils in metals or ceramic wafers used in semiconductor fabrication.

The laser radiation from the fiberoptic conduit (1) passes through the ruby rod (2) and becomes incident on the lens (3), which focuses the laser beam within the coniform chamber (4) of the stylus. The beam diverges from the focus as it leaves the exit aperture (5) at the tip of the stylus; when the stylus tip is placed on an error character and pressed against it, the laser is triggered and the character receives the divergent beam (at point 5) and is instantly vaporized. For microwelding, the lens is internally adjusted to move the focus of the beam to the plane of the stylus tip. Thus, when the tip is placed on the workpiece to be welded, the beam is in focus at the center of the thickness of the material being welded. For drilling, the lens or the laser power is adjusted prior to triggering the laser.

Other modifications of the laser eraser are also available, one of which resembles a ballpoint pen and can be carried in a pocket. The laser generator has a pulse-forming circuit operating from a rechargeable battery. The laser element is an array of semiconductors that furnish the radiant energy within a variable energy level; this energy is adjustable within the device. Triggering of the laser beam can occur only when the stylus is perpendicularly positioned on the error character and pressed against it. This feature, which is also found in the other eraser devices already described, makes the laser eraser safe to use. The user can never see the laser radiation or its reflections.

Color Television Projection System

The Japanese firm Hitachi has recently developed a laser television projection system that can project a color television picture composed of 1125 scanning lines per frame on a screen that can be from 10 to over 90 feet square. The three primary colors of the picture are obtained from ion laser beams: red from a krypton ion laser and green and blue from an argon laser of tunable wavelengths. The system is said to have a polygon mirror driven by an 81,000-rpm motor, which causes a horizontal resolution of 800 lines and a vertical resolution of 700 lines. A wideband video amplifier operating at 30 MHz is used.

Hitachi has also developed a laser videodisk system that uses a modulated argon laser beam for recording color pictures on a disk, which is played back using a He-Ne laser beam. The disk is made of sheet glass with a reflective metallic coating on its surface and is 30 cm in diameter. The system is capable of recording as many as 50,000 still color pictures from a television camera or a videotape recorder. The argon laser beam is electro-optically modulated and optically focused to a point of 6000-Å diameter, the length of one wavelength of orange light. The disk is provided with 8000-Å wide tracks separated from one another by 20,000 Å.

In the playback operation, the He-Ne laser beam is focused to a point of 10,000 Å on the track and the reflection of the beam from the track is recorded in the data storage bank of the system. The definition of the reproduced picture is claimed to be as high as that of a color television image. A signal-to-noise ratio of 40 dB or more has been achieved. The potential applications for the system are cited to be in broadcasting still pictures and in recorded storage of radiographs and fingerprints.

A similar system has been developed by RCA in the United States; this system can take 10,000 still color pictures on a 250-mm-wide stripe disposed on a disk 30 cm in diameter. For playback, the system uses a laser beam from a He-Cd laser source.

Laser Printing System

The laser printing system supplants the traditional stereotyping and engraving procedures. The pasted-up copies of a newspaper page are scanned by computer-controlled laser beams to produce a relief image on thin metal or plastic plates by vaporizing the scanned areas. This procedure eliminates the necessity of making photographic negatives or acid-etched plates and speeds up production.

In the process, two laser systems are used. One of the laser systems scans and reads the pasted-up copies and makes a negative film, from which a positive image is transmitted by a cable or microwave system to the second laser system at the receiving end. At the second laser system, which may be located anywhere nearby or at a remote facility, the image is exposed with an ultraviolet laser beam to make a printing plate. The method is said to have reduced overall production costs by about 40% and the time to about one-third that of the conventional plate-making process.

Several other laser companies are also manufacturing laser printers. Among them, the Hewlett-Packard laser printer is of special interest because of its sophisticated system design. This system employs an electrophotographic process for producing printed pages at a rapid rate. It uses six steps in producing printed pages. In the first step, a pattern of latent laser signals corresponding to the information to be printed is generated by the digital electronic circuit in response to computer controller command. In the next step, the laser beam produces an electrostatic image that consists of charged and uncharged areas on a cylindrical drum. Following this, an ink toner consisting of fine black particles that are electrically charged is projected to the drum, where the particles adhere electrostatically to the charged areas on the drum. The paper to be printed on is brought adjacent the drum, and an electrostatic charge is applied to the back of the paper and fixes the toner to the paper permanently. The last step is to restore the drum to its quiescent stage, ready for the next operation.

Laser Power Converter-Transmitter

A laser-assisted power transmission system for transmitting power through space has been developed by NASA's Lewis Research Center. The device consists of a vacuum discharge tube with a thermionic filament as one electrode and an electron-capturing element as the other electrode. Between the electrodes is a low-pressure ionizable gas (possibly metal vapor) that provides a plasma in the tube when a relatively low voltage is impressed across the tube. When the operating discharge tube is illuminated with a laser beam, the energy of the laser beam is converted directly into a direct current, thus increasing the output of the plasma current from the tube. It is calculated that the efficiency of conversion of the laser beam energy into electricity could be as high as 50%. Recent work on this concept has indicated that up to 10 kW of laser beam input can produce at least 2 kW of direct-current output. The goal is to obtain up to 10 MW density in a space vehicle by projecting a high-power laser beam from the ground to the plasma converter in the traveling vehicle; in such an arrangement, the laser beam is locked on the vehicle.

Rocket Propulsion by Laser

Work is in progress on the concept of propelling rockets into earth orbit by means of CO_2 laser energy. The idea was first advanced several years ago by AVCO Everett Research Laboratories, which had already developed the most powerful laser systems for metalworking in the world. To accomplish the purpose of the concept, the laser energy must overcome difficulties posed by atmospheric conditions, such as absorption of laser radiation by the water vapor in the atmosphere, scattering by aerial particles, and thermal blooming. However, because the energy attenuation by these obstacles is small, the laser energy that reaches its destination would be sufficient to produce propulsion in a rocket. The propulsion is achieved by focusing the laser beam on a small, lightweight propellant in the rocket engine and vaporizing the propellant to provide thrust for lifting the rocket to orbit the earth. Compared with the weight of chemical fuel now used in rocket engines, the weight of the fusion propellant would be significantly smaller, so the payload would be greater than in a chemical fuel–propelled space vehicle.

Phase-Locked Semiconductor Laser Diode

The development of high-power continuous-wave phase-locked laser diodes has been reported by Xerox Palo Alto Research Center. This accomplishment in laser design consists of an array of 10 closely spaced injection diode lasers grown on a single GaAs chip. The achievement is akin to the author's first attempt, in 1965, to produce an array of diode lasers for use in his laser eraser development. However, the new development transcends the original idea by its production of higher power in a phase-locked format, which results from optical coupling between the laser elements. Operating as a single optical source, the device emits about 0.6 W per mirror with an optical conversion efficiency of 23%. The principal application is in optically pumping a Nd-YAG laser rod or ruby rod. Because of its good affinity for isotopic selection, the feasibility of this development for an isotope separation process has been demonstrated under room temperature conditions.

Gold Vapor Laser

The first commercially successful gold vapor laser has been developed by Plasma Kinetics, Inc. The device operates in pulsed mode, emitting at 6280 Å with 30-nsec pulses at a 5-kHz repetition rate. The beam diameter of the radiation is 3 cm with an exit port divergence of 3 mrad. The device has also operated with a peak power of 30 kW, the average power being 4.5 W. The gold vapor plasma operates at 1450°C, a temperature achieved by using the impressed electrical power on the tube containing neon as a buffer. The arrangement eliminates use of an external heating means. Unlike an argon-pumped dye laser, which is costly, the new development is projected for applications in radiotherapy for cancerous cells, avoiding contiguous healthy cells by the use of hematoporphyrin dye. (See Chapter 11.)

14-2 NEEDED LASER SYSTEMS

Moving Vehicle Proximity Indicator

An automobile proximity indicator using a GaAlAs diode to indicate the distance of a car ahead as well as the approach of another car from the rear is needed. This would be especially useful for foggy days. For front view, when an automobile is detected within 100 feet ahead, a yellow flashing light should appear at the rear of the car moving in the front to warn other cars following it that any sudden slowing down by it may cause a rear collision. A flashing red light should indicate that there is a car following in the rear. The approach distance either in the front or in the rear may be determined by the frequency of the flashing light—the closer the two light flashes are, the faster the indicator light will flash. The indication can be accomplished by modulating the detector cell at 100 MHz, providing a certain frequency threshold that is received by the device receiver, and converting this frequency into distance in feet or yards. The principle is based on target designators.

Nasal Passage Treatment Device

Since an argon laser furnishes copious amounts of ultraviolet rays, which are used in medical treatment of nasal passages to kill bacteria, a small, compact, and lightweight laser system for medical office use would be a useful instrumentation tool. The device should have a flexible waveguide with a variable-focus lens system at its distal end to vary the amount, direction, and even the wavelength of the beam directly from the waveguide stylus. A tunable dye laser emitting at ultraviolet wavelengths may also be suitable for the purpose. Streptococci and staphylococci find the rhinal cavities to be good breeding cites, and at times they produce suppuration and other foul discharges which are vulnerable to the action of ultraviolet light. The device could be fitted with a timing device to avoid excessive exposure of the mucous sites, but this is not entirely needed if the density of the radiation is reduced to a low level.

Otological Treatment Instrument

Nasal sinus inflammation due to allergy or rhinal infection causes the eustachian tube to constrict or close up to the extent of interfering with the passage of air to and from the inner ear. This condition causes distention of the tympanum, causing pain and pressure in the ear. To relieve the pressure, the physician pneumatizes the eustachian tube; if this cannot be achieved, the otologist punctures a hole in the tympanum and inserts a grommet for free passage of air into and out of the inner ear. This relieves the pressure and the earache, and the patient can hear almost normally. The treatment is usually carried out in a hospital.

A clinical argon laser instrument or a frequency-doubled Nd-YAG laser of 5 to 10 J output can be used for vaporizing a tiny hole in the tympanic membrane, using a long-focal-length lens. This procedure will further cauterize the periphery of the hole so that no grommet will be required. The technique can be carried out at the otologist's own clinic, without resorting to hospital care. No local or general anesthesia will be required, since the treatment is rapid and painless. Such an instrument is direly needed by otologists who perform this type of operation.

Laser Medical Imaging System

Thermal imaging methods now in use employ either an infrared device or a spray of liquid crystals in solution. In the process, the colors thus produced are changeable in rapid progression as the patient's nervous condition changes. A thermal imaging system using complementary laser beams compatible with the texture of the examination site would yield improved imaging performance, making the medical diagnosis more reliable. A television screen can be used for visualization and diagnosis of the patient. Subsurface breast or dermal carcinomatous conditions can be detected by the use of a television color camera mechanism transmitting signals from the patient's cutaneous areas to a black-and-white television screen. The author's preliminary tests with such instrumentation have shown high-definition anomalies both on the surface and immediately below the surface layer of the skin. The system should be constructed as one unit, with the patient positioned on one side of the system and the diagnostician examining the affected sites from the opposite side of the diagnostic instrumentation system.

Optical Fiber for 3000 to 106,000 Å Wavelengths

A new fiberoptic material that is compatible with wavelengths from the ultraviolet to the infrared is needed immediately for medical as well as optical communication purposes. The material should combine the spectroscopic properties with physical properties, such as hardness, coefficient of expansion, high thermal conductivity, and flexibility. The material could possibly be selected from rare-earth ions contained in a durable host material. The fiber diameter should be approximately 2 mm; it should be capable of being employed as a single fiber or in bundles.

Metrological Instrument

Present metrological instruments to conduct quality assurance manipulations for controlling the accuracy of manufactured product dimensions generally employ micrometers used manually. The procedure is slow and cumbersome when demand for rapid production work is critical. A He-Ne laser system of about 2-mW power can be made to project a beam in the path of moving manufactured parts and automatically expel any part that does not meet the dimensional tolerances as set forth on the machine. Such a system will eliminate human errors and increase production. The cost of such a machine should be kept low, possibly not to exceed $10,000.

Meteorological Sounding System

A meteorological sounding system for shipboard application would be welcomed by various military agencies because the present telemetering methods cannot be used under emission-control conditions. The system should have unlimited altitude capabilities, should employ a completely passive means of obtaining data, and should operate reliably in all climatic conditions and at all temperatures. The laser beam, being highly directional and insensitive to jamming, would be an admirable candidate for this purpose.

Indicator for Aircraft-Wing Icing Condition

The present indicators used for sensing the wing icing condition of an aircraft integrate the differential pressure of a heated impact tube in relation to the icing condition. Such an instrument is said to be subject to excessive lag in indication. The needed instrument to be developed must indicate the approach of the icing condition rather than indicating the icing condition after it is predominant. In addition, the device should indicate to the pilot at all times the ambient conditions tending to form icing and at the same time should automatically deice the surfaces as the icing begins. Such an indicating instrument should be independent of specific atmospheric characteristics not directly contributing to a condition of icing. The sensor for such an instrument could be one of the photosensors, such as GaAlAs, combined with a temperature- and humidity-sensing crystal to produce beat frequencies to differentiate the conditions between the icing and melting temperatures.

Aircraft Collision-Preventive System

Several collision-preventive systems are commercially produced at the present time, but apparently none of them is effective enough to become standard equipment on every aircraft manufactured. A system is needed that will be reliable, effective, and reasonable in cost so that it can be installed on private planes as well as on commercial planes. Two different-color visible laser beams can be used for this purpose. The beam should be projected on the oncoming plane, detected by scanning it, and then should be locked on it. This will create a reference point that will be displayed on the instrument panel in the cockpit. As the two planes are moving, the reference point will also move with respect to the image of the aircraft displayed on the same instrument, so that the pilot can maneuver his plane with reference to the original reference point and take evasive action. A corresponding tactical system on the oncoming plane should not be required.

Portable Security Sentinel

A portable security-surveillance and alerting laser system is needed that can be portable from one place to another and set up easily. The setup should be invisible to intruders so that the device will not be vulnerable to artificial implementation or destruction by an intruder. There are several types of surveillance laser systems, but none appears to be effective in performing the function required. They are also prone to detection by the intruder, and optical jamming is not impossible. A small and compact device that can be produced at low cost and easily set up at any location should be attempted for the development work. It could employ an invisible beam from a laser diode or from a Nd-YAG laser if the installation area is extensive and a more powerful laser beam is required to make its rounds from the transmitter to the receiver.

Underwater-Target Detection System

When a submarine detects an unknown vessel in the ocean, the only technique for identification of the target appears to be by tracking the vessel for an extended period and attempting communication. In such a situation, the ambient sea noise complicates the acquisition of acoustic transmission, causing confusion in identification of the vessel. Long-range detection techniques and equipment are needed for this purpose. Identification of the target appears to be the primary requirement of such a system. Therefore, a deep-sea penetrating beam (such as a green laser beam) for scanning and acquisition and a double beam for identification are required. The system should be easily and conveniently installed on any underwater craft.

Instrument for Referred Pain

A medical instrument for alleviating pain due to an organic disorder in the body by external treatment would be a useful medical tool for physicians practicing internal medicine. Pain due to disorders of many organs in the body travels remotely from the source and appears in an area adjacent the skin. Since pain manifests itself by irritation of a nerve ending, nerve ganglion, or nerve trunk, there are several hundred areas on the skin where pain from different organs appears. Such a condition is known as referred pain or reflex pain. This subject has been fully taken up in Chapter 10 and should be studied before attempting to develop the instrument suggested in this section.

The laser system developed should be small and compact so that it is easily portable in hand from place to place as the physician travels from house to house, as in the old days (which procedure some doctors have already started). The instrument should consist of a low-level visible laser source, such as He-Ne or frequency-doubled Nd-YAG. The beam diameter should be 1 or 2 mm at power of 1 or 2 mW. The beam should be directed through a fiberoptic waveguide having a stylus at its distal end provided with a means of laser power control. An exposure timer may also be added to the power supply section so that the power administered to the skin does not exceed that tolerable in each dosage. The instrument should be used only by a physician specializing in organic disorders and in the use of laser radiation as therapeutic treatment. No other person, including those involved in anatomical and physiological treatments, should use this laser instrument unless licensed by the medical profession.

Gynecological Laser Instrument

There are a number of ailments in women that require treatment at a hospital. The treatment at times is time-consuming and costly, because the same techniques that are used in general surgery are also used in minor female problems because of the requirements of the conventional techniques. With the advent of lasers, many female internal disorders that can be reached externally without an incision can be treated with a laser beam effectively and rapidly. An example is the opening of blocked fallopian tubes to aid in fertilization of an infertile woman. A laser waveguide could be developed that can transmit CO_2 laser radiation through a controllable, elongated stylus with a long-focal-length lens system to the site of disorder and can be used to perform the work rapidly in a doctor's office. The waveguide may also be provided with an observing optical eyepiece connected to a conduit made of a coherent fiberoptic bundle, whereby the operating physician can examine the blocked area and administer the surgery effectively and accurately. The cost of such a system could be kept low enough to make it attractive to many practicing physicians.

Meteorological Instrument

A meteorological instrument is needed for measuring accurately and instantaneously the atmospheric characteristics at altitudes from 100,000 to 250,000 feet. This will be an advantageous and distinct improvement over the prevalent instruments. The physical considerations should include smallness of size, about 20 to 30 cm^3, a minimum life span of 1 hour, a maximum response time of 10 msec, output adaptable to telemetering technique, a calibration accuracy of ± 1 to $\pm 3\%$, and a power capability of several milliwatts. The device is to be used on a missile or other high-altitude vehicle, which is to be returned to earth by parachute after collecting information.

In the rapidly changing environment at high altitude, where the atmospheric medium is of extremely low density, the time lag to reach equilibrium is critical in obtaining accuracy. The rarefaction of air molecules at high altitude must be considered. The greater mean free path of the molecules may exceed the size of the sensor, and thus counting the molecules and measuring their mean kinetic energy would be a problem. In addition, the absorption of this energy in the projected laser beam and translation into a color-change device in a light-sensitive material would be further problems to be resolved.

Printed Circuit Board Welder

In conducting failure analysis on printed circuit boards, occasionally one or more leads are found to be either broken or loose from the board, and when this occurs the loose leads are usually soldered with a soldering iron. If these leads are close to active devices, such as transistors, diodes, or resistors formed in the thin-film surface, the solder may run on them and spoil them unless special care is taken with the technique; the procedure is also somewhat cumbersome. A small laser unit with a Nd-YAG laser source transmitting a narrow beam through a fiberoptic conduit will be a useful device for the microelectronics laboratory. The Nd-YAG must be either frequency-doubled or used with a He-Ne laser beam. A ruby laser may also be used in the pulsing mode. The cost of such a device should be kept within $5000.

Underwater Torpedo Timing Device

Develop a rugged, inexpensive, and highly reliable timing device for use with a torpedo or any underwater missile, with an accuracy of $\pm 1\%$ without significantly affecting the short-period missile travel, which may be a maximum of 5 minutes. The existing electronic and mechanical timers are vulnerable to vibration, shock, and temperature. A diode laser beam–aided timer, using the usual clock frequency and operable from the selected frequency termination, would not seem to be influenced adversely by environmental conditions in deep water.

Laser-Aided Rocket Engine

Develop a compact, high-power laser source that can be carried in a cruising spacecraft and has means for vaporizing a quantity of mercury, sodium, or potassium metal. In passing through an electric field, the vapor is ionized into positive ions, which are repelled by the positively charged electrodes into space at an increasing velocity because of the absence of drag. This produces a forward thrust. The system may become an ion-driven engine of the future for rocket propulsion. The rocket must be shuttled into space by conventional methods.

Laser Sewing Machine

A laser sewing machine that makes use of the melting and bonding capabilities of laser radiation may be developed for sewing materials such as nylon, acetates, and other synthetic clothing materials. As in a mechanical or electric sewing machine, the material should be conveyed automatically under the laser head, and the sewing path should be variable from a seam bond to a zigzag pattern. The system may employ a Nd-YAG laser generator that operates in the pulsing mode and uses a closed-circuit refrigerated deionized water system for cooling. Provision must also be made for an instrument panel for selection of the sewing pattern and speed.

Nuclear Warhead–Destroying Laser System

A laser system that will destroy a traveling nuclear rocket is urgently needed. The laser system must provide both a pulsed and a CW laser beam of a power density exceeding 100 MW/cm^2. This could be made in two units using a common optical lens system, or it could be a single unit developing and transmitting sequentially a pulsed and a CW laser beam. The beam must be capable of searching, detecting, acquiring, and then locking on the flying rocket. To accomplish this, a conical laser beam should scan the sky, find the target, and lock on it. The beam then should be instantly collimated to a diameter of 1 cm.

In operation, the system projects a conical beam with about 1-mrad divergence, seeks the target, acquires it, and locks on it, and simultaneously the beam becomes collimated. A high-power pulsing beam starts to work on the target, followed by a burst of CW beam for one-tenth of a second. The pulsing beam penetrates the surface, and the CW beam cleans the drilled hole and permits the pulsing beam to penetrate further. The sequentially pulsing and CW beams, after penetrating the target in less than 1 second, ignite the nuclear fuel in the warhead to explode the rocket in space before its synchro system directs the rocket's course toward the earth. The laser beam may also be directed to the driving section of the rocket to incapacitate its motion; the rocket then falls on earth without exploding.

QUESTIONS AND PROBLEMS

14-1 How does the laser system for personnel verification operate? Describe its construction and the function of each part. How does the fingerprint identifier operate? What is the image enhancement technique? What type of laser beam do you think the fingerprint identifier should use? Why?

14-2 Why is an argon laser beam used in Xerox Company's fingerprint identifier? Describe the technique used in the Xerox fingerprint identifier. How can one make a permanent record of the Xerox fingerprint image? How old a fingerprint can the Xerox fingerprint identifier identify? Can the date of the print be determined?

14-3 How does a tire-mold roundness checkout system operate? What type of laser does it use? How is the error in roundness determined? Can this machine measure any other manufactured parts? What is a laser microphone? What characteristic features of the microphone make it operable? Under what conditions can it not operate?

14-4 Why should a gyroscope be balanced? How is it balanced? What types of laser sources can be used to balance a gyroscope? Are the laser beams used pulsed or in the CW mode? What pulsewidth is used? How does the gyroscope balancer automatically cut off the laser beam? Explain the method.

14-5 How does the algae mapping system developed by NASA operate? What type of laser beam has a 5900-Å wavelength? What color is a 6850-Å-wavelength beam? How is the amount of algae recorded? Why should salmon be branded? What is the advantage of branding cattle with a laser beam? What type of laser is used for branding? How does the branding in fish occur?

14-6 What is the function of a laser-aided cane? How are any objects located on the right and left determined by the blind? What type of laser is used in the cane? Why is that laser chosen? What is a stenographic laser eraser? How is it made? How does it operate? Does the beam operate focused or defocused? Explain why. How many joules of laser energy are used in the eraser? Is the laser beam used in pulsing mode or in continuous-wave mode?

14-7 A certain typewritten letter is to be erased. A biconvex lens is used, with a focal length of 14 mm and optical plane of 8 mm. If the size of the letter to be erased is 3 mm, what will be the distance between the focus of the beam and the letter to be erased? If for each 2-mm-size letter a laser energy of 1.234 J is used, what energy laser beam is necessary to erase the typewritten error?

14-8 In the microwelder shown in Fig. 14-3, if the distance between the focus of the beam and the workpiece at the exit port is 10 mm and the optical plane of the lens is 14 mm, what is the focal length of the lens (biconvex)?

14-9 Does the microwelder shown in Fig. 14-3 operate CW or in the pulsed mode? Why? How is the microwelder triggered if no external switch is used? How is the power to the laser source controlled? If the microwelder is used as a laser eraser, describe the technique used in erasing an error letter. What safety feature does the microwelder possess as a laser eraser? Is this feature necessary for proper operation of the device? Can the device work without the feature? Explain.

14-10 If the laser microwelder uses a fiberoptic waveguide and operates with a Nd-YAG laser, explain how one can determine where the focus on the workpiece will be. How is the focus of the beam adjusted? Is the beam for microwelding focused or defocused? How can this be adjusted? Explain.

14-11 How does the color projection system of Hitachi operate? How large can the screen be for projection? What are the vertical and horizontal resolutions of the television system? What kind of video amplifier is used in this system? Describe the construction and operation of the Hitachi laser videodisk system? How is the playback operation performed? How does the RCA system for the same purpose operate? Describe.

14-12 What is a laser printing system? How does it operate? Why should one employ a laser printing system? What is Hewlett-Packard's electrophotographic process? What does it do? Describe the technique of its operation. What is the function of providing an electrostatic charge during the operation of the system?

14-13 How does NASA's laser power transmitter convert the laser beam into electric power? What efficiency can be obtained from the conversion of a laser beam into electric power? What laser is used in propelling a rocket into earth orbit? How is the propulsion achieved? What provides thrust in the rocket engine? Explain.

14-14 What is a phase-locked semiconductor diode? Describe the laser system that operates as an optical pump. What optical conversion can be obtained from this system? What is a gold vapor laser? At what wavelength does the gold vapor laser operate? How is the plasma temperature of 1450°C achieved in the gold vapor laser? Why is the gold vapor laser useful in radiotherapy of cancerous cells in medical work?

14-15 Describe one method of constructing a laser system that will indicate the proximity of one car to another. Draw a diagram to show how an ultraviolet laser device can be constructed for use in the treatment of infected nasal passages. Draw a diagram for a laser treatment instrument for otological work, as described in Section 14-3. Give the type of optical fibers that can be used to transmit 106,000- to 3000-Å wavelengths. What type of laser beam does a 3000-Å wavelength designate?

14-16 How does an indicator for aircraft-wing icing operate? Give at least two methods. Describe a method for construction and operation of an aircraft collision-preventive system. How should an underwater target-detection system work? Give your own method.

14-17 Draw a diagram showing the plan of designing an instrument for treatment of referred pain on the body. Could you use a CO_2 or Nd-YAG laser? Explain why. How would you construct a printed circuit board welder? Explain. In building a meteorological instrument as described in Section 14-3, why should you consider the mean free path of the atmospheric molecules?

14-18 Construct diagrammatically a device for underwater torpedo timing, using a diode laser. Why should the device using a laser beam not be affected by ocean turbulence?

14-19 Design a laser-aided rocket engine and describe its operation and function, indicating how forward thrust is produced in the engine. Determine how much thrust (force exerted) will be required to push a mass of 5 kg through space.

14-20 Draw a simple circuit diagram for a laser system that can produce both a pulsed beam and a continuous-wave beam, using a power of 10 kW. The operation of switching from a pulsed to a CW mode should be automatic. Choose any pulsed duration and CW duration in the circuit; each duration should not exceed a quarter of a second. After the circuit is constructed in accordance with your design, determine how much power density will be required to drill a hole through 4-mm-thick steel, provided no divergence has occurred during the beam's travel to the target. Assume the beam diameter to be 8 mm. Use the parameters already learned in previous chapters. Use the equation given in one of the chapters.

SELECTED SOURCES OF RELATED DATA

Bhatti, M. S., and A. Marzocchi. 1986. Owens-Corning. Method and Apparatus for Forming Optical Fibers. U.S. Patent 4,430,105.

Channell, P. J., ed. 1982. *Laser Acceleration of Particles*. New York: American Institute of Physics.

Christensen, C. P. 1982. Emerging laser applications. *Science* 218:115.

Churnside, J. H. 1982. Three-dimensional laser vector velocimetry. *Appl. Opt.* 21(5):845.

Clarady, J. F., and K. L. Jessee. 1984. Holographic Inspection of Bonded Joints. U.S. Patent No. 4,408,881.

Cunningham, R. 1984. Erasable optical storage of data. *Lasers and Applications* May, pp. 109–112.

Demokan, M. S. 1982. *Mode-Locking in Solid-State and Semiconductor Lasers*. New York: Wiley.

Hosaka, Y., et al. 1982. Laser Copier/Facsimile Machine. U.S. Patent 4,358,793.

Hutley, M. C. 1982. *Diffraction Gratings*. New York: Academic Press.

Infrared technology fights contraband at border. 1984. *Photonics Spectra* March, pp. 36–38.

Javan, A., 1983. Transient Regenerative Amplification (Laser Beam). U.S. Patent No. 4,410,992.

Kaplan, H. 1984. Thermal sensing instruments. *Laser Focus/Electro-Optics* March, pp. 57–69.

Kompa, K. L., et al., eds. 1984. *Laser Applications in Chemistry*. New York: Plenum.

Laser Focus Staff. 1983. Selected research and applications. *Laser Focus/Electro-Optics* January, pp. 33–44.

Lerner, J. M., and A. Thevenon. 1984. Diffraction grating's anamorphic properties. *Lasers and Applications* January, pp. 89–93.

Leskovar, B. 1984. Recent advances in high-speed photon detectors. *Electro-Optics* February, pp. 73–81.

Marathay, A. S. 1982. *Elements of Optical Coherence Theory*. New York: Wiley.

McCartney, E. J. 1983. *Absorption and Emission by Atmospheric Gases*. New York: Wiley.

McMahan, R. K. 1982. Laserbeam Directional Stabilization Device. U.S. Patent No. 4,361,383.

Muscatell, R. P. 1983. Laser Microphone (Distortion-Free). U.S. Patent 4,412,105.

Ohara, Y., and M. Ohnishi. 1983. Laser Recorder. U.S. Patent No. 4,370,175.

Presta, P. S., et al. 1982. Laser System and Its Method of Use. U.S. Patent No. 4,396,285.

Simons, B. S. 1982. Laser Light Detection System with Gain Integrator. U.S. Patent No. 4,366,378.

Solem, J. C., and G. C. Baldwin. 1982. X-ray holography of living organisms. *Science* 218:229.

Taylor, L. H. 1982. Techniques for Amplifying Two or More Laser Beams in a Single Laser Amplifier. U.S. Patent No. 4,397,024.

Tew, C. E., and F. E. Ellis. 1983. Heterodyne Laser Ranging System. U.S. Patent No. 4,405,230.

Willardson, R. K., and A. C. Beer, eds. 1981. *Semiconductors and Semimetals*. New York: Academic Press.

APPENDIX

I

ELEMENTS AND THEIR CHARACTERISTICS

Element	Symbol	At. no.	At. wt.	Melting point, °C	Specific gravity
Actinium	Ac	89	227.00	—	—
Alabamine*	Ab	85	221.0	—	—
Aluminum	Al	13	26.97	660	2.70
Antimony	Sb	51	121.76	630.5	6.68
Argon	A	18	39.94	−189.2	1.27 g/l
Arsenic	As	33	74.93	814.0	5.73
Barium	Ba	56	137.36	850.0	3.50
Beryllium	Be	4	9.02	1350.0	1.85
Bismuth	Bi	83	209.0	271.0	9.78
Boron	B	5	10.82	2300.0	2.54
Bromine	Br	35	79.91	−7.2	3.12 liq.
Cadmium	Cd	48	112.41	320.9	8.65
Calcium	Ca	20	40.08	810.0	1.54
Carbon	C	6	12.00	>3500.0	2.25 graph.
Cerium	Ce	58	140.13	640.0	6.90
Cesium	Cs	55	132.81	26.4	1.87
Chlorine	Cl	17	35.45	−101.6	3.2 g/l
Chromium	Cr	24	52.01	1615.0	7.1
Cobalt	Co	27	58.94	1480.0	8.9
Columbium†	Cb	41	93.30	1950.0	8.4
Copper	Cu	29	63.57	1083.0	8.95
Dysprosium	Dy	66	162.46	—	—
Erbium	Er	68	167.64	—	4.77
Europium	Eu	63	152.00	—	—
Fluorine	F	9	19.0	−223.0	1.69 g/l
Gadolinium	Gd	64	157.30	—	—

Element	Symbol	At. no.	At. wt.	Melting point, °C	Specific gravity
Gallium	Ga	31	69.72	29.75	5.91
Germanium	Ge	32	72.60	958.5	5.36
Gold	Au	79	197.20	1063.0	19.32
Hafnium	Hf	72	178.60	1700.?	13.30
Helium	He	2	4.00	<−272.2	0.177 g/l
Holmium	Ho	67	163.50	—	—
Hydrogen	H	1	1.0078	−259.1	0.0899 g/l
Illinium	Il	61	146.00	—	—
Indium	In	49	114.80	155.0	7.28
Iodine	I	53	126.93	113.5	4.93 sol.
Iridium	Ir	77	193.10	2350.0	22.42
Iron (cast)	Fe	26	55.84	1275.0	7.85
Krypton	Kr	36	82.90	−169.0	3.7 g/l
Lanthanum	La	57	138.90	826.0	6.15
Lead	Pb	82	207.22	327.5	11.35
Lithium	Li	3	6.94	186.0	0.534
Lutecium	Lu	71	175.00	—	—
Magnesium	Mg	12	24.32	651.0	1.74
Manganese	Mn	25	54.93	1260.0	7.20
Masurium	Ma	43	—	—	—
Mercury	Hg	80	200.61	−38.87	13.59
Molybdenum	Mo	42	96.00	2620.0	10.20
Neodymium	Nd	60	144.27	840.0	6.95
Neon	Ne	10	20.18	−248.67	0.9 g/l
Nickel	Ni	28	58.69	1452.0	8.90
Nitrogen	N	7	14.008	−209.86	1.25 g/l
Osmium	Os	76	190.80	2700.00	22.48
Oxygen	O	8	16.00	−218.4	1.42 g/l
Palladium	Pd	46	106.70	1555.0	11.40
Phosphorus	P	15	31.02	44.1	1.8–2.2
Platinum	Pt	78	195.23	1755.0	21.45
Polonium	Po	84	210.00	—	—
Potassium	K	19	39.10	62.3	0.86
Proseodymium	Pr	59	140.92	940.0	6.50
Protactinium	Pa	91	—	—	—
Radium	Ra	88	225.97	960.0	5.0
Radon	Rn	86	222.00	—	9.73 g/l
Rhenium	Re	75	186.31	3000.?	20.53
Rhodium	Rh	45	102.91	1955.0	12.50
Rubidium	Rb	37	85.44	38.5	1.53
Ruthenium	Ru	44	101.70	—	12.20
Samarium	Sm	62	150.43	>1300.0	7.7
Scandium	Sc	21	45.10	1200.0	2.5
Selenium	Se	34	79.20	220.0	4.5
Silicon	Si	14	28.06	—	2.42
Silver	Ag	47	107.88	960.5	10.50
Sodium	Na	11	22.997	97.5	0.97
Strontium	Sr	38	87.63	800.0	2.6
Sulphur	S	16	32.06	120.0	2.07
Tantalum	Ta	73	181.40	2850.0	16.6
Tellurium	Te	52	127.50	452.0	6.24

Element	Symbol	At. no.	At. wt.	Melting point, °C	Specific gravity
Terbium	Tb	65	159.20	—	—
Thallium	Tl	81	204.39	303.5	11.85
Thorium	Th	90	232.12	1845.0	11.20
Thulium	Tm	69	169.40	—	—
Tin	Sn	50	118.70	170.0	5.7–6.5
Titanium	Ti	22	47.90	1800.0	4.5
Tungsten	W	74	184.0	3370.0	19.3
Uranium	U	92	238.14	<1850.0	18.68
Vanadium	V	23	50.95	1710.0	5.87
Virginium	Vi	87	224.00	—	—
Xenon	Xe	54	131.30	−140.0	5.85 g/l
Ytterbium	Yb	70	173.50	—	—
Yttrium	Y	39	88.92	1490.0	5.51
Zinc	Zn	30	65.38	419.4	7.14
Zirconium	Zr	40	91.22	1700.0	6.40

* Astatine

† Niobium

APPENDIX

II

IMPORTANT PHYSICAL CONSTANTS

Electrical units

Charge (quantity)

Electronic charge (e) .... = 4.802×10^{-10} e.s.u.
= 1.592×10^{-20} e.m.u
= 1.592×10^{-19} coulomb

Coulomb (Q)................ = 10^{-1} e.m.u.
= 3×10^{9} e.s.u.
= 6.28×10^{18} electronic charges

Microcoulomb (μQ) .... = 10^{-6} coulomb

Faraday .... = 96,500 coulombs

Current (I)

Ampere (A) .... = 10^{-1} e.m.u.
= 3×10^{9} e.s.u.
= 1 coulomb per second
= 1.036×10^{-5} faraday per second

Milliampere (mA)........... = 10^{-3} ampere

Microampere (μA).......... = 10^{-6} ampere

Potential (E or V)

Volt (V)....................... = 10^{8} e.m.u.
= 10^{8} gauss
= $\frac{1}{3} \times 10^{-2}$ e.s.u.

Kilovolt (kV) .... = 1000 volts

Million volts (MV) .... = 10^{6} volts

Millivolt (mV)............... = 10^{-3} volt

Microvolt (μV)............. = 10^{-6} volt

Resistance (R or r)

Ohm (Ω)...................... = The resistance of a uniform column of mercury 106.3 cm long at 0°C, having a mass of 14.452 grams
= 10^9 e.m.u.
= $1/9 \times 10^{-11}$ e.s.u.

Megohm (MΩ) .... = 10^6 ohms

Microhm ($\mu\Omega$).............. = 10^{-6} ohm
= 10^{-12} megohm

Capacity (C)

Farad (F)...................... = 10^{-9} e.m.u.
= 9×10^{11} e.s.u.
= 10^6 microfarads

Microfarad (μF)............ = 10^{-6} farad

Picofarad (pF) .... = 10^{-12} farad

Inductance (L)

Henry (H) .... = 10^9 e.m.u.
= $1/9 \times 10^{-11}$ e.s.u.

Millihenry (mH)............ = 10^{-3} henry

Magnetic units

1 line of force = 1 maxwell = 1 e.m.u. = 10^{-8} volt-second
1 gauss = 1 maxwell/sq. cm. = 1 line/sq. cm.
1 gilbert = 1 gauss-centimeter
Magneton (Bohr) = 9.22×10^{-21} erg per gauss

Other constants

$$\text{Velocity of light} \ldots = \frac{\text{electrostatic unit}}{\text{electromagnetic unit}}$$

$$= \frac{\text{e.s.u.}}{\text{e.m.u.}} = 3 \times 10^{10} \text{ cm/second}$$

Positron charge (positive electron) = 4.802×10^{-10} e.s.u.
Positron mass = 9.107×10^{-28} gram
Electron mass = 9.107×10^{-28} gram
Mass of proton = 1.672×10^{-24} gram
e/m = 5.2741×10^{17} e.s.u. per gram

Electron volt = 1.59×10^{-12} erg
1 erg = 0.629×10^{-12} electron volts

Gas constant (R) = 1.9864 calories/degree/mole
= 8.3136×10^7 ergs/degree/mole
Avogadro's number = 6.06×10^{23} per mole

Planck's constant = 6.55×10^{-27} erg-second
Boltzmann's constant = 1.3708×10^{-16} erg/degree

Gravitational constant (G) = 6.66×10^{-8} C.G.S. units
Radius of earth (equatorial) = 6.3784×10^{8} cm
Mean density of earth = 5.52 grams per cc
Mean distance from earth to sun = 1.495×10^{13} cm
Mean distance from earth to moon = 3.844×10^{10} cm

APPENDIX

III

INTERNATIONAL WEIGHTS AND MEASUREMENTS

Units of Length

Angstrom units	Micrometers	Millimeters	Centimeters	Inches	Meters
1	10^{-4}	10^{-7}	10^{-8}	2.5×10^{-8}	10^{-10}
10^4	1	10^{-3}	10^{-4}	2.5×10^{-4}	10^{-6}
10^7	10^3	1	$1/10$	$1/25$	10^{-3}
10^8	10^4	10	1	$2/5$	10^{-2}
2.54×10^8	2.54×10^4	25.4	2.54	1	$1/39$
10^{10}	10^6	10^3	10^2	39.4	1

1 light year = 5.9 10^{12} miles = 9.5 × 10^{12} kilometers
1 kilometer = 10^3 meters = 10^5 centimeters = 10^6 millimeters
1 mile = 1760 yards = 5280 feet = 63,360 inches
1 yard = 3 feet = 36 inches = .9144 meter
1 foot = 12 inches = .3048 meter
1 inch = 2.54 centimeters
1 meter = 1.093 yards = 3.279 feet = 39.37 inches

Units of Area

1 square meter = 100 sq. decimeters = 10^4 sq. cm. = 10^6 sq. mm.
1 square mile = 640 acres = 3,097,600 sq. yd. = 27,878,400 sq. ft.
1 acre = 4840 sq. yd. = 43,560 sq. ft.
1 sq. yd. = 9 sq. ft.
1 sq. ft. = 144 sq. in.

Units of Volume

1 cubic meter = 10^3 cubic decimeters = 10^6 cu. cm. = 10^9 cu. mm.
1 cubic decimeter = 10^3 cu. cm. = 10^6 cu. mm.
1 cc = 10^3 cu. mm.

The weight of 1 cc of water at 4°C is 1 gram.

Units of Capacity and Equivalents

1 liter = 1000 cc = 1000 milliliters
1 cc = 1 milliliter = 16.2 minims (drops)
1 cc distilled water = 1 milliliter distilled water = 1 gram
1 gallon = 4 quarts = 8 pints = 128 fl. oz.
1 quart = 2 pints = 32 fl. oz.
1 pint = 16 fl. oz.
1 fl. oz. = 8 fl. drams = 29.57 mils = 29.57 cc
1 fl. dram = 3.696 mils = 3.696 cc

Avoirdupois

1 ton = 2000 pounds = 32,000 ounces = 907.2 kilograms
1 pound = 16 ounces = 453.59 grams
1 ounce = 28.34 grams
1 gram = 100 centigrams = 1000 milligrams = 15.43 grains
1 grain = .0648 gram = 6.48 centigrams = 64.8 milligrams

Units of Angle

1 circumference = 360 degrees = 2π radians = ($2\pi r$; r = radius)
½ circumference = 180 degrees = π radians
¼ circumference = 90 degrees = $\pi/2$ radians
⅛ circumference = 45 degrees = $\pi/4$ radians
(π = 3.1416)
1 degree = 60 minutes = 3600 seconds = 0.01745 radian
1 minute = 60 seconds
1 radian = 57° 17′ 44.8″ = 57.2958° (degrees)
1 knot (nautical mile) = length of 1′ (minute) of arc on the earth's equatorial surface.

Other Units Used in Book

Nanosecond = 10^{-9} second
Nanometer = 10^{-9} meter
Microfarad = 10^{-6} farad
Picofarad = 10^{-12} farad
Femtosecond = 10^{-15} second

Megawatt = 10^6 watts
Gigawatt = 10^9 watts
Terawatt = 10^{12} watts

APPENDIX

IV

TRIGONOMETRIC FUNCTIONS USED IN THIS BOOK

At times, the reader has occasion to use trigonometric functions in solving certain laser-related problems in this book. The problems involve laser beam propagation in space, reflection from reflecting surfaces, refraction in media in which the beam travels, and determining the intensity of a given beam from the source to the area of incidence. The reader who is not familiar with these functions (or has forgotten after leaving the university) will find them easy to review and apply in cases where other methods would not be as simple to use. The trigonometric functions are generally given in sine, cosine, tangent, cotangent, secant, and cosecant. Since only the first four functions are used in this book occasionally, only they will be explained in the ensuing discussion.

In the illustration, we may take up a case in measuring distance with a laser beam. A narrow-diameter laser beam is projected from the source and focused at C of the target. To measure the distance between the source and C, we may use one of the

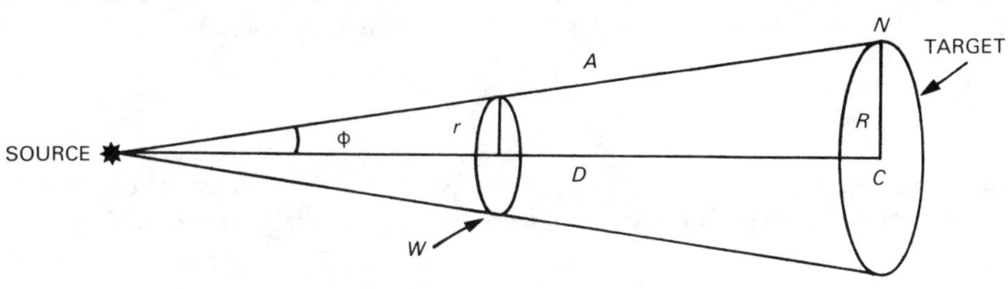

Figure IV-1 Laser source-to-target relations.

trigonometric functions, such as the tangent function. Depending on the length, a second mark N at a distance of, for instance, 2 meters from the point C may be made perpendicular to the source–C, or at right angles to the laser beam projection. First the beam is focused on C and then on point N. The angle θ through which the beam is rotated is measured in degrees. Then, by using a relationship selected from the following trigonometric functions, one can calculate the distance D between the source and the target point C. The trigonometric functions may be given as follows:

(a) Sine	$\sin\theta$	$= R/A$	$= \cos(90° - \theta)$
(b) Cosine	$\cos\theta$	$= D/A$	$= \sin(90° - \theta)$
(c) Tangent	$\tan\theta$	$= R/D$	$= \cot(90° - \theta)$
(d) Cotangent	$\cot\theta$	$= D/R$	$= \tan(90° - \theta)$

In the triangle source–N–C, shown in the figure, we may use the tangent function of the triangle, since one side of the triangle is given as $R = 2$ meters and the angle θ has been measured after the laser beam is rotated. Thus, we have

$$\tan\theta = \frac{R}{D}$$

and

$$D = \frac{R}{\tan\theta}$$

$$= \frac{2}{\tan\theta}$$

Assume that the measured angle θ was 20°. Then tan 20° will be, as looked up and found from trigonometric function tables, 0.3640. Therefore, the distance will be

$$D = \frac{200\ \text{cm}}{0.3640}$$

$$= 550\ \text{cm approximately}$$

$$= 5.5\ \text{meters}$$

For laser beam intensity measurement, assume that the intensity of a given laser radiation at distance r from the source is 80 W/cm^2. What will be the intensity of the projected radiation at R, if r is located midway between the source and C?

$$\tan\theta = \frac{r}{\frac{1}{2}D} = \frac{2r}{D}$$

Since r is equal to ½R, its numerical value is 1 meter or 100 centimeters, and the distance D is 5.5 meters or 550 centimeters. Equating a relation for r, we have

$$r = \frac{D\tan\theta}{2} = 100\ \text{cm}$$

Since the laser intensity at the target T to that at W is inversely proportional to the numerical values of these respective areas, we may equate the proportion as follows:

$$\frac{\pi R^2}{\pi r^2} = \frac{80 \text{ W/cm}^2}{X}$$

where X is the unknown intensity and 80 W/cm^2 is the given intensity at position r. Thus,

$$X = \frac{80 \text{ W/cm}^2 \times \pi r^2}{\pi R^2}$$

$$= \frac{80(100)^2}{(200)^2}$$

$$= 20 \text{ W/cm}^2$$

The above relations prove that when the distance from a reference point is doubled, the intensity of radiation becomes one-fourth, and when the distance is tripled, the intensity falls to one-ninth of that at the reference point, which is the point r in this case.

APPENDIX

V

LASER-ORIENTED MANUFACTURERS

Advanced Kinetics, Inc.
1231 Victoria Street — Phone (714) 646-7165
Costa Mesa, California 92627
Mfg—pulsed lasers, detectors, power supplies, interferometers, spectrometers.

American Laser Systems, Inc.
106 James Fowler Road — Phone (805) 967-0423
Goleta, California 93017
Mfg—Laser communication systems, modulators, silicon detectors, power supplies.

American Optical Corporation
10 Optical Avenue — Phone (603) 352-3202
Keene, New Hampshire 03431
Mfg—laser systems, optical components, beam expanders, beam splitters, lenses.

Apollo Lasers, Inc.
6357 Arizona Circle — Phone (213) 776-3343
Los Angeles, California 90045
Mfg—ruby, glass, YAG, CO_2 lasers; photodiode detectors, radiometers, holography.

Ardel Kinamatic Corporation
125-20 18th Avenue — Phone (212) 353-3600
College Point, New York 11356
Mfg—optical mounts, translators, optical benches, He-Ne lasers.

Atomergic Chemetals Corporation
100 Fairchild Avenue — Phone (516) 822-8800
Plainview, New York 11803
Mfg—Birefringent windows, infrared windows, ultraviolet windows, laser crystals.

AVCO Everett-Metalworking Lasers
32 Cobble Hill Road — Phone (617) 389-3000
Somerville, MA 02143
Mfg—Carbon dioxide lasers, metalworking systems, metal surface treatment, etc.

Barnes Engineering Company
30 Commerce Road — Phone (203) 348-5381
Stamford, Ct 06904
Mfg—photon detectors, collimators, infrared radiometers, spectrum analyzers

Bausch & Lomb, Inc.
820 Linden Avenue — Phone (716) 385-1000
Rochester, New York 14625
Mfg—diffraction gratings, grating monochromators, laser safety goggles.

BOC Ltd, Industrial Power Beams
7 Royal Oak Way South — Phone (03272) 4813
Daventry, Northants, UK
Mfg—CO_2 lasers, Nd-YAG lasers, beam expanders, radiometers, Q-switches

Bond Optics, Inc.
Etna Road — Phone (603) 448-2300
Lebanon, New Hampshire 03766
Mfg—beam splitters, etalons, filters, lenses, mirrors, prisms, windows, lasser rods.

Broomer Research Corporation
23 Sheer Plaza — Phone (516) 249-1544
Plainview, New York 11803
Mfg—thin-film coatings, optical filters, etalons, lenses, mirrors, polarizers.

Burleigh Instruments, Inc.
Burleigh Park — Phone (716) 924-9355
Fishers, New York 14453
Mfg—Fabry-Perot interferometers, mirror mounts, power supplies, piezoelectric/ translators.

Cleveland Crystals, Inc.
19306 Redwood Avenue — Phone (216) 486-6100
Cleveland, Ohio 44110
Mfg—optical components, KDP group, Q-switches, polarizers, waveplate retarders.

Coherent, Inc.
3210 Porter Drive Phone (415) 493-2111
Palo Alto, California 94304
Mfg—pulsed CO_2 lasers, He-Ne lasers, radiometers, detectors, collimators, dye lasers.

Control Laser Corporation
11222 Astronaut Boulevard Phone (305) 851-2540
Orlando, Florida 32809
Mfg—Nd-YAG lasers, holographic equipment, collimators, etalons, expanders.

Daedal, Inc.
P.O. Box G Phone (412) 744-4451
Harrison City, PA 15636
Mfg—beam positioners, optical benches, optical mounts, motor-driven stages.

The Ealing Corporation
22 Pleasant Street Phone (617) 655-7000
St. Natick, MA 01760
Mfg—laser monitors, safety goggles, collimators, detectors, attenuators, filters.

Eastman Kodak Company
343 State Street Phone (716) 325-2000
Rochester, New York 14650
Mfg—laser dyes, Q-switch for ruby and Nd lasers, optical materials, lenses, phosphors.

EG&G Inc.
35 Congress Street Phone (617) 745-3200
Salem, MA 01970
Mfg—detectors, radiometers, flashlamps, krytons, thyratrons, pulse selectors.

Edmund Scientific Company
1776 Edscorp Building Phone (609) 547-3488
Barrington, New Jersey 08007
Mfg—Lenses, mirrors, gratings, fiberoptics, optical benches, flats, mirrors, goggles.

General Photonics Corporation
2255 F Martin Avenue Phone (408) 496-6040
Santa Clara, California 95050
Mfg—CW and pulsed YAG lasers for science, medical field, and industrial applications.

Hughes Aircraft Company
6155 El Camino Real Phone (714) 438-9191
Carlsbad, California 92008
Mfg—He-Ne lasers, power supplies, collimators, interferometry, CO_2 and military lasers.

Infrared Optics, Inc.
5 Jean Court Phone (516) 694-2977
Farmingdale, New York 11735
Mfg—beam expanders, splitters, filters, coatings, lenses, mirrors, wedges, domes.

IntraAction Corporation
766 Foster Avenue Phone (312) 595-3770
Bensenville, Illinois 60106
Mfg—acousto-optics modulators, mode lockers, Q-switches, deflectors, beam generators.

ITT Corporation
7635 Plantation Road Phone (703) 563-0371
Roanoke, Virginia 24019
Mfg—all types fiberoptic components, communication systems, connectors.

Korad/Hadron, Inc.
2520 Colorado Avenue Phone (213) 829-3377
Santa Monica, California 90404
Mfg—pulsed lasers, holography, mode lockers, mirrors, Q-switches, photodiodes.

Karl Lambrecht Corporation
4204 North Lincoln Avenue Phone (312) 472-5442
Chicago, Illinois 60618
Mfg—beam splitters, collimators, filters, coatings, polarizers, waveplates, radiometers.

Laserkinetics Corporation
1735 N. Morningside Street Phone (714) 637-4683
Orange, California 92667
Mfg—developers of medical and dental laser systems, communication systems, military.

Lasermetrics, Inc.
111 Galway Place Phone (201) 837-9090
Teaneck, New Jersey 07666
Mfg—modulators, Q-switches, polarizing prisms, optical elements, translators.

Lexel Corporation
928 East Meadow Drive Phone (415) 494-3241
Palo Alto, California 94303
Mfg—argon and krypton lasers, holographic systems, power meters, mode lockers.

Lumonics Incorporated
105 Schneider Road, Kanata Phone (613) 592-1460
Ontario, Canada K2K 1Y3
Mfg—gas lasers, markers, detectors, power meters, energy meters.

Metrologic Instrument, Inc.
143 Harding Avenue Phone (609) 933-0100
Bellmawr, New Jersey 08033
Mfg—He-Ne lasers, collimators, filters, optical benches, holography, radiometers.

Molecron Corporation
177 North Wolfe Road Phone (408) 738-2661
Sunnyvale, California 94086
Mfg—pyroelectric detectors, Nd-YAG and tunable dye lasers, radiometers, etc.

Newport Corporation
18235 Mount Baldy Circle Phone (714) 963-9811
Fountain Valley, California 92708
Mfg—optical tables, vibration isolators, beam positioners, laser kits.

Optical Coating Laboratory, Inc.
15251 E Don Julian Road Phone (213) 968-6581
City of Industry, California 91761
Mfg—silicon photodiodes, beam splitters, beam expanders, lenses, polarizers.

Owen-Illinois, Inc.
P.O. Box 1035 Phone (419) 247-8364
Toldeo, Ohio 43666
Mfg—Nd-glass rods, disks, Faraday rotators, samarium filter tubes, clad rods.

Phase-R Corporation
Old Bay Road Phone (603) 859-3800
New Durham, NH 03855
Mfg—tunable dye and nitrogen lasers, expanders, optical benches, flashlamps.

RCA Corporation
New Holland Avenue Phone (717) 397-7661
Lancaster, PA 17604
Mfg—helium-cadmium lasers, semiconductor lasers, detectors, transceivers.

Space Optics Research Laboratories
7 Stuart Road Phone (617) 256-4511
Chelmsford, MA 01824
Mfg—Fourier lenses, infrared components, parabolic mirrors, mounts, interferometers.

Spectra-Physics, Inc.
1250 West Middlefied Road Phone (415) 961-2550
Mountain View, California 94042
Mfg—He-Ne lasers, Ar lasers, krypton lasers, radiometers, splitters, expanders, etalons.

United Detector Technology, Inc.
2644 30th Street Phone (213) 396-3175
Santa Monica, California 90405
Mfg—silicon photodiodes, radiometers, positioning devices, photometers.

Valtec Corporation
99 Hartwell Street Phone (617) 835-6082
West Boylston, MA 01583
Mfg—custom fiberoptic cables, telecommunications, industrial and medical illuminators.

WEC Engineering, Inc.
86 Woodbridge Road Phone (617) 369-4183
Carlisle, MA 01741
Mfg—laser systems, amplifiers, beam expanders, modulators, mounts, radiometers.

INDEX